Evolutionary Computation in Scheduling

Evolutionary Computation in Scheduling

Edited by

Amir H. Gandomi, Ali Emrouznejad,
Mo M. Jamshidi, Kalyanmoy Deb, and Iman Rahimi

This edition first published 2020

Registered Office
John Wiley & Sons, Inc., 111 River Street, Hoboken, NJ 07030, USA

Editorial Office
111 River Street, Hoboken, NJ 07030, USA

For details of our global editorial offices, customer services, and more information about Wiley products visit us at www.wiley.com.

Wiley also publishes its books in a variety of electronic formats and by print-on-demand. Some content that appears in standard print versions of this book may not be available in other formats.

Library of Congress Cataloging-in-Publication data applied for

ISBN: 9781119573845

Cover Design: Wiley
Cover Image: © NicoElNino/Shutterstock

Set in 9.5/12.5pt STIXTwoText by SPi Global, Pondicherry, India

Printed in United States of America

10 9 8 7 6 5 4 3 2 1

Contents

List of Contributors

Abbas Ahmadi
Department of Industrial Engineering and Management Systems, Amirkabir University of Technology, Tehran, Iran

Abdollah Ahmadi
School of Electrical Engineering and Telecommunications, University of New South Wales, Sydney, NSW, Australia

Reza Behmanesh
Young Researchers and Elite Club, Isfahan (Khorasgan) Branch, Islamic Azad University, Isfahan, Iran

João P.S. Catalão
INESC TEC and Faculty of Engineering of the University of Porto, Porto, Portugal

P. Deepalakshmi
Department of Computer Science and Engineering, Kalasalingam Academy of Research and Education, Krishnankoil, Tamilnadu, India

Ali Emrouznejad
Aston Business School, Aston University, Birmingham, UK

Ali Esmaeel Nezhad
Department of Electrical, Electronic, and Information Engineering, University of Bologna, Bologna, Italy

Amir Mohammad Fathollahi-Fard
Department of Industrial Engineering and Management Systems, Amirkabir University of Technology, Tehran, Iran

Amir H. Gandomi
Faculty of Engineering and IT, University of Technology Sydney, Ultimo, Australia

Mohammad Sadegh Javadi
Department of Electrical Engineering, Shiraz Branch, Islamic Azad University, Shiraz, Iran

Javidan Kazemi Kordestani
Department of Computer Engineering, Science and Research Branch, Islamic Azad University, Tehran, Iran

Preetam Kumar
Department of Electrical Engineering, Indian Institute of Technology Patna, Bihar, Patna, India

Jiayi Liu
School of Information Engineering, Wuhan University of Technology, Wuhan, China

Mohammad Reza Meybodi
Soft Computing Laboratory, Computer Engineering and Information Technology Department, Amirkabir University of Technology (Tehran Polytechnic), Tehran, Iran

Mohsen Naderpour
Centre for Artificial Intelligence, Faculty of Engineering and Information Technology, University of Technology Sydney (UTS), Sydney, NSW, Australia

Prabina Pattanayak
Department of Electronics and Communication Engineering, National Institute of Technology Silchar, Assam, Silchar, India

Duc Truong Pham
Department of Mechanical Engineering, University of Birmingham, Birmingham, UK

Iman Rahimi
Young Researchers and Elite Club, Isfahan (Khorasgan) Branch, Islamic Azad University, Isfahan, Iran

Fahimeh Ramezani
Centre for Artificial Intelligence, Faculty of Engineering and Information Technology, University of Technology Sydney (UTS), Sydney, NSW, Australia

Seyed-Ehsan Razavi
School of engineering and IT, Murdoch University, Perth, Australia

Jack Romanous
Centre for Artificial Intelligence, Faculty of Engineering and Information Technology, University of Technology Sydney (UTS), Sydney, NSW, Australia

Mohsen S. Sajadieh
Department of Industrial Engineering and Management Systems, Amirkabir University of Technology, Tehran, Iran

S. Sarathambekai
Department of Information Technology, PSG College of Technology, Coimbatore, Tamilnadu, India

K. Shankar
Department of Computer Applications, Alagappa University, Karaikudi, Tamilnadu, India

Javid Taheri
Department of Computer Science, Karlstad University, Karlstad, Sweden

K. Umamaheswari
Department of Information Technology, PSG College of Technology, Coimbatore, Tamilnadu, India

Wenjun Xu
School of Information Engineering, Wuhan University of Technology, Wuhan, China

Mostafa Zandieh
Department of Industrial Management, Management and Accounting Faculty, Shahid Beheshti University, G.C., Tehran, Iran

Zude Zhou
School of Information Engineering, Wuhan University of Technology, Wuhan, China

Albert Y. Zomaya
Centre for Distributed and High Performance Computing, School of Computer Science, University of Sydney, Sydney, NSW, Australia

Editors' Biographies

Amir H. Gandomi is a Professor of Data Science at the Faculty of Engineering & Information Technology, University of Technology Sydney. Prior to joining UTS, Prof. Gandomi was an Assistant Professor at the School of Business, Stevens Institute of Technology, USA and a distinguished research fellow in BEACON center, Michigan State University, USA. Prof. Gandomi has published over 160 journal papers and five books. He has been named as one of the most influential scientific minds (top 1%) and a Highly Cited Researcher (H-index = 56) for three consecutive years, 2017 to 2019. He also ranked 19th in GP bibliography among more than 12 000 researchers. He has served as associate editor, editor, and guest editor in several prestigious journals, such as AE of SWEVO and IEEE TBD. Prof. Gandomi is active in delivering keynote and invited talks. His research interests are global optimization and (big) data mining using machine learning and evolutionary computations in particular.

Ali Emrouznejad is a Professor and Chair in Business Analytics at Aston Business School, UK. His areas of research interest include performance measurement and management, efficiency and productivity analysis, as well as data mining and big data. He holds an MSc in applied mathematics and received his PhD in operational research and systems from Warwick Business School, UK. Prof Emrouznejad is editor, associate editor, and guest editor to several respected journals including: *Annals of Operations Research*, *European Journal of Operational Research*, *Socio-Economic Planning Sciences*, and *Journal of Operational Research Society*. He has published over 150 articles in top-ranked journals and authored/edited several books including *Applied Operational Research with SAS* (CRC Taylor & Francis), *Big Data Optimization* (Springer), *Performance Measurement with Fuzzy Data Envelopment Analysis* (Springer), *Big Data for Greater Good* (Springer), *Managing Service Productivity* (Springer), *Fuzzy Analytics Hierarchy Process* (CRC Taylor & Francis), and *Handbook of Research on Strategic Performance Management and Measurement* (IGI Global). See www.emrouznejad.com.

Mo M. Jamshidi is a F-IEEE, F-ASME, AF-AIAA, F-AAAS, F-TWAS, F-NYAS. He received BSEE (Cum Laud) at Oregon State University in 1967, and MS and PhD in EE from the University of Illinois at Urbana-Champaign in June 1969 and February 1971, respectively. He holds honorary doctorate degrees from the University of Waterloo, Canada, 2004, Technical University of Crete, Greece, 2004, and Odlar Yourdu University, Baku, Azerbaijan in 1999. Currently, he is the Lutcher Brown Endowed Distinguished Chaired Professor at the University of Texas, San Antonio, TX, USA.

He was an advisor to NASA for 10 years (including with 1st MARS Mission and 7 years with NASA HQR), spent 9 years with US AFRL, 8 years with USDOE, and 1 year with the EC/EU. Currently he is a consultant on the Army Science Board. He has close to 800 technical publications, including 75 books (11 textbooks), research volumes, and edited volumes in English and five foreign languages. He is the Founding Editor or co-founding editor or Editor-in-Chief of five journals, including *IEEE Control Systems Magazine* and the *IEEE Systems Journal*. He has graduated or advised 65 PhD and 85 MS students. Moreover, he has advised over 120 US ethnic minority students at MS and PhD and over 850 undergraduate students. Among them are 4 Native American PhDs, 10 Hispanic PhDs and 8 African American PhDs. His former students are successful professionals in 22 nations around the world. Six of his edited and authored books are on System of Systems Engineering in English and Mandarin.

He is the recipient of IEEE Centennial Medal 1984 and WAC Medal of Honor 2014, among many other awards and honors. He is a member of the UTX System Chancellor's Council. He is currently involved in research on system of systems engineering with an emphasis on robotics, drones, biological and sustainable energy systems. He has over 10 940 citations on Google Scholar.

Kalyanmoy Deb is Koenig Endowed Chair Professor at the Department of Electrical and Computer Engineering in Michigan State University, USA. Prof. Deb's research interests are in evolutionary optimization and their application in multi-criterion optimization, modeling, and machine learning. He has been a visiting professor at various universities across the world including IITs in India, Aalto University in Finland, University of Skovde in Sweden, and Nanyang Technological University in Singapore. He has been awarded the IEEE Evolutionary Computation Pioneer Award, Infosys Prize, TWAS Prize in Engineering Sciences, CajAstur Mamdani Prize, Distinguished Alumni Award from IIT Kharagpur, Edgeworth-Pareto award, Bhatnagar Prize in Engineering Sciences, and Bessel Research award from Germany. He is fellow of IEEE, ASME, and three Indian science and engineering academies. He has published over 520 research papers with over 132 000 Google Scholar citations with an h-index of 114.

He is on the editorial board of 18 major international journals. More information about his research contribution can be found at http://www.coin-lab.org.

Iman Rahimi, BSc. (Applied Mathematics), MSc (Applied Mathematics – Operations Research) received his PhD in the Department of Mechanical and Manufacturing Engineering, Faculty of Engineering, Universiti Putra Malaysia, Malaysia in 2017. His research interests include supply chain management, data mining, and optimization.

Preface

Scheduling problems are devoted to allocating tasks to resources. When the number of tasks is increased, the scheduling and planning problems become complex, large-scale, and involve numerous constraints. To catch a real solution, most real-world problems must be formulated as discrete or mixed variable optimization problems. Moreover, finding efficient and lower-cost procedures for the common use of the structure is critically important. Although several solutions are suggested to solve the issues mentioned above, there is still an urgent need for more efficient methods. By cause of their complexity, real-world scheduling problems are challenging to solve using derivative-based and local optimization algorithms. Evolutionary Computation (EC) approaches are known as the more effective approaches to cope with this limitation. Evolution can be viewed as a method for searching through enormous numbers of possibilities in parallel, in order to find better solutions to computational problems. It is a way to find solutions that, if not necessarily optimal, are still good.

This book intends to show a variety of single- and multi-objective problems which have been solved using ECs including evolutionary algorithms and swarm intelligence. Because of clear space constraints, the set of presentations included in the book is relatively small. However, we trust that such a set is illustrative of the existing trends among both scholars and practitioners across several disciplines.

This book aims to display a representative sampling of real-world problems as well as to to offer some visions into the diverse features related to the use of ECs in real-world applications. The reader might find the material mainly useful in studying the realistic opinion of each contributor concerning how to choose a specific EC and how to validate the results which have been found using metrics and statistics.

This edited book provides an indication of several of the state-of-the-art developments in the field of evolutionary scheduling and reveals the applicability of evolutionary computational approaches to tackle real-world scheduling problems.

This edited book will emphasize the audiences of engineers in industries, research scholars, students (advanced undergraduates and graduate students), and faculty teaching and conducting research in operations research and industrial engineering from academia, who work mainly on evolutionary computations in scheduling problems (ECSP). Many scientists from operations research labs will also benefit from this book. There is a scarcity of quality books on ECSP which deal with the practicability of this technology. We hope that this book is a great source of research material for enthusiastic people who deal with ECSP.

To facilitate this goal, Chapter 1 presents scientometric analysis to analyze scientific literature in EC in Scheduling. Chapter 2 presents the implementation of Ant Colony Optimization (ACO) in the Job Shop Scheduling Problem with makespan. This is followed by Chapter 3, which describes the application of ACO algorithm in healthcare scheduling.

The focus of the next two chapters is on swarm optimization. Chapter 4 introduces the significance of neighborhood structure in discrete particle swarm optimization algorithm for meta-tasks scheduling problem in heterogeneous computing systems while Chapter 5 addresses genetic algorithm and particle swarm optimization for multi-antenna systems with various carries.

Chapter 6 presents a new variant of the truck scheduling problem in the cross-docking system. This application has been explained along with introducing a new modified version of the Red Deer Algorithm. Chapter 7 presents the management function evaluations (FEs) for intelligent distribution of FEs among sub-populations.

A comprehensive review of the recent literature on models that use evolutionary algorithms to optimize task scheduling in cloud environments is given in Chapter 8. Chapter 9 addresses scheduling of robotic disassembly using the Bees algorithm. Finally, Chapter 10 investigates the state of the art of power generation scheduling problem using a modified Fireworks algorithm.

This book is suitable for research students at all levels, and we hope it will be used as a supplemental textbook for several type of courses, including operations research, computer science, statistics, and many fields of science and engineering related to scheduling problems/ECs.

Amir H. Gandomi
Faculty of Engineering and IT, University of Technology Sydney, Australia

Ali Emrouznejad
Aston Business School, Aston University, Birmingham, UK

Mo M. Jamshidi
Department of Electrical and Computer Engr., University of Texas San Antonio, USA

Kalyanmoy Deb
Department of Electrical and Computer Engineering, Michigan State University, USA

Iman Rahimi
Young Researchers and Elite Club, Islamic Azad University, Iran

Acknowledgments

This book would not have been possible without help of a number of people. First of all, we would like to thank all the authors who contributed in this book. We would also like to extend our appreciation to all the reviewers for their critical review of the chapters and their insightful comments and suggestions provided in several rounds.

The editors also would like to thank the project team in Wiley, for supporting us to finish this project. Special thanks to Kathleen Santoloci, Mindy Okura-Marszycki, and Linda Christina E, project editors in Wiley, for their continuous support to complete this book. The editors hope the readers will find this book to be a valuable contribution to the body of knowledge in Evolutionary Computation in Scheduling.

Amir H. Gandomi
Ali Emrouznejad
Mo M. Jamshidi
Kalyanmoy Deb
Iman Rahimi

1

Evolutionary Computation in Scheduling

A Scientometric Analysis

Amir H. Gandomi[1], *Ali Emrouznejad*[2], *and Iman Rahimi*[3]

[1] *Faculty of Engineering and IT, University of Technology Sydney, Ultimo, Australia*
[2] *Aston Business School, Aston University, Birmingham, UK*
[3] *Young Researchers and Elite Club, Isfahan (Khorasgan) Branch, Islamic Azad University, Isfahan, Iran*

1.1 Introduction

Evolutionary computation (EC) is known as a powerful tool for global optimization-inspired nature. Technically, EC is also known as a family of population-based algorithms which could be addressed as metaheuristic or stochastic optimization approaches. The term "stochastic" is used because of the nature of these algorithms, such that a primary set of potential solutions (initial population) is produced and updated, iteratively. Another generation is made by eliminating the less desired solutions stochastically. Increasing the fitness function of the algorithm resulted from evolving the population. A metaheuristic term refers to the fact that these algorithms are defined as higher-level procedures or heuristics considered to discover, produce, or choose a heuristic which is an adequately good solution for an optimization problem [1, 2]. Applications of metaheuristics can be found in the literature, largely [3–9]. Swarm intelligence algorithms are also a family of EC, based on a population of simple agents which are interacting with each other in an environment. The inspiration for these algorithms often comes from nature, while these algorithms behave stochastically and the agents possess a high level of intelligence as a colony. The most common used algorithms reported in literature are: particle swarm optimization, ant colony optimization, the Bees algorithm, and the artificial fish swarm algorithm [10–15].

Scheduling and planning problems are generally complex, large-scale, challenging issues, and involve several constraints [16–19]. To find a real solution,

Evolutionary Computation in Scheduling, First Edition. Edited by Amir H. Gandomi, Ali Emrouznejad, Mo M. Jamshidi, Kalyanmoy Deb, and Iman Rahimi.

most real-world problems must be formulated as discrete or mixed-variable optimization problems [16, 20]. Moreover, finding efficient and lower-cost procedures for frequent use of the system is crucially important. Although several solutions are suggested to solve the problems mentioned above, there is still a severe need for more cost-effective methods. As a result of their complexity, real-world scheduling problems are challenging to solve using derivative-based and local optimization algorithms. A possible solution to cope with this limitation is to use global optimization algorithms, such as EC techniques [21]. Lately, EC and its branches have been used to solve large, complex real-world problems which cannot be solved using classical methods [22–24]. Another critical problem is that several aspects can be considered to optimize systems simultaneously, such as time, cost, quality, risk, and efficiency. Therefore, several objectives should usually be considered for optimizing a real-world scheduling problem.

This is while there are usually conflicts between the considered objectives, such as cost-quality, cost-efficiency, and quality-cost-time. In this case, the multi-objective optimization concept offers key advantages over the traditional mathematical algorithms. In particular, evolutionary multi-objective computations (EMC) is known as a reliable way to handle these problems in the industrial domain [22, 25–27].

With the advent of computation intelligence, there is renewed interest in solving scheduling problems using evolutionary computational techniques. The spectrum of real-world optimization problems dealt with the application of EC in industry and service organizations, such as healthcare scheduling, aircraft industry, school timetabling, manufacturing systems, and transportation scheduling in the supply chain. This chapter gives a general analysis of many of the current developments in the growing field of evolutionary scheduling using scientometrics and charts.

1.2 Analysis

1.2.1 Data Collection

For this scientific literature review, we use a scientometric mapping technique to find the most common keywords used among research articles. First, we searched for the topics "evolutionary scheduling," "metaheuristic scheduling," and "swarm intelligence scheduling" in the SCOPUS database between 2000 and the present. We identified 1107 scientific articles. Figure 1.1 presents the distribution of papers from 2000 (articles in the area of the multi-objective vs. total number of documents).

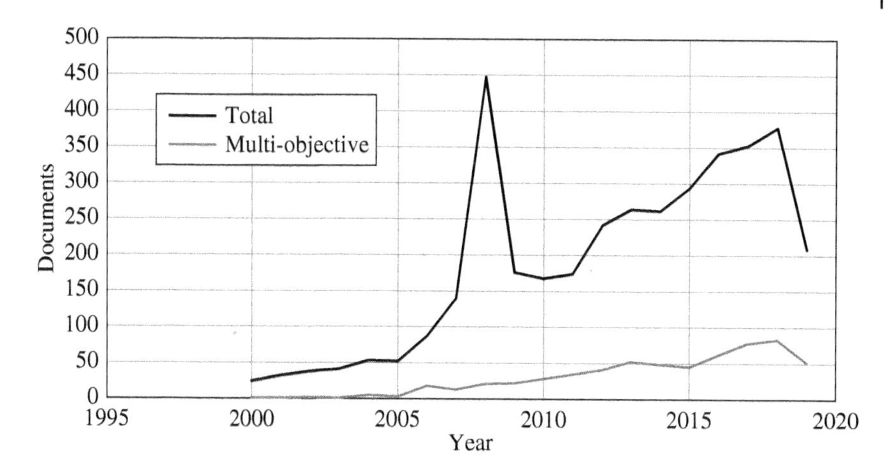

Figure 1.1 Number of documents on "Evolutionary Computation in Scheduling" (multi-objective in total).

Most of the analysis in this part has been carried out by VOSviewer, which is known as a powerful tool for scientometric analysis [28–30]).

1.3 Scientometric Analysis

1.3.1 Keywords Analysis

Figure 1.2 shows a cognitive map where the node size is comparable with a number of documents in the specified scientific discipline. Links among disciplines are presented by a line whose thickness is proportional to the extent to which two subjects are employed in one paper.

Top keywords and the number of occurrences found in the analysis are presented in Table 1.1.

1.3.2 An Analysis on Countries/Organizations

Figure 1.3 presents an organization ranking indicating the top 10 organizations which have the most contribution in the field. As is observed from Figure 1.3, the Huazhong University of Science and Technology is the most active organization in this area with 107 published documents, the Ministry of Education China and Tsinghua University are in second and third places, respectively.

Figure 1.4 illustrates the ranking of countries by number of documents. As shown, China, with almost 1100 published articles, possesses the first rank, followed by India, United States, Iran, respectively.

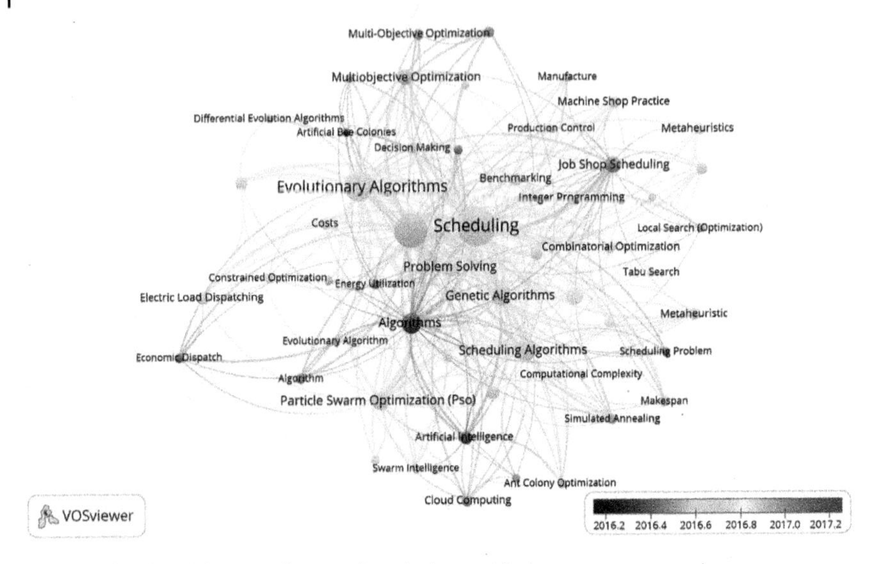

Figure 1.2 Cognitive map (keyword analysis considering co-occurrences).

Table 1.1 Top 10 keywords.

No.	Keyword	Occurrences
1	Scheduling	1185
2	Optimization	840
3	Evolutionary algorithms	698
4	Scheduling algorithms	345
5	Genetic algorithms	335
6	Algorithms	321
7	Particle swarm optimization	256
8	Problem solving	248
9	Heuristic algorithms	224
10	Job shop scheduling	208

1.3.3 Co-Author Analysis

In Figure 1.5, the analysis of co-authors and networks shows the robust and fruitful connections among collaborating scholars. The links across the networks show channels of knowledge, and networks which highlight the scientific communities engaged in research on the EC in scheduling.

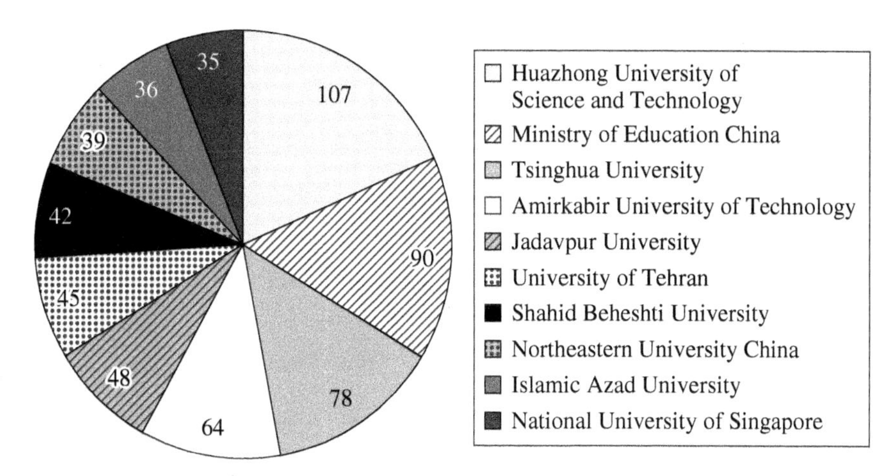

Figure 1.3 Top 10 organizations ranking by number of documents.

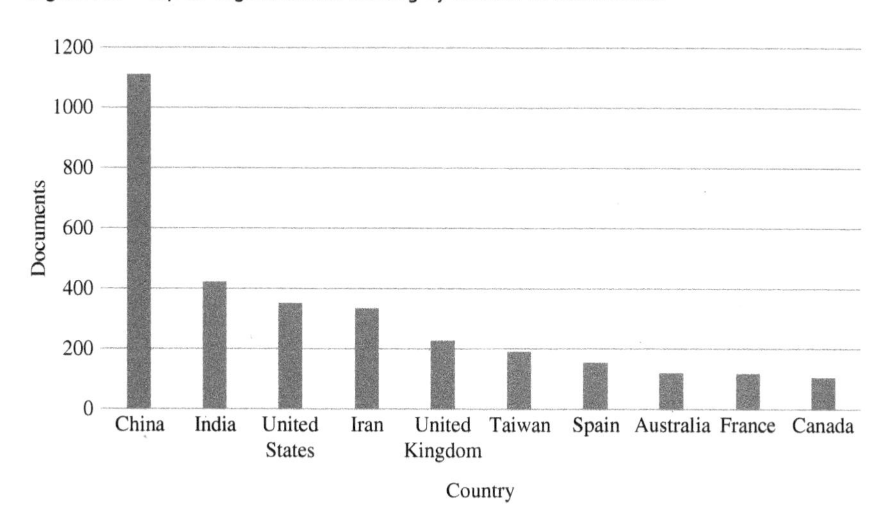

Figure 1.4 Ranking of countries by number of documents.

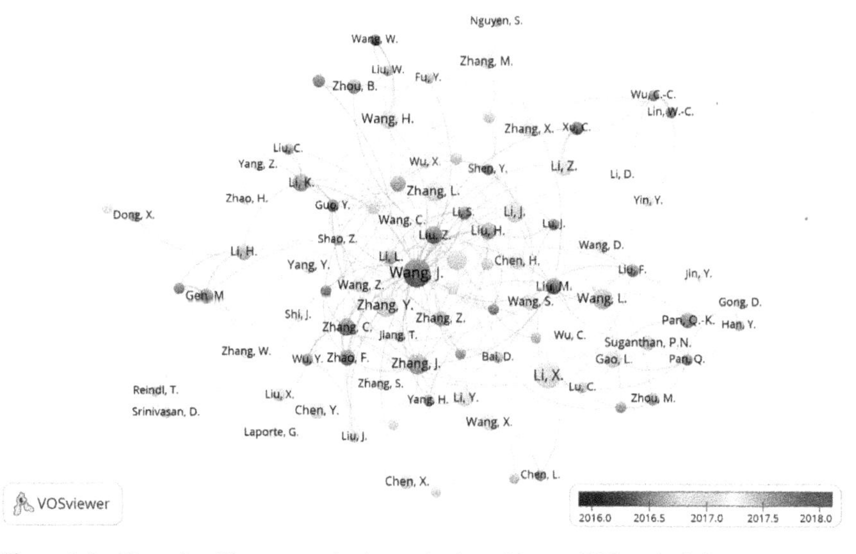

Figure 1.5 The scientific community (co-author) working on EC in scheduling.

Figure 1.6 Bibliographic coupling (title).

1.3.4 Journal Network Analysis

Figures 1.6 and 1.7 show bibliographic coupling and a density map of the active sources (journals) of EC in scheduling, respectively. Figure 1.6 shows the journals aggregated by network visualization. For Figure 1.6, a bibliographic coupling analysis for sources has been used. Considering a minimum number of one document of a source, a total of 585 sources have been found. The most frequent active journals are *Applied Soft Computing*, *Computers and Industrial Engineering*, *International Journal of Advanced Manufacturing Technology*, *European Journal of Operational Research*, and *International Journal of Production Research*. The colors/shadings in Figure 1.6 represent clusters, indicating five clusters for all the items.

In Figure 1.7, the color/shading of each node in the map is related to the density of the nodes at the point. The shading ranges from high density of journals (*Applied Soft Computing*) to low density (e.g. *Neurocomputing*).

1.3.5 Co-Citation Analysis

Figure 1.8 displays the co-citation analysis of cited authors (first author only) who have a minimum of one citation for each author, resulting in 28 203 authors with strength co-citation links. In Figure 1.8, the full strength of co-citation links to other authors has been considered. The top-cited authors in the field are Kalyanmoy Deb, Eckart Zitzler, David E. Goldberg, and Edmund Kieran Burke.

1.4 Conclusion and Direction for Future Research

In this chapter, scientometric analysis of "Evolutionary Computation in Scheduling" for a time zone between 2000 and 2019 has been investigated. Keywords analysis and citation analysis fractional counting with VOSviewer software was used,

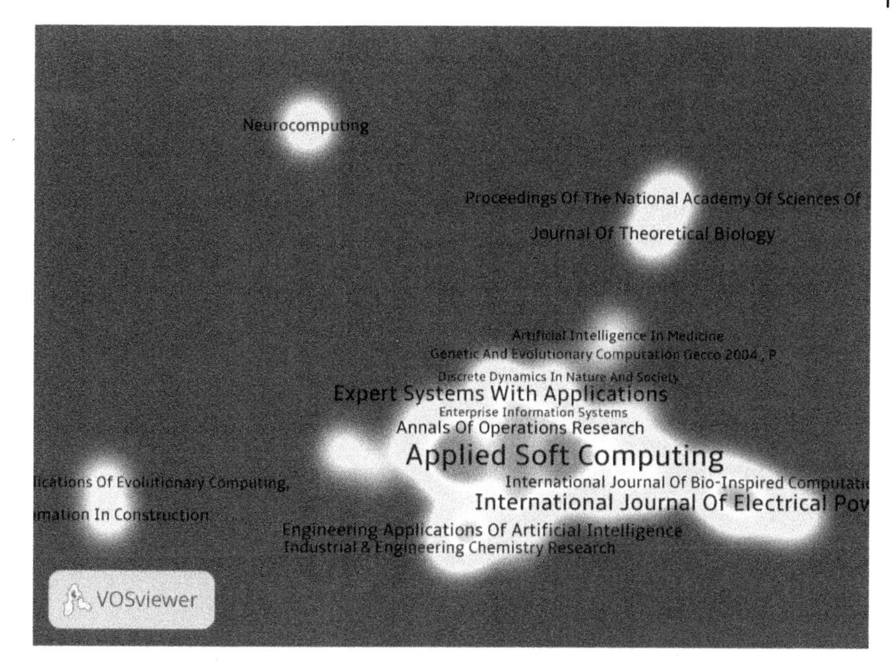

Figure 1.7 Density map.

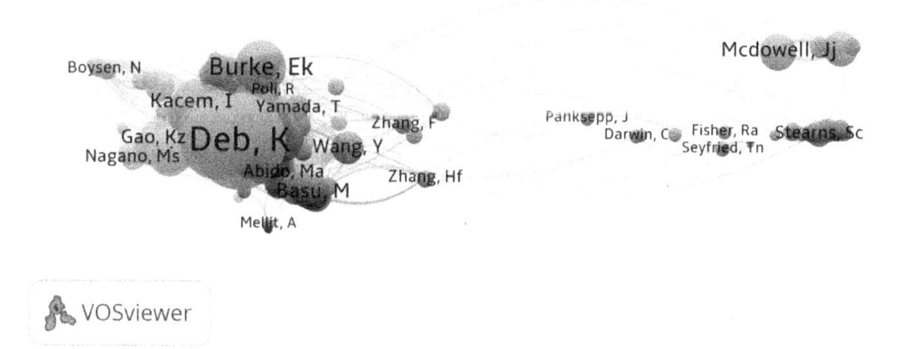

Figure 1.8 Co-citation analysis (cited authors).

and commonly used keywords were identified. The analysis includes different parts such as keyword, organization, country, bibliographic coupling, co-author, and co-citation. Keyword analysis shows that genetic algorithms and particle swarm optimization have been used frequently among other metaheuristic approaches, and job shop scheduling is the most challenging problem in the field of scheduling. China, India, United States, and Iran have the most active organizations in the research area. Moreover, co-authorship and co-citation analysis have been addressed. From a source analysis point of view, *Applied Soft Computing*,

Expert System with Application, *European Journal of Operational Research*, and *Annals of Operations Research* have been found as well-known active journals in the field.

As a future study, it is recommended to do other analyses with more detail, such as co-citation analysis on cited references and cited sources and/or bibliographic coupling analysis on authors. The analysis also shows how interdisciplinary works have been focused recently; more research in this area is recommended. There are also some other gaps which need to be focused on: robustness is one of these urgent needs, where practitioners should focus more on solving real problems. Scalability is another issue in the field; applicability of EC algorithms in the case of big data is an important matter for practical problems. Comparative works which try to compare evolutionary algorithms and other methods could be also a promising area for more research in the future, especially when tackling these problems in different sources of benchmark data.

References

1 Balamurugan, R., Natarajan, A., and Premalatha, K. (2015). Stellar-mass black hole optimization for biclustering microarray gene expression data. *J. Appl. Artif. Intell.* 29: 353–381.

2 Bianchi, L., Dorigo, M., Gambardella, L.M., and Gutjahr, W.J. (2009). A survey on metaheuristics for stochastic combinatorial optimization. *J. Nat. Comput.* 8: 239–287.

3 Blum, C. and Roli, A. (2003). Metaheuristics in combinatorial optimization: overview and conceptual comparison. *J. ACM Comput. Surveys* 35: 268–308.

4 Blum, C., Puchinger, J., Raidl, G.R., and Roli, A. (2011). Hybrid metaheuristics in combinatorial optimization: a survey. *J. ACM Comput. Surveys* 11: 4135–4151.

5 Ganesan, T., Elamvazuthi, I., and Vasant, P. (2011). Evolutionary normal-boundary intersection (ENBI) method for multi-objective optimization of green sand mould system. 2011 IEEE International Conference on Control System, Computing and Engineering, ICCSCE, Penang, Malaysia, 86–91.

6 Ganesan, T., Elamvazuthi, I., Shaari, K.Z.K., and Vasant, P. (2013). Swarm intelligence and gravitational search algorithm for multi-objective optimization of synthesis gas production. *J. Appl. Energy* 103: 368–374.

7 Tomoiagă, B., Chindriş, M., Sumper, A. et al. (2013). Pareto optimal reconfiguration of power distribution systems using a genetic algorithm based on NSGA-II. *Energies* 6: 1439–1455.

8 Yang, X.-S. (2011). Metaheuristic optimization. *Scholarpedia* 6: 11472.

9 Zarei, M., Davvari, M., Kolahan, F., and Wong, K. (2016). Simultaneous selection and scheduling with sequence-dependent setup times, lateness penalties, and

machine availability constraint: heuristic approaches. *Int. J. Ind. Eng. Comput.* 7: 147–160.

10 Banks, A., Vincent, J., and Anyakoha, C. (2007). A review of particle swarm optimization. Part I: background and development. *J. Nat. Comput.* 6: 467–484.

11 Bonabeau, E., Dorigo, M., and Theraulaz, G. (1999). *Swarm Intelligence: From Natural to Artificial Systems*. Oxford University Press.

12 Lones, M.A. (2014). Metaheuristics in nature-inspired algorithms. Proceedings of the Companion Publication of the 2014 Annual Conference on Genetic and Evolutionary Computation, 1419–1422.

13 Miller, P. (2010). *The Smart Swarm: How Understanding Flocks, Schools, and Colonies Can Make us Better at Communicating, Decision Making, and Getting Things Done*. Avery Publishing Group, Inc.

14 Neshat, M., Sepidnam, G., Sargolzaei, M., and Toosi, A.N. (2014). Artificial fish swarm algorithm: a survey of the state-of-the-art, hybridization, combinatorial and indicative applications. *J. Artif. Intell. Rev.* 42: 965–997.

15 Pham, D.T., Ghanbarzadeh, A., Koç, E. et al. (2006). The bees algorithm – a novel tool for complex optimisation problems. In: *Intelligent Production Machines and Systems* (eds. D.T. Pham, E.E. Eldukhri and A.J. Soroka), 454–459. Elsevier.

16 Jia, Z. and Ierapetritou, M. (2004). Efficient short-term scheduling of refinery operations based on a continuous time formulation. *J. Comput. Chem. Eng.* 28: 1001–1019.

17 Magalhaes, M.V. and Shah, N. (2003). Crude oil scheduling. Proceedings of the 4th Conference on Foundations of Computer-Aided Process Operations, 323–326.

18 Pinedo, M. (2012). *Scheduling*. Springer.

19 Toumi, A., Jurgens, C., Jungo, C. et al. (2010). Design and optimization of a large scale biopharmaceutical facility using process simulation and scheduling tools. *J. Pharm. Eng.* 30: 1–9.

20 Papavasileiou, V., Koulouris, A., Siletti, C., and Petrides, D. (2007). Optimize manufacturing of pharmaceutical products with process simulation and production scheduling tools. *J. Chem. Eng. Res. Des.* 85: 1086–1097.

21 Gandomi, A.H., Yang, X.-S., Talatahari, S., and Alavi, A.H. (2013). Metaheuristic algorithms in modeling and optimization. In: *Metaheuristic Applications in Structures and Infrastructures* (eds. A.H. Gandomi, X.-S. Yang, S. Talatahari, and A.H. Alavi), 1–24. Elsevier.

22 Coello, C.A.C., Lamont, G.B., and Van Veldhuizen, D.A. (2007). *Evolutionary Algorithms for Solving Multi-Objective Problems*. Springer.

23 Liu, J., Abbass, H.A., and Tan, K.C. (2019). Evolutionary computation. In: *Evolutionary Computation and Complex Networks* (eds. J. Liu, H.A. Abbass, and K.C. Tan), 3–22. Springer.

24 Simon, D. (2013). *Evolutionary Optimization Algorithms*. Wiley.

25 Deb, K. (2014). Multi-objective optimization. In: *Search Methodologies* (eds. E.K. Burke and G. Kendall), 403–449. Springer.

26 Deb, K. (2015). Multi-objective evolutionary algorithms. In: *Springer Handbook of Computational Intelligence* (eds. J. Kacprzyk and W. Pedrycz), 995–1015. Springer.

27 Ishibuchi, H., Tsukamoto, N., and Nojima, Y. (2008). Evolutionary many-objective optimization: A short review. 2008 IEEE Congress on Evolutionary Computation (IEEE World Congress on Computational Intelligence), IEEE, 2419–2426.

28 Van Eck, N. and Waltman, L. (2009). Software survey: VOSviewer, a computer program for bibliometric mapping. *Scientometrics* 84: 523–538.

29 Van Eck, N.J. and Waltman, L.J.A.P.A. (2011). Text mining and visualization using VOSviewer.

30 Van Eck, N.J. and Waltman, L.J.L.U.L. (2013) VOSviewer manual. 1.

2

Role and Impacts of Ant Colony Optimization in Job Shop Scheduling Problems

A Detailed Analysis

P. Deepalakshmi[1] *and K. Shankar*[2]

[1] *Department of Computer Science and Engineering, Kalasalingam Academy of Research and Education, Krishnankoil, Tamilnadu, India*
[2] *Department of Computer Applications, Alagappa University, Karaikudi, Tamilnadu, India*

2.1 Introduction

In manufacturing units, the word "assignment" generally indicates a methodology step related to the demand to change unrefined materials into finished merchandise [1]. The Job Shop Scheduling Problem (JSP) is familiar, involving making a calendar for the instrument by exhausting a static reason series in multi-machine surroundings [2]. At the point when there is a machine breakdown, the real game plan might be moved from a standard program [3]. The effect of starts and finishing up of all work that must be near its standard program is assigned as the outcome durability and is generally controlled as a trustworthiness metric of the arrangement [4]. JSP is an NP-hard, difficult, combinatorial optimization problem in which finding the problem's optimal solution increments with the intricacy [5]. The present reality is that dynamic JSPs are challenging. Simplified dispatching rules which depend on experience and basic rationale have been broadly received in the industry [6]. The traditional JSP is likewise a static scheduling scenario in which all activities having a place with the equivalent task should be prepared in a singular request and there should be considered to be no gap between one task completing and its successor beginning [7, 8]. The dispatch rules are not commonly ideal timetables. The manufacturing condition is characterized by the accompanying features, which rapidly result in a significantly more mind-boggling scheduling issue. To settle this issue in scheduling problems, distinctive metaheuristics optimization – that is, genetic algorithms (GA), particle swarm optimization (PSO),

Evolutionary Computation in Scheduling, First Edition. Edited by Amir H. Gandomi, Ali Emrouznejad, Mo M. Jamshidi, Kalyanmoy Deb, and Iman Rahimi.

and ant colony optimization (ACO) – are considered to solve the JSP [9]. There are four variables to depict the JSP: Arrival Patterns, Number of Work Stations (machines), Work Sequence, and Performance on Evaluation Criterion. There are two sorts of arrival examples: static and dynamic. In the event that the JSP is less unpredictable, a simple graphical display technique (Gantt diagram) is progressively appropriate for introducing the problem with no tenets and showing and assessing paradigm results [10].

2.1.1 Process of JSP

JSP is a working area where n employments need to be set up on m sets of machines with various endeavors [11, 12]. In most amassing conditions, especially in those with a wide combination of things, strategies, and age levels, the advancement of improvement designs is seen as fundamental to achieving this goal. The detailed process of the JSP model appears in Figure 2.1. It incorporates some essential parts:

Job: A portion of work that goes throughout a series of operations.
Shop: A place for manufacturing or renovation of goods or machinery.
Scheduling: Decision procedure aiming to construe the order of processing.

Each processing time of every movement on each machine must be known early, whatever the progression of jobs to be readied [7]. The primary machine is believed to be arranged continually; any action is to be set up on it first. Machines may be latent or, exceptionally, still. A branch and bound algorithm was industrialized in [13] to solve JSP with availability constraints in the case of a single machine. Each machine is accessible for maintenance once in a while, without great distribution of the scale into developments or days and without considering brief detachment, for instance breakdown or support [14].

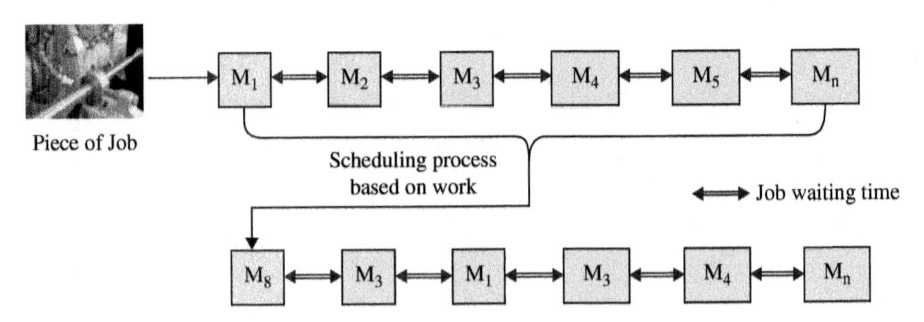

Figure 2.1 Process of Job Shop Scheduling Model.

2.1.2 Problem Description of JSP

Algorithms like GA and PSO in JSP treat great optimal solution. However, they accomplish a major issue as the absence of optimal solution and computational expense. The activities of each job must be handled in the given grouping. Each machine $M \Rightarrow \{i = 1, 2, \dots n\}$ can process at most one activity at once, and at most one task of each job $Job \Rightarrow \{j = 1, 2, \dots n\}$ can be processed at once. A schedule is an allotment of a solitary time period for every operation. Each job has four tasks to be processed and every activity can be processed by any two of the three machines; the JSP can be seen as a collection of 220 JSPs. While JSP comprises just a sequencing problem in light of the fact that the task of activities to the machines is given ahead of time, the Flexible JSP (FJSP) transforms into a routing just as much as a sequencing problem: allotting every task R_{ij} to a machine chosen from the set M_{ij} and requesting activities on the machines so that makespan is minimized.

2.1.3 Assumption of JSP

To solve the JSP objective function, the following assumptions are considered, based on the job completion and delay time.

- All jobs must be prepared in a sole machine.
- Each job must be processed in the due time.
- Each job can be prepared with the immediate time to such an extent that it finishes with direct time.
- Completion time.
- Every single task should be appointed to a machine to systematize the capacities on the machines.
- This JSP does not have parallel processing.
- The request for processing is not the equivalent and the operations cannot be intruded upon.
- Setup times of machines and move time between activities are unimportant.
- After a job is handled on a machine, it is transported to the next machine quickly and the transportation time is insignificant or incorporated into the activity time.
- Each activity should be handled as a continuous time of a given length on a given machine. The purpose is to discover a calendar; that is, an allotment of the tasks to time interims to machines that have negligible length.

2.1.4 Constraints of JSP

It suggests the processing times of jobs are extending the limit of their starting time. In most of the examination related to planning falling apart jobs, a straightforward direct decay work is acknowledged. Each job J_i contains a lot of functions $L_i = \{L^i, \ldots L^n\}$ which must be performed between a ready time (r_{ti}) and a due time (d_{ti}). The implementation of each function (L_{ik}) requires the utilization of a lot of resources ($P_{ik} \subseteq P$) during a time interval (du_k^i) [15]. The start time St_{ik} of function L_{ik} demonstrates when the function may begin to utilize the assets L_{ik}. The constraints of job-shop setting up problems can be depicted as binary, disjunctive, metric, and point-based. JSP with machine accessibility constraint is an exceptional issue. The considerable optimization algorithm to calculate the job complete time with machine accessibility is imperative to JSP. The optimization to compute the complete time of employment by considering the machine accessibility requirement means the working time calculation. This constraint $K_i(q_j, q_k) = \{[a_1^- a_1^+][a_2^- a_2^+]...[a_n^- a_n^+]\}$ disjunctively confines the sequential separation between q_j and q_k. Two unique functions L^i_k and L^j_l cannot be connected unless they utilize distinctive assets. Capacity constraints give rise to disjunctive linear inequalities.

2.1.5 Objective Functions for JSP

JSP in the manufacturing industry has a distinctive objective function, in the light of which a job gets a short time and least energy in a machine. Many objective functions are considered for the exploration work. Some significant functions are investigated below:

i) *Makespan time:* The least makespan, as a rule, derives a decent usage of the machine(s). Most of the writing on the job shop issue considers makespan as the scheduling criterion.

$$MT_{\max} = \min(MT_{ij}) \tag{2.1}$$

ii) *Total weighted tardiness:* The distinction between a delay and a deferment lies in the fact that the delay is never negative. This objective is in like manner a broader cost work than the total weighted completion time.

$$TWT_j = \max\{MT_{ij} - A_j, 0\} \tag{2.2}$$

iii) *Total workload:* Considering machines (aggregate of the working occasions over all machines), the aim is minimization of the maximal machine workload (total of the handling times of activities). This implies the total time on every one of the machines working in the production.

$$TWL = \sum_{i=1}^{n} \text{work}_i \tag{2.3}$$

From the above equations, the documentation is shown by $i \in 1,2,\ldots n$, is completion time of jobs if $O_{ij}(\text{work}_i)$ is a summation of the handling time of tasks that are processed on the machine. To give a brisk reaction to the market prerequisites, a smaller makespan is desired, which makes the generation quicker. The total workload implies the total time on every one of the machines working in the process of production. Finally, the most extreme workload is the machine with the most noteworthy working time.

2.1.6 Types of Job Shop Scheduling

JSP has three main types of scheduling models: Flexible Job Shop Scheduling (FJS), Flow Shop Scheduling (FSS), and Hybrid Job Shop Scheduling (HJSP).

FJS: The jobs are affirmed with the goal that at least one execution technique might be advanced. It contains a single machine to process n jobs. The goal of the setting-up procedure is to plan these n jobs on a single machine with the end goal that a given measure of performance is limited [14, 16].

FSS: Distinctive machines given with the objective that most of the items being processed through this machine go in a comparative request. The machines are normally set up a course of action [17], which includes a flow shop, and the scheduling of the jobs in this condition is typically alluded to as FSS [18].

HJSP: HJSP is normally utilized for industrialized surroundings in which a lot of n jobs are to be handled in a progression of stages, every one of which has a few machines working in a practically identical way [18]. A few phases may have just a single machine, yet somewhere around one phase must have various machines.

2.1.7 Advantages and Application of JSP

This section describes the merits of JSP and discusses some of the real-time applications.

- By using all machines completely and successfully, fewer machines are required to fabricate a wide assortment of products. Along these lines, job shop manufacturing needs lower speculation as a result of the relatively smaller number of machines.
- Improved adaptability in progress planning and augmentation flexibility.
- Low obsolete quality and decreased time usage of formation of machines and versatile nature of machines.
- High power to machine disappointments.

Applications

- Textile industry, bioprocess industry, vehicle manufacturing, parallel figuring, distributed frameworks, ongoing machine vision frameworks, transportation problems, compartment designation of holder terminal, and workforce task, genuine enterprises.
- Semiconductor manufacturing, food handling, metalworking along with ship scheduling at route locks.
- Exhausting machines.
- Operations of CNC cutting mechanical assembly machines.
- Control of true proportions of execution of machines.

2.2 Ant Colony Optimization

The ACO algorithm created by Colorni et al. got its motivation from genuine ant colony behavior in finding the shortest path from the home to a food source [19]. The ACO metaheuristic created by Dorigo et al. presents the principal features of artificial ants and these highlights have enlivened diverse ant algorithms to tackle hard optimization problems. ACO utilizes artificial ants so as to probabilistically develop an answer by iteratively adding solution segments to fractional solutions. Tasks need to recognize an appropriate machine to process them [20]. Just as ants search for the shortest path to achieve a food source, activities need to scan for the briefest way to achieve machines. The ants' home and the food source are comparable to the activity beginning and end of JSP.

2.2.1 Inspiration of ACO

Swarm knowledge, as discussed in [21], is the order that bargains with regular and counterfeit frameworks made out of numerous people that coordinate utilizing decentralized control and self-association. Specifically, the control centers on the aggregate practices that emerge from the local communications of the people with one another and with their condition. Instances of systems considered to display swarm insight are provinces of ants and termites, schools of fish, groups of birds, and crowds of land creatures.

Ant colonies, and social insect societies more generally, are conveyed frameworks that, disregarding their individual straightforwardness, present an exceedingly organized social association. Because of this association, the ant colony can achieve complex undertakings. Diverse parts of the conduct of ant colonies such as scrounging, division of labor, brood arranging and helpful

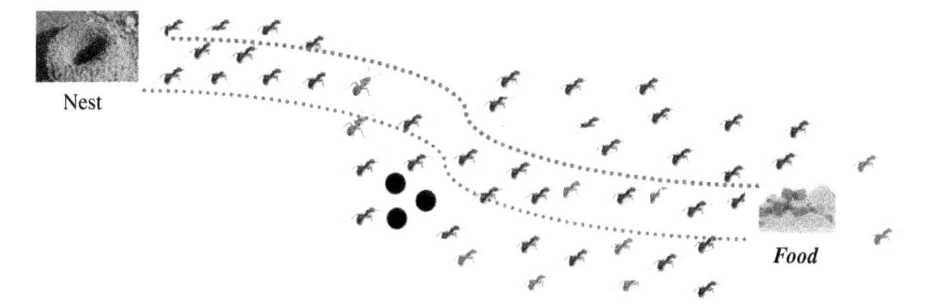

Figure 2.2 Inspiration of ACO.

transport have motivated various types of subterranean ant algorithms. At the position, ants need to choose whether to turn right or left. Under this circumstance, they may turn right or left. The main ant turning right will achieve the food first and the position appears in Figure 2.2.

2.2.2 ACO Metaheuristic

Artificial ants as utilized in ACO are a stochastic solution development methodology that probabilistically fabricates an answer by iteratively including arrangement segments. A stochastic component in ACO enables the ants to assemble a wide range of arrangements and thus investigate a much larger number of arrangements than greedy heuristics.

- The way the ants in ACO develop solutions for the problem is explained by moving simultaneously and non-concurrently on a properly characterized development diagram.
- It characterizes the way the solution development, the pheromone refresh, and conceivable activities are communicated in the solution procedure.
- An ACO metaheuristic is enlivened by the searching behavior of ant colonies and targets discrete optimization problems. It tends to be utilized to tackle static and dynamic optimization problems. For static problems, the qualities are given when the problem is characterized and does not change while the problem is being unraveled.
- The traveling salesman problem is a static problem in which the urban areas and relative separations are static.
- In the case of dynamic problems such as system directing, powerful scheduling problem occasion characterized by the capacity of a few amounts changes at run time, the optimization must be equipped for adjusting on the web to the evolving condition.

2.2.3 Characteristics of Real Ants

Here we will discuss the real ant characteristics in ACO, an algorithm for solving the JSP problems, namely:

- Each operator is an independent development process using a stochastic policy and aiming to build a solitary answer for the current problem, conceivably in a computationally light way.
- The thought behind this sub-division of specialists is to enable the retrogressive ants to use the helpful data accumulated by the forward ants on their trek from source to goal. In light of this rule, no node routing refreshes are performed by the forward ants [21].
- The foraging behavior of the ant colony can be mimicked to make fake ants to adopt this idea to tackle some realistic designing optimization problems.

2.2.4 Applications

ACO has been connected to numerous combinatorial enhancement issues. The general algorithm is moderately straightforward and dependent on a lot of ants, each creating one of the conceivable round-trips within the urban cities.

- In recent years, the enthusiasm of established researchers in ACO has risen dramatically. A few fruitful utilizations of ACO for a wide range of discrete optimization problems are currently accessible.
- By far the majority of these applications are to NP-difficult problems; that is, to problems for which the best-known algorithms that can definitely reach an ideal solution require time that increases exponentially for more pessimistic scenarios of a complex nature.
- The utilization of such calculations is frequently infeasible, and ACO calculations can be valuable for rapidly discovering outstanding solutions.
- ACO is broadly utilized in applications such as successive requesting problems, job scheduling problems, vehicle directing problems, quadratic task problems, traveling salesman problems, scheduling problems, and system routing problems to discover an ideal solution.
- Sharing the constructive phase with local search to acquire a novel ACO calculation that utilizes a heterogeneous colony of ants is very successful in finding the best-known values in all cases of a generally utilized set of benchmark problems.

2.2.5 Suitability of ACO for JSP

The fitness evaluation or objective solving by ACO described as follows makes ACO suitable for JSP.

- Problem portrayal which enables ants to steadily manufacture/alter solutions.
- A constraint satisfaction strategy which powers the development of attainable solutions and the pheromone refreshing standard which determines how to adjust pheromone trail τ on the edges of the diagram.
- A probabilistic progress principle regarding heuristic popularity and the pheromone trail.
- Positive feedback represents fast disclosure of good arrangements.
- The specialists' experience impacts the fabrication of the solutions in the next cycles: utilizing a whole colony of agents offers power to the solution, and the aggregate connection of the agents makes it more effective.

At the same time, we do have the following restrictions on ACO to be used for JSP also:

- The ACO solution for JSP reaches the solution in an acceptable time. Be that as it may, when the number of jobs and machines are expanded to the stochastic qualities and the measure turns out to be multiple, an ACO solution can be futile.
- The extensive scale problem of jobs and machines in JSSP means that multi-threading systems can be actualized by methods for agent innovation.
- A probability distribution can change for every emphasis and it has a troublesome hypothetical investigation and also dependent groupings of irregular choices.

2.2.6 Implementation Procedure of ACO

The ant state computation is a method for discovering perfect ways that rely on the example of ants chasing down a food source. The major typical for this estimation is that the pheromone worth is updated at all cycles itself by all of the ants included. Applying ant estimation to deal with a couple of issues, the pheromone is used as a medium to trade messages, so the assortment of pheromone has a basic effect to deal with issues. This ACO strategy, with two important procedures which update the pheromone as well as the probability computation, and the flow graph of ACO appear in Figure 2.3.

2.2.7 Implementation Procedure of ACO

2.2.7.1 Initialization

Table 2.1 demonstrates the processing time for 7×7 machines and the processing time in each job comparing machines. In this procedure, starting with one generation then onto the next generation, the crossover and mutation process is rehashed until the most extreme number of generations is fulfilled.

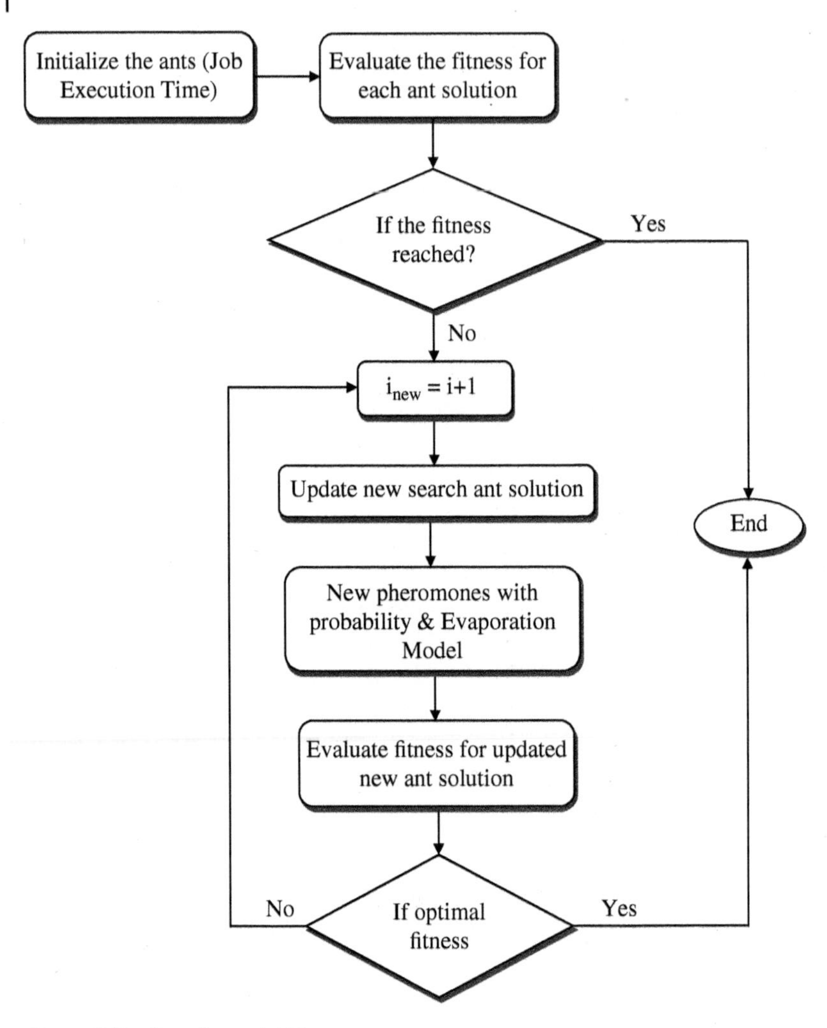

Figure 2.3 Flowchart of ACO.

2.2.7.2 Probability Transition Matrix

At every single configuration stage, ant *c* applies a probabilistic activity decision rule, known as an irregular corresponding principle, to figure out which area must be visited next. In particular, the probability with which ant (finish time) *c*, now in area *i*, wants to go to area *j* is:

$$Pro_{ij}{}^{c} = \frac{(\beta_{ij})^{\alpha}\,(\delta_{ij})^{\beta}}{\Sigma(\beta_{ij})^{\alpha}\,(\delta_{ij})^{\beta}} \tag{2.4}$$

Table 2.1 Job completion time allocation of 7×7 JSP.

	Jobs						
Machines	**M1**	**M2**	**M3**	**M4**	**M5**	**M6**	**M7**
J1	P1	P2	P4	P3	P6	P5	P7
J2	P2	P4	P3	P7	P6	P5	P1
J3	P7	P5	P2	P4	P1	P2	P6
J4	P6	P4	P1	P5	P2	P7	P3
J5	P2	P3	P7	P6	P4	P5	P6
J6	P3	P1	P4	P5	P2	P6	P7
J7	P1	P3	P4	P6	P7	P2	P5

According to this probability direction, the probability of picking a particular circular segment (i, j) upgrades the estimation of the connected pheromone trail β_{ij} and of the heuristic information $\eta_{ij,}$ and the parameters α and β categorize the similar importance of the pheromone versus the heuristic information, which is denoted by δ_{ij}, the coterminous urban communities further liable to be picked; while if $\delta_{ij} = 0$, pheromone enhancement is working correctly.

2.2.7.3 New Pheromone Updating Model

While searching for food, pheromones are conveyed by ants and they store them on trails. It is a method used to change the proportion of pheromone based on where the ants went in the midst of the improvement of the game plan. To update the numerical values, the sigmoid limit is induced as a limit which is melded with the fake neural framework.

2.2.7.4 Evaporation Model

The pheromone esteems are refreshed when all the s ants have framed solution after their emphasis. The measure of pheromone is equivalent to $\Delta\beta_{ij} = 0$. Over time, the pheromone trail is changed due to new pheromone being connected and old pheromone vanishing. ρ is set as the evaporation coefficient of the pheromone. The pheromone force connected with the edge joining areas i and j is depicted by

$$\beta_{ij} = (1-\rho) * \beta_{ij} + \sum_{C=1}^{S} \Delta\delta_{ij}^{\ C} \tag{2.5}$$

The locations of ACO are recorded in the tabu list, followed by the memory of ant n is assumed and the term ρ indicates pheromone evaporation rate. Finally, δ_{ij} is the quantity of pheromone laid on the edge by the nth ant.

$$\Delta\delta_{ij} = \begin{cases} Q/L_n & \text{the path from picking } i \text{ to picking } j \text{ by the ant } k \\ 0 & \text{otherwise} \end{cases} \tag{2.6}$$

The speed of this decay is symbolized by ρ, which is the evaporation limitation. The correct section upgrades the pheromone on every one of the edges visited by ants. The measure of pheromone an ant stores on an edge is demonstrated by L_n, which indicates the length of the visit created by the related ant. At long last, the evaporation is gauge until the minimization.

2.2.7.5 Termination Model

Until reaching the optimal solution for the JSP process, the model runs to get least makespan time, most extreme tardiness, and average workload. If ants examine different routes, at that point there is a higher probability that one of them will find an improved course of action. From the ant settlement motivated strategy, a perfect course of action will be found by using differing accentuations.

2.3 Review of Recent Articles

This section in Table 2.2 reviews many JSP articles with multi-objective functions and solved using optimization models. Here, we will prefer ACO techniques for solving the manufacturing problem.

2.3.1 Articles Related to ACO to Solve JSP Objective Function

In 2017, Elmi and Topaloglu [30] recommended the ACO-based algorithm while deciding the ideal height of jobs in the cyclic timetable, the robot assignments for transportation activities, and the ideal sequencing of the robot's moves, which in return maximizes the throughput rate. The proficiency of the proposed model is analyzed by a computational report on many arbitrarily created problem cases.

In 2018, Engin and Güçlü [17] proposed the hybrid ant colony algorithm (ACA), which is connected to the 192 benchmark occurrences from the literature so as to limit makespan. The execution of the proposed algorithm is contrasted with the Adaptive Learning Approach and the Genetic Heuristic algorithm which were utilized in past investigations to illuminate a similar arrangement of benchmark problems.

A dynamic schedule technique is connected to Dynamic JSP (DJSP) by Meilin et al. [31]. Manufacturing Execution System is an equipment stage sent in the shop floor with the point of constant and remote manufacturing. This approach has been put into genuine practice in a few manufacturing undertakings, as indicated by its

Table 2.2 Review of recent articles in JSP manufacturing.

Author/Ref	Technique	Objective functions	Description
Reddy and Padmanabhan [22]	Teaching Learning-based Optimization (TLBO)	Makespan Time	Teacher stage involves learning something from a teacher and Learner phase involves learning without input from another person. TLBO execution can be studied by solving with 10 Taillard benchmark problems.
Ling Wang and Ye Xu [23]	Hybrid Biogeography-Based Optimization (HBBO)	Makespan Time	In the relocation stage, the way relinking heuristic is used as an item nearby search procedure to upgrade the gathering.
Akram et al. [24].	Fast Simulated Annealing (FSA)	Makespan Time	This algorithm was tested by attempting to solve 88 benchmark problems taken from previously published research. The proposed algorithm solved 45 problem within a reasonable time and to the best known values. The algorithm also measured against 18 other published works.
Muthiah and Rajkumar [25]	Artificial Bee Colony (ABC)	Makespan Time	ABC algorithm was skilled to achieve staggering outcomes. Similarly, they have explored and separated the run time of the two systems according to the general taking care of the CPU for finishing the total number of accentuation.
Kumar et al. [26]	Krill Herd Optimization (KHO)	Makespan Time	A couple of earlier standards are given to build the underlying population with an unusual value state. This examination used four unique optimization systems to find the best outcomes using the KHO process.

(Continued)

Table 2.2 (Continued)

Author/Ref	Technique	Objective functions	Description
Akram et al. [24]	Fast Simulated Annealing (FSA)	Makespan Time	FSA for worldwide search and stifling, for bounded solutions in the territory of the current solution, while a tabu list is used to prevent the search from coming back to previously examined solutions.
Salido et al. [27]	Genetic algorithm	Makespan Time	Shows an augmentation of the traditional JSP, where every task must be executed by one machine and this machine can work at various rates.
Yazdani et al. [28]	Imperialist Competitive Algorithm and Neighborhood Search	maximum earliness and maximum tardiness	Motivated by various advances in solving this famously troublesome problem, another surmised optimization approach is built up, which depends on the radical aggressive algorithm hybridized with an effective neighborhood search.
Mokhtari and Hasani [29]	Enhanced Evolutionary Algorithm	minimizing total completion time, total energy cost and maximizing the total availability of the system	To adapt to this multi-target optimization problem, an improved evolutionary algorithm joined with the global criterion is proposed as a method of dealing with a multi-target system, and execution assessment is performed dependent on a broad numerical analysis.

all-inclusiveness. It has proved astoundingly effective in relation to constant scheduling and arranging JIT (Just-In-Time) manufacturing.

In numerous applications to hard optimization problems, ACO performs best when coordinated with neighborhood seek schedules since they move the ants' answers to a nearby space [18]. If the ant can do that, it will almost certainly develop increasingly exact nearby ideal states in the forthcoming visits by rehashing the entire procedure again and again and hence improving its misuse capacities.

Computational outcomes confirm the enhancements accomplished by the proposed method and demonstrate the prevalence of the proposed algorithm; more than seven of those looked at are effective as regards arrangement quality.

A scientific model which is made out of JSP and Carbon Fiber Reinforced Polymer (CFRP) at the same time is a two-phase ACA presented by Saidi-Mehrabad et al. [32]. The problem under examination is NP-hard. The outcomes demonstrate that ACA is an effective metaheuristic for this problem, particularly for the substantial measured problems. Moreover, the ideal number of both AGVs, as well as rail-routes in the generation condition, is dictated by the financial investigation.

The traditional ACO algorithm along with a load balancing was introduced for JSP in 2013 by Chaukwale and Kamath [33]. They presented the attained outcomes and examined them with reference to traditional ACO. The proposed algorithm gave better outcomes when contrasted with traditional ACO. To enhance the ACA, inadequacy of poor union and simple to fall in neighborhood optima, a genetic ACA with a master–slave structure was planned by Wu et al. [34]. In this algorithm, ACO is viewed as a master and a GA as a slave. Finally, computational tests dependent on the notable benchmark suites in the paper are directed, and then the computational outcomes demonstrate that the displayed algorithm is powerful in finding optimal as well as near-optimal solutions.

The objectives that are utilized to quantify the nature of the produced timetables are weighted-total of makespan, the tardiness of jobs, and splitting the cost. The created algorithms are investigated broadly on genuine information acquired from a printing organization and recreated information by Huang and Yu [35]. A scientific programming model is produced and combined examples *t*-tests are performed between attained solutions of the numerical programming model and proposed algorithms so as to check the viability of proposed algorithms.

In 2014, Turguner and Sahingoz [36] expressed the advantages of utilizing ACO for JSP. Since ACO is a memoryless algorithm, it tends to be valuable for frameworks that have memory limitations; the iterative idea of ACO can prompt a worldwide optimum solution. The creators considered N×M JSP in which the number of ants will be the same as a number of jobs, N. However, to all intents and purposes, it should be constrained. The creators likewise expressed that CPU utilization rate and getting stuck in neighborhood optima in some cases have become serious problems. To deal with these problems, static max and min estimation of pheromones is utilized. In the event that the problem occasion of JSP scales up as far as jobs and machines, the writers propose using a multi-threading method with multi-agent innovation to decrease the number of cycles and the time for handling.

The quantity of ant clusters is set haphazardly. At each emphasis, each ant beginning at a source vertex chooses one of the unvisited vertices as indicated by the standards of progress that consolidate pheromone level data and heuristic separation between activities [37]. To enhance the time attributes of the ACA and to evade the neighborhood ideal, Chernigovskiy et al. presented elitist ants as 1% of the ants overall. Elitist ants improve the circular segments of best courses found so far. Their test results showed that Elite-ACO and ACOA+ lessens the general generation life cycle by 5%, notwithstanding the decrease in count time contrasted with traditional ACO and considerably better execution contrasted with branch and bound procedure.

The heuristic value is any of the normal turnaround time, normal response, and waiting time. Clearly, an arrangement with a higher likelihood of higher esteem will get more space in the roulette wheel, leading to a higher decision of determination [38]. The creators have tested their algorithm with 10 jobs with 5 non-primitive procedures, each requiring a static time and 30 ants, and two emphases as a constant setup to discover normal response and waiting time. The number of ants is decreased to 10 to gauge normal turnaround time in just a single cycle. One can also plan to enhance other CPU-based parameters like CPU usage and CPU throughput as well as to to consider crude procedures with various entry times that typically happen continuously.

In 2013, Tawfeeket al. [39] used ACO for ideal asset assignment for undertakings in a unique cloud framework with the aim of limiting the makespan. A colossal number of cloud clients – in the millions – offer cloud assets for finishing their assignment. To make the best use of these assets, a productive task scheduling component is in reality required to apportion approaching requests/jobs to virtual machines (VMs). Ants begin their search from a VM at each emphasis, and fabricate answers for the cloud assignment scheduling problem by moving to start with one VM then onto the next for the next task until they get an assignment for all undertakings.

In 2017, Imen Chaouch et al. [40] aimed to limit the worldwide makespan over every one of the plants. This paper was an initial step to manage the DJSP utilizing three variants of a bio-motivated algorithm: the Ant System (AS), the Ant Colony System (ACS), and a Modified Ant Colony Optimization (MACO), with the intent of investigating more hunt space and accordingly guaranteeing better solutions for the problem.

The ACO algorithm has been produced to unravel the proposed numerical model of coordinated procedure arranging and scheduling. The optimization algorithm is separated into two phases. The scheduling plan optimization algorithm and dynamic crisis circumstance were dealt with by Xiaojun Liu et al. [41]. The scheduling plan optimization algorithm is utilized to get a plausible and upgrade scheduling plan, and the dynamic crisis circumstance taking care of component is utilized to deal with dynamic crisis circumstance; for example, embeddings new parts. The proposed strategy establishes a system of incorporation of simulation and heuristic optimization. Simulation was used to assess the local fitness function for ants by Korytkowski et al. [42]. The technique, dependent on ACO, was used to decide the imperfect distribution of dynamic multi-attribute dispatching rules to boost job shop framework execution (four measures were examined: mean flow time, max flow time, mean tardiness, and max tardiness).

In 2017, Lei Wang et al. [43] proposed enhancing the makespan for FJSP. The accompanying aspects are done on their enhanced ACO algorithm: select machine rule problems, introduce uniform disseminated component for ants, change pheromone's directing instrument, select node strategy, and refresh the pheromone's system. A real generation instance and two sets of common benchmark occurrences were tried, and correlations with some different methodologies confirmed the adequacy of the proposed improved ACO (IACO). The outcomes demonstrate that that the proposed IACO can give a better arrangement in a sensible computational time.

The computer simulations on a lot of benchmark problems have led to evaluating the value of the proposed algorithm in contrast with different heuristics in the existing literature. The solutions found by Habibeh Nazif [44] were of good quality and exhibited the adequacy of the proposed algorithm. A fairly good solution was created in negligible computation time and, after that, the trail powers were started dependent on this solution.

We can summarize the observations and findings of this literature survey as follows. The majority of research in scheduling job shops has kept to fixed-course single-arrange job shop problems. From the literature investigation, the seminal research is the use of new tools, for example ACO algorithms and particle swarm algorithms for JSP with great coding structures [15]. The objective here is to limit the makespan, which is polynomially feasible for two multi-processor groups, regardless of whether seizures are confined to indispensable time [24, 26]. Researchers have developed an uncommon whole number program in two measurements and demonstrated that in the region of the ideal arrangement of its unwinding, there are constantly necessary cross-section focuses which structure an ideal arrangement of the number program. In one hybrid JSP each machine had indistinguishable parallel machines with an objective of limiting the makespan time [17].

2.4 Results Analysis

This section analyzes the JSP with different optimization and the ACO model with objective function results with implementation platforms. Generally, a JSP problem can be implemented in Matlab software. Here, the datasets have been taken from http://bach.istc.kobe-u.ac.jp/csp2sat/jss.

Table 2.3 examines the makespan time of various cases and distinctive research papers; the optimizations are Krill Herd Optimization (KHO) [26], Fast Simulated Annealing (FSA) [24], ACO [17], MACO [35], ACS [40], MACO [40], IACO [43], and ACA [44], with Best Known Solutions (BKS). Benchmark problems were run on various occasions by the techniques using similar parameters, which included worst case, best, and normal makespan. Based on the optimization behavior as well as refreshing procedure, the objective function is tackled and the span of the examples is expanded with a settled time-out. It simulates the behavior of the two algorithms on genuine problems when the time is confined.

Figure 2.4 and Table 2.4 explain the job consummation time and Gannt chart of case 5 and case 8 respectively. Every machine processed its jobs for a specific time and made note of the minimum completion time. In the benchmark problem FT10, the processing time for jobs in machine 1 and 4 are lower than for other machines. The makespan time for ACO is almost equivalent to the BKS time. Additionally, for all the benchmark problem measures, the hybrid methods give the least makespan time contrasted with independent algorithms. The varieties are for the most part reliant on the progressions of an emphasis value and the makespan period.

Figure 2.5a and b demonstrates that with the maximum tardiness of various cases with optimization problems, the optimal solution can be attained. It decreases the working time of the considerable number of jobs processing in every one of the machines. Job waiting time for Matlab simulation process is low compared to the string evaluation process. Matlab gets the best outcome compared to the others. On average, the waiting time difference is 10–12.56% in the existing technique. In job 5, the Matlab procedure gets the most extreme best tardiness from the MACO and ACO method. In light of the cycle, only the objective functions are illuminated; the varieties mostly rely upon changes of an iteration value and the makespan time period. Figure 2.6 shows the computational complexity of all cases with different optimization techniques, for example in case 10, the CC of MACO [35] is 12 485, its minimum value compared to other techniques, and other cases are similarly represented.

2.5 Conclusion

The ACO algorithms presented here functioned admirably and found an allotment of dispatching rules that gave preferable outcomes for all criteria for only one standard in a whole system. The attained outcomes support the adequacy of using the proposed algorithms for settling JSP. Distinctive arrangements can be made for this problem – for instance, given a makespan threshold, an answer that less energy consumption utilization can be achieved and, vice versa, given an energy consumption threshold, a solution that minimizes makespan can be

Table 2.3 Benchmark problems with objective function.

Cases	Size	Makespan Time								
		BKS	KHO [26]	FSA [24]	ACO [17]	MACO [35]	ACS [40]	MACO [40]	IACO [43]	ACA [44]
Case 1	10×5	666	665	666	666	659	659	688	666	667
Case 2	10×5	677	672	655	678	578	679	678	623	688
Case 3	10×5	597	592	597	560	586	590	593	598	560
Case 4	10×5	590	585	590	585	586	588	589	585	588
Case 5	10×10	593	588	593	575	605	605	590	583	596
Case 6	10×10	945	940	926	942	945	948	942	942	595
Case 7	10×10	784	779	890	786	788	786	783	783	789
Case 8	15×10	842	837	820	845	846	849	840	845	843
Case 9	10×10	902	909	863	905	918	916	910	906	930
Case 10	10×10	1263	1221	1260	1269	1279	1270	1259	1369	1270
Case 11	10×10	951	928	956	953	960	956	950	950	956

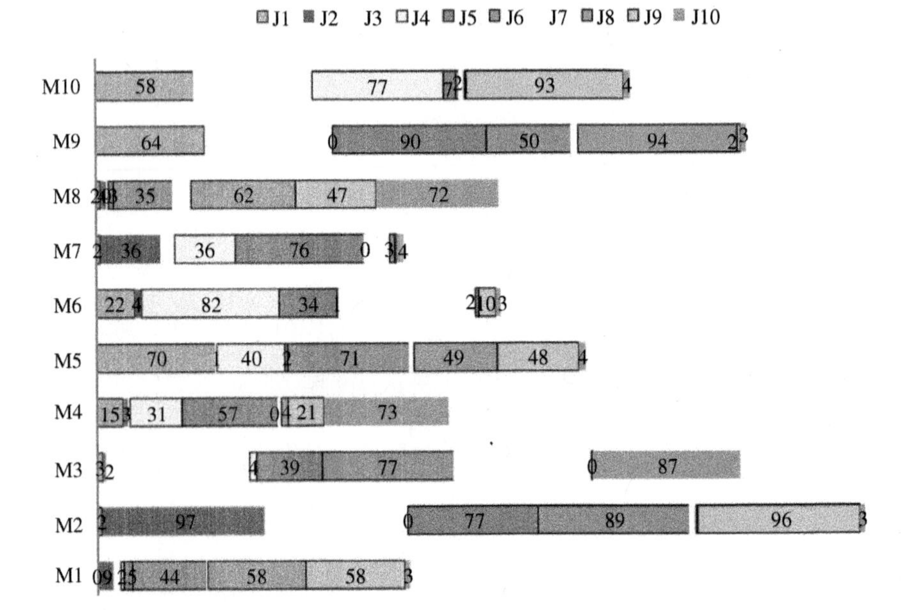

Figure 2.4 Model Gantt chart for Case 5 (10×10).

Table 2.4 Job completion time of case 8 (15×10): sample model.

Job/Machine	M1	M2	M3	M4	M5	M6	M7	M8	M9	M10
J1	1	35	37	58	111	115	210	213	213	268
J2	4	87	91	117	133	204	211	215	215	289
J3	35	87	103	120	135	205	215	257	355	394
J4	112	167	168	171	250	254	254	334	421	423
J5	116	201	204	241	251	273	275	334	504	568
J6	118	201	205	295	338	400	403	495	583	587
J7	211	214	214	364	365	404	491	582	660	662
J8	211	215	275	405	407	445	494	606	664	747
J9	213	313	317	405	408	462	497	650	689	796
J10	309	313	320	409	488	565	648	652	732	797
J11	313	341	343	504	504	600	643	728	733	804
J12	374	378	378	506	601	611	644	763	766	820
J13	378	381	440	565	692	693	709	765	812	813
J14	379	385	483	617	692	721	723	815	818	845
J15	466	511	513	617	696	722	762	856	885	843

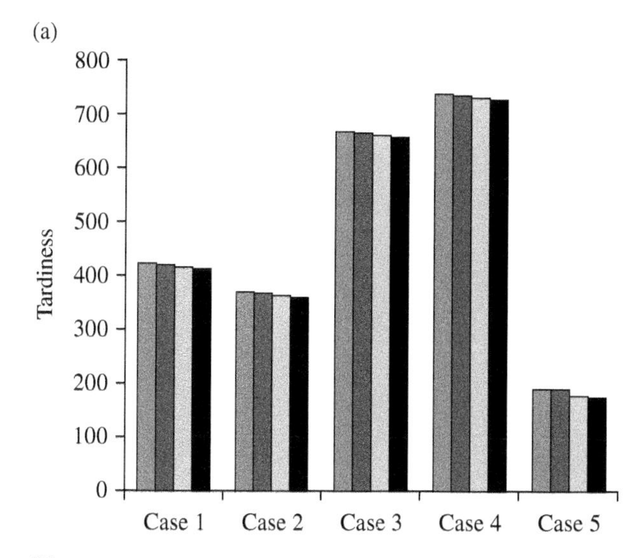

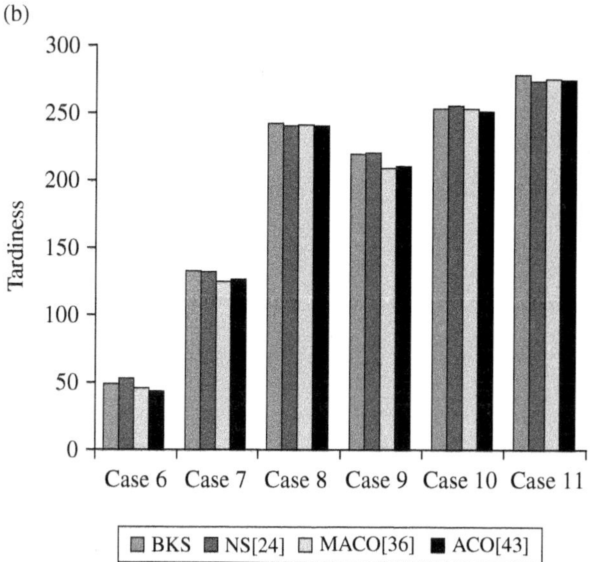

Figure 2.5 Comparative analysis of Maximum Tardiness.

accomplished by an ACO pheromone and probability updating model. In future research, an ACO hybrid with other swarm optimizations could be considered for the JSP process, with multiple goals such as improving delay, workload, earliness, etc. To consolidate the scheduling hypothesis with every profession characteristic to use to manage the genuine generation, the assistance could be taken from any manufacturing industry case investigation.

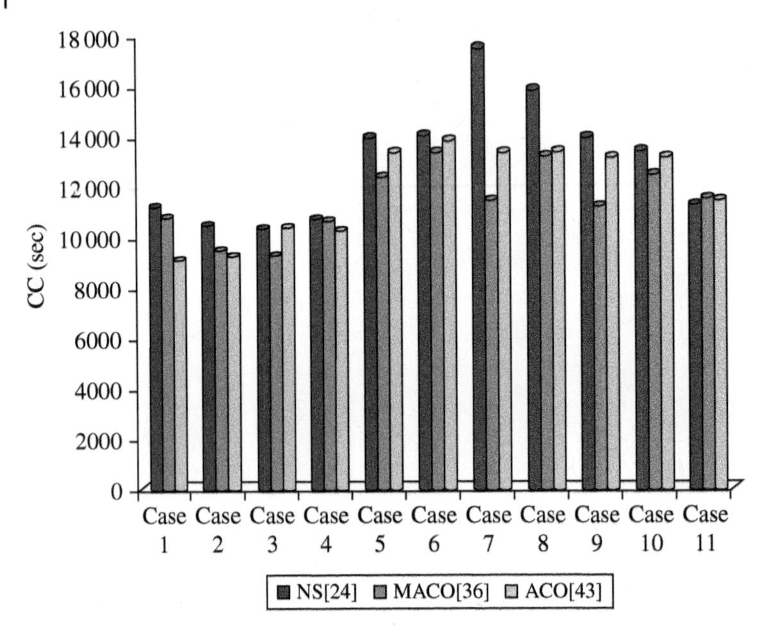

Figure 2.6 Computational complexity (CC) analysis.

References

1. Sun, L., Lin, L., Wang, Y. et al. (2015). A Bayesian optimization-based evolutionary algorithm for flexible job shop scheduling. *Procedia Comput. Sci.* 61: 521–526.
2. Pang, H. and Jiang, X. (2017). A solution of Flexible Job Shop Scheduling Problem Based on Ant Colony Algorithm and Complex Network. In Computational Intelligence and Design (ISCID), 2017 10th International Symposium on (Vol. 2: 463–466). IEEE.
3. Li, L., Keqi, W., and Chunnan, Z. (2010). An improved ant colony algorithm combined with particle swarm optimization algorithm for multi-objective flexible job shop scheduling problem. In Machine Vision and Human-Machine Interface (MVHI), 2010 International Conference on (pp. 88–91). IEEE.
4. Fox, B., Xiang, W., and Lee, H.P. (2007). Industrial applications of the ant colony optimization algorithm. *Int. J. Adv. Manuf. Technol.* 31 (7–8): 805–814.
5. Lin, J., Zhu, L., and Wang, Z.J. (2018). A hybrid multi-verse optimization for the fuzzy flexible job-shop scheduling problem. *Comput. Ind. Eng.* 127: 1089–1100.
6. Xing, L.N., Chen, Y.W., Wang, P. et al. (2010). A knowledge-based ant colony optimization for flexible job shop scheduling problems. *Appl. Soft Comput.* 10 (3): 888–896.

7 Huang, K.L. and Liao, C.J. (2008). Ant colony optimization combined with taboo search for the job shop scheduling problem. *Comput. Oper. Res.* 35 (4): 1030–1046.

8 Huang, R.H. and Yang, C.L. (2008). Ant colony system for job shop scheduling with time windows. *Int. J. Adv. Manuf. Technol.* 39 (1–2): 151–157.

9 Wen-Xia, W., Yan-Hong, W., Hong-Xia, Y., and Cong-Yi, Z. (2013). Dynamic-Balance-Adaptive Ant Colony Optimization Algorithm for Job-Shop Scheduling. 2013 Fifth International Conference on Measuring Technology and Mechatronics Automation (ICMTMA), 496–499.

10 Kis, T. (2003). Job-shop scheduling with processing alternatives. *Eur. J. Oper. Res.* 151 (2): 307–332.

11 Ho, N.B., Tay, J.C., and Lai, E.M.K. (2007). An effective architecture for learning and evolving flexible job-shop schedules. *Eur. J. Oper. Res.* 179 (2): 316–333.

12 Zhang, C., Li, P., Guan, Z., and Rao, Y. (2007). A tabu search algorithm with a new neighborhood structure for the job shop scheduling problem. *Comput. Oper. Res.* 34 (11): 3229–3242.

13 Mauguière, P., Billaut, J.C., and Bouquard, J.L. (2005). New single machine and job-shop scheduling problems with availability constraints. *J. Sched.* 8 (3): 211–231.

14 Jensen, M.T. (2003). Generating robust and flexible job shop schedules using genetic algorithms. *IEEE Trans. Evol. Comput.* 7 (3): 275–288.

15 Chan, F.T.S., Wong, T.C., and Chan, L.Y. (2006). Flexible job-shop scheduling problem under resource constraints. *Int. J. Prod. Res.* 44 (11): 2071–2089.

16 Kavitha, S., Venkumar, P., Rajini, N., and Pitchipoo, P. (2018). An efficient social spider optimization for flexible job shop scheduling problem. *J. Adv. Manuf. Syst.* 17 (2): 181–196.

17 Engin, O. and Güçlü, A. (2018). A new hybrid ant colony optimization algorithm for solving the no-wait flow shop scheduling problems. *Appl. Soft Comput.* 72: 166–176.

18 Kurdi, M. (2019). Ant colony system with a novel Non-DaemonActions procedure for multiprocessor task scheduling in multistage hybrid flow shop. *Swarm Evol. Comput.* 44: 987–1002.

19 Comuzzi, M. (2019). Optimal directed hypergraph traversal with ant-colony optimisation. *Inf. Sci.* 471: 132–148.

20 Jin, H., Wang, W., Cai, M. et al. (2017). Ant colony optimization model with characterization-based speed and multi-driver for the refilling system in hospital. *Adv. Mech. Eng.* 9 (8): 1–18.

21 Wu, Y., Gong, M., Ma, W., and Wang, S. (2019). High-order graph matching based on ant colony optimization. *Neurocomputing* 328: 97–104.

22 Reddy, K.N. and Padmanabhan, G. (2016). Teaching learning based optimization (TLBO) for job shop scheduling problems. First International Conference on

Productivity, Efficiency and Competitiveness in Design and Manufacturing, Coimbatore, Tamilnadu, India.

23 Wang, L. and Xu, Y. (2011). An effective hybrid biogeography-based optimization algorithm for parameter estimation of chaotic systems. *Expert Syst. Appl.* 38 (12): 15103–15109.

24 Akram, K., Kamal, K., and Zeb, A. (2016). Fast simulated annealing hybridized with quenching for solving job shop scheduling problem. *Appl. Soft Comput.* 49: 510–523.

25 Muthiah, A. and Rajkumar, R. (2014). A comparison of artificial bee colony algorithm and genetic algorithm to minimize the makespan for job shop scheduling. *Procedia Eng.* 97: 1745–1754.

26 Kumar, V., Singh, O., and Mishra, R.S. (2018). Krill herd based optimization for job shop scheduling problems (JSSP) to minimize make span time. *Int. J. Appl. Eng. Res.* 13 (11): 9627–9635.

27 Salido, M.A., Escamilla, J., Giret, A., and Barber, F. (2016). A genetic algorithm for energy-efficiency in job-shop scheduling. *Int. J. Adv. Manuf. Technol.* 85 (5–8): 1303–1314.

28 Yazdani, M., Aleti, A., Khalili, S.M., and Jolai, F. (2017). Optimizing the sum of maximum earliness and tardiness of the job shop scheduling problem. *Comput. Ind. Eng.* 107: 12–24.

29 Mokhtari, H. and Hasani, A. (2017). An energy-efficient multi-objective optimization for flexible job-shop scheduling problem. *Comput. Chem. Eng.* 104: 339–352.

30 Elmi, A. and Topaloglu, S. (2017). Cyclic job shop robotic cell scheduling problem: ant colony optimization. *Comput. Ind. Eng.* 111: 417–432.

31 Meilin, W., Xiangwei, Z., Qingyun, D., and Jinbin, H. (2010). A dynamic schedule methodology for discrete job shop problem based on ant colony optimization. 2nd IEEE International Conference on Information Management and Engineering (ICIME), 306–309.

32 Saidi-Mehrabad, M., Dehnavi-Arani, S., Evazabadian, F., and Mahmoodian, V. (2015). An ant colony algorithm (ACA) for solving the new integrated model of job shop scheduling and conflict-free routing of AGVs. *Comput. Ind. Eng.* 86: 2–13.

33 Chaukwale, R. and Kamath, S.S. (2013). A modified ant colony optimization algorithm with load balancing for job shop scheduling. 15th International Conference on Advanced Computing Technologies (ICACT), 1–5.

34 Wu, Z.J., Zhang, L.P., Wang, W., and Wang, K. (2009). Research on Job-Shop Scheduling Problem Based on Genetic Ant Colony Algorithms. In Computational Intelligence and Security, 2009. CIS'09. International Conference on (Vol. 2: 114–118). IEEE.

35 Huang, R.H. and Yu, T.H. (2017). An effective ant colony optimization algorithm for multi-objective job-shop scheduling with equal-size lot-splitting. *Appl. Soft Comput.* 57: 642–656.

36 Turguner, C. and Sahingoz, O.K. (2014). Solving job shop scheduling problem with Ant Colony Optimization. 15th International Symposium on Computational Intelligence and Informatics (CINTI), 385–389.

37 Chernigovskiy, A.S., Kapulin, D.V., Noskova, E.E. et al. (2017, October). Production scheduling with ant colony optimization. *IOP Conf. Ser.: Earth Environ. Sci.* 87 (6): 062002.

38 Nosheen, F., Bibi, S., and Khan, S. (2013). Ant Colony Optimization based scheduling algorithm. International Conference on Open Source Systems and Technologies (ICOSST), 18–22.

39 Tawfeek, M.A., El-Sisi, A., Keshk, A.E., and Torkey, F.A. (2013). Cloud task scheduling based on ant colony optimization. 8th International Conference on Computer Engineering & Systems (ICCES), 64–69.

40 Chaouch, I., Driss, O.B., and Ghedira, K. (2017). A modified ant colony optimization algorithm for the distributed job shop scheduling problem. *Procedia Comput. Sci.* 112: 296–305.

41 Liu, X., Ni, Z., and Qiu, X. (2016). Application of ant colony optimization algorithm in integrated process planning and scheduling. *Int. J. Adv. Manuf. Technol.* 84 (1–4): 393–404.

42 Korytkowski, P., Rymaszewski, S., and Wiśniewski, T. (2013). Ant colony optimization for job shop scheduling using multi-attribute dispatching rules. *Int. J. Adv. Manuf. Technol.* 67 (1–4): 231–241.

43 Wang, L., Cai, J., Li, M., and Liu, Z. (2017). Flexible job shop scheduling problem using an improved ant colony optimization. *Sci. Prog.*: 9016303.

44 Nazif, H. (2015). Solving job shop scheduling problem using an ant colony algorithm. *J. Asian Sci. Res.* 5 (5): 261–268.

3

Advanced Ant Colony Optimization in Healthcare Scheduling

Reza Behmanesh[1], Iman Rahimi[1], Mostafa Zandieh[2], and Amir H. Gandomi[3]

[1] *Young Researchers and Elite Club, Isfahan (Khorasgan) Branch, Islamic Azad University, Isfahan, Iran*
[2] *Department of Industrial Management, Management and Accounting Faculty, Shahid Beheshti University, G.C., Tehran, Iran*
[3] *Faculty of Engineering and IT, University of Technology Sydney, Ultimo, Australia*

3.1 History of Ant Colony Optimization

3.1.1 Introduction to Ant Colony Optimization

The first ant colony optimization (ACO) algorithm, ant system (AS), was introduced by [1] as an optimizer, learning, natural algorithm and meta-heuristic; and also it was applied to tackle the Traveling Salesman Problem (TSP) by [2]. It must be noted that the researcher introduced the basic algorithm and some extended versions of the basic method in his dissertation in 1992. This algorithm was designed to find an optimized route or vertices through the graph according to an ant's behavior between the nest and food resource. Since this method is inspired by the nature of ant colony behavior, it is categorized by bio-inspired computation methods to solve problems. In nature, the ants move from their nest to find food along a random path. While moving along the route, the ants secrete pheromone on the way back. The smell of the pheromone helps other ants to tail the path for finding food. The pheromone is liquid, and it will evaporate after a short time, but more pheromone on the shorter routes remains and more ants are attracted by sniffing, and as a consequence, the ant colony on short routes is higher than on long routes because more pheromone is trailed before evaporation.

Evolutionary Computation in Scheduling, First Edition. Edited by Amir H. Gandomi, Ali Emrouznejad, Mo M. Jamshidi, Kalyanmoy Deb, and Iman Rahimi.

To continue, we categorize some essential and basic characteristics of ACO to ascertain this methodology. ACO includes the following special components:

1) This approach is a metaheuristic algorithm.
2) This method is classified as a stochastic local search (SLS) method.
3) ACO is inspired by an ant's behavior in the colony, and so this is classified as a biologically-inspired algorithm.
4) ACO is a population-based method because the algorithm applies a swarm of ants to find the objective.
5) The algorithm's strategy for finding a solution is constructive, i.e. decision variables of the problem are constructed in each iteration, and it is different from the strategy of other meta-heuristic algorithms such as genetic algorithms, which use an improvement.
6) ACO is a prominent swarm intelligence technique because the passed path by each ant helps to improve the objective function in the next iteration.

On the other hand, the ACO is missing some characteristics:

1) It is not a single algorithm.
2) It is not an evolutionary algorithm, because there is no evolution path in the ACO approach.
3) It is not a model of the behavior of real ants in nature, because the equations of ACO have been extracted according to pheromone trail foraging and probability rules. These formulations are not by nature exactly the same as an ant's movement, which passes along the shortest path.

3.1.2 The Nature of Ants' Behavior in a Colony

As stated previously, ACO was inspired by the movement of swarm ants from the nest to the food resource. In other words, it was inspired by the indirect communication of some ant species through pheromone trails. All ants secrete pheromone when coming back from a food resource to the nest. Each ant forages the higher probability paths that are determined by stronger pheromone concentrations. It was demonstrated that a big swarm of ants passes along the shortest path to find food. It was experienced by [3] in a double bridge experiment, and therefore they indicate that more ants choose the shortest path to find food. This famous experiment is indicated in Figure 3.1 and [3] presented a stochastic model and confirmed it by simulation as follows:

$$P_{i,b} = \frac{(k+\tau_{i,b})^{\alpha}}{(k+\tau_{i,b})^{\alpha} + (k+\tau'_{i,b})^{\alpha}} \tag{3.1}$$

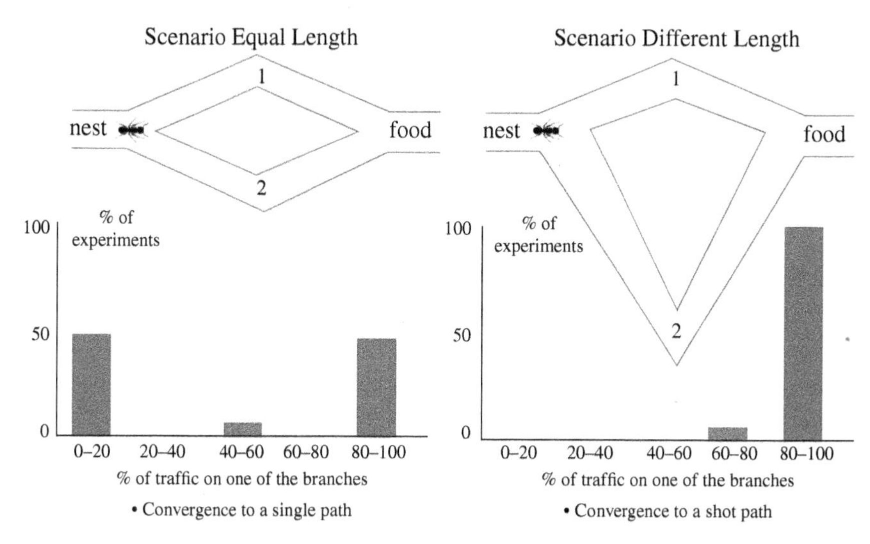

Figure 3.1 Double bridge experiment.

Where, $P_{i,b}$ denotes the probability of choosing branch #*b* when ant #*i* decides to choose and $\tau_{i,b}$ denotes pheromone concentration corresponding to ant #*i* in branch #*b*.

As it was notated, real ants pass the shortest way between nest and food, so they solve the shortest path problem. However, artificial ants in swarm intelligence problems are taken into account as stochastic solution construction procedures, which can be considered as a searcher on a construction graph network, as indicated in Figure 3.2.

Artificial ants' constructed paths (solutions) are based on segregated pheromone probabilistically and then these ants record the solutions in memory.

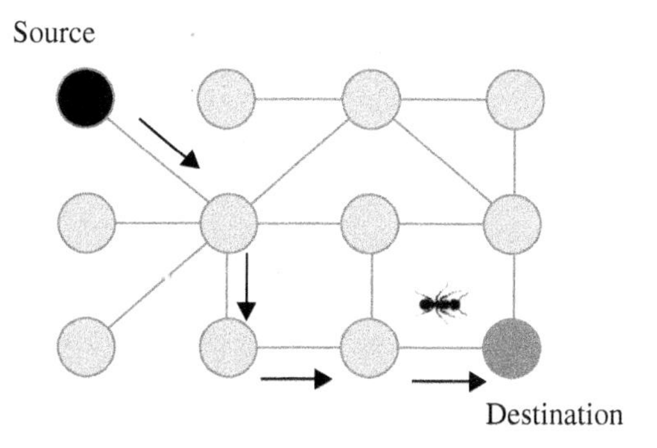

Figure 3.2 Artificial ants searching on a graph.

Therefore, an artificial memory is built to be applied in the next iterations. These artificial ants deposit pheromones on the path that they walk on. It is possible that the pheromone evaporates after some time. Finally, the paths represent better solutions that have more pheromone trails.

3.2 Introduction to ACO as a Metaheuristic

3.2.1 Ant Colony Optimization Approach

This approach consists of two substantial parts. The first is to construct the solution with a probability rule and the second is to update the pheromone in order to search for better solutions. Artificial ants construct routes (solutions) probabilistically by pheromones, and these memorize the solution. On the other hand, agents secrete pheromone on the route and as a consequence there is more pheromone on the routes that present better solutions. The first ACO is called AS, in which each ant (agent) with tag #*k* builds a complete tour (a solution) in the grap h iteratively, and from the last visited node *i* chooses an unvisited node *j* to visit next (Figure 3.3).

It is noted that unvisited nodes are feasible neighborhood (N_i^k). The probability of choosing node *j* after node *i* by ant *k* is determined as shown in the following equation (probabilistic choice rule):

$$P_{ij}^k(t) = \frac{\left[\tau_{ij}(t)\right]^\alpha \cdot \left[\mu_{ij}\right]^\beta}{\sum_{l \in N_i^k} \left[\tau_{il}(t)\right]^\alpha \cdot \left[\mu_{il}\right]^\beta} \qquad if\ j \in N_i^k \tag{3.2}$$

Where $\tau_{ij}(t)$ notates pheromone information that changes in any iteration, μ_{ij} is heuristic information, which corresponds to the problem, α controls the relative importance of pheromone, and β controls the relative importance of heuristic information. In the second part, the pheromone trail updating strategies are presented. The pheromone update procedure in AS is formulated as follows:

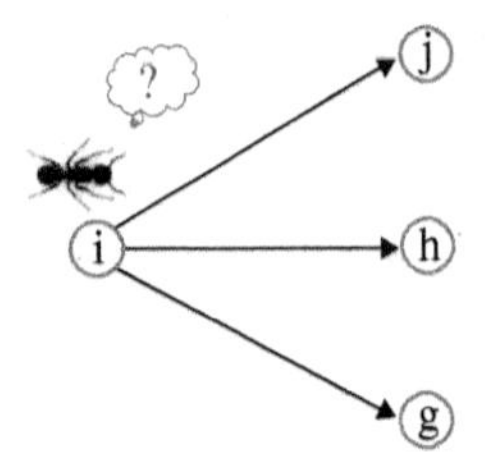

Figure 3.3 Probability of choosing a model by ant *k*.

$$\tau_{ij}(t+1) = (1-\rho).\tau_{ij}(t) + \sum_{k=1}^{m} \Delta\tau_{ij}^{k} 0 < \rho \leq 1 \tag{3.3}$$

$$where \ \ \Delta\tau_{ij}^{k} = \begin{cases} \dfrac{Q}{f(s_k)} & if\ ant\ k\ goes\ through\ (i,j)\ in\ this\ iteration \\ 0, & o.w. \end{cases} \tag{3.4}$$

where s_k is the solution of the kth ant and $f(s_k)$ is its cost function. It is notated by L_k in some literature, which means it found the tour's length by ant k.

The pseudo code of ACO is presented in Figure 3.4, in which the stages of the algorithm are articulated.

For example, the AS algorithm is applied to solve TSP for the first time by [2], and we describe details of the algorithm as follows:

1) The ants are set on nodes (cities) randomly so that each agent memorizes the path recorded (Figure 3.5). In other words, a solution consisting of all decision variables is saved in the memory of agent. As indicated in Figure 3.6, there are five cities and the matrix of D denotes the distances between all cities. On the other hand, a graph with prime pheromone is constructed.
2) As indicated in Figure 3.6, ant #1 has several choices (C, E, D, and B) and according to transition rule Eq. (3.2), the probability choosing is determined and the ant can move to the next node (city). If setting ants on each city is done to construct the first partial solution (a decision variable of completed solution) in Figure 3.6, this stage can be repeated to the complete solution that is shown in Figure 3.7.

Algorithm 1. Ant Colony Optimization

Input: an instance P of a combinatorial problem P
Initialize Pheromone Values (τ_0)
While termination conditions not met **do**
S_{iter}: = ∅
for $k = 1, \ldots, m$ **do**
s_k: = { }
Repeat
C_j: = Probabilistic_Choice ($N(s_k)$, τ)
s_k: = $s_k \cup \{c_j\}$
until s_k is a complete solution
S_{iter}: = $S_{iter} \cup \{s_k\}$
τ: = Pheromone Update (τ)
Return the best solution found

Figure 3.4 Pseudo code of ant colony optimization.

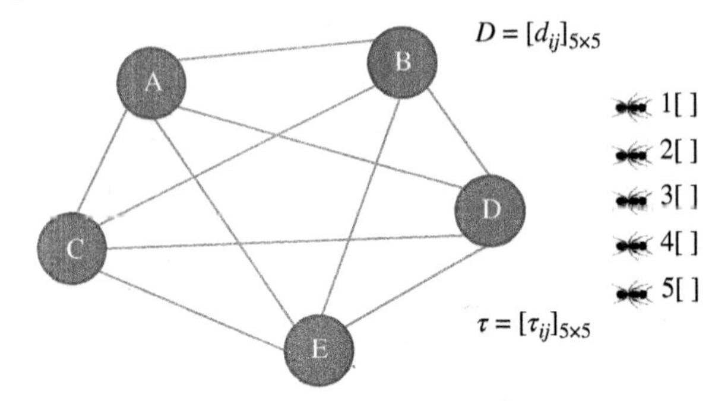

Figure 3.5 Inputting an instance and initializing prime pheromone.

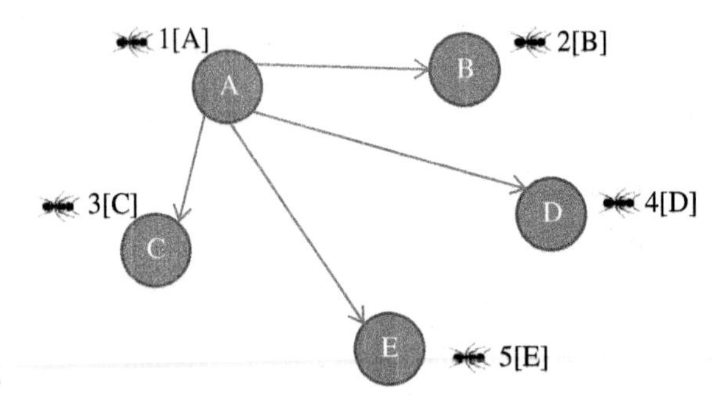

Figure 3.6 Setting ants on cities.

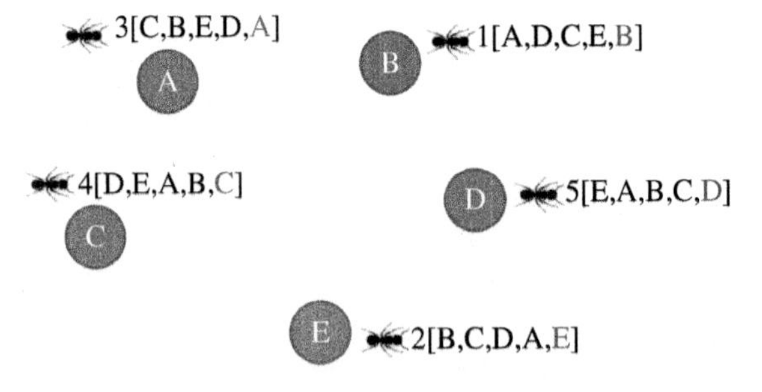

Figure 3.7 Moving ants and completing solution.

3) After obtaining a solution for each agent, the length of tour for each ant is calculated, and then the segregated pheromone value of the arcs (the path between two cities) is evaluated based on Eq. (3.4). Finally, the pheromone value of each arc is updated according to the second part of Eq. (3.3). The results are displayed in Figure 3.8. For example, we can observe that ants #4

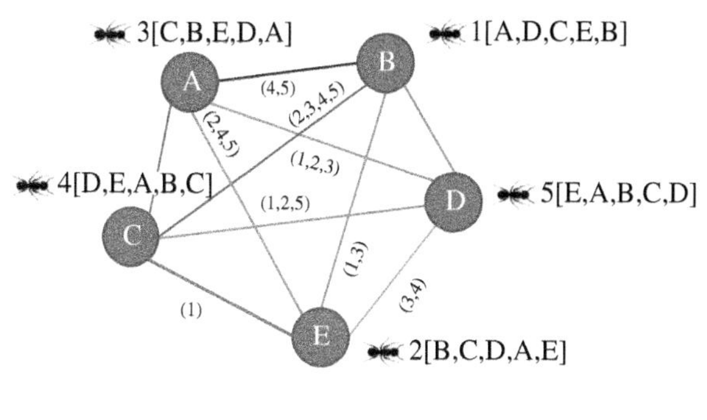

1[A,D,C,E,B] $L_1 = 30$
2[B,C,D,A,E] $L_2 = 45$
3[C,B,E,D,A] $L_3 = 28$
4[D,E,A,B,C] $L_4 = 28$
5[E,A,B,C,D] $L_5 = 42$

$$\Delta\tau_{AB}^{total} = \Delta\tau_{AB}^{1} + \Delta\tau_{AB}^{2} + \Delta\tau_{AB}^{3} + \Delta\tau_{AB}^{4} + \Delta\tau_{AB}^{5}$$

$$\Delta\tau_{AB}^{total} = 0 + 0 + 0 + Q.\left(\frac{1}{28} + \frac{1}{42}\right)$$

Figure 3.8 Updating pheromone trail value for paths between cities.

and #5 passed the arc AB and deposited pheromone on this arc. Thus, for assessing the pheromone value of arc AB, only ants #4 and #5 are involved in calculating the increased pheromone. It must be noted that there is no difference between AB or BA, i.e. the reverse path for updating the pheromone value in the path. As shown in the graph, some arcs are colored, which means agent tags inside parentheses update the pheromone value of these arcs. For example, the pheromone of arc BC or CB is updated by ants #(2, 3, 4, 5).

4) Repeat stages 1 to 3 until termination conditions are met, or the algorithm reaches the convergent condition.

3.2.2 Intensification and Diversification Mechanism in ACO

As we know, there are two major components in the literature of metaheuristic algorithms: intensification and diversification, known as exploitation and exploration [4, 5]. To make a qualified and robust algorithm metaheuristic, a balanced and adjusted combination of the exploration and exploitation mechanism is vitally important because their strategies are in contradiction with each other [6].

A good diversification or exploration strategy guarantees that an algorithm can search as many regions as possible inside the solution space in an efficient manner.

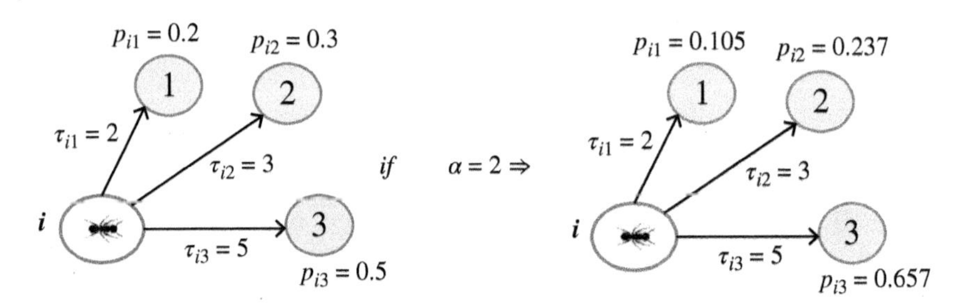

Figure 3.9 Compare two probabilities of choosing nodes for alpha = 1 and alpha = 2.

Moreover, this component prevents premature convergence or being trapped in local optima, and, as a consequence, the stronger diversification leads to a slowdown in the convergence of the algorithm since this strategy implements randomization or stochastic procedures in the algorithm and the weaker exploration leads to premature convergence. In the first term of the transition rule equation of ACO, i.e. the pheromone-based probability of choosing a node, the diversification is focused. So, more alpha as an important factor of pheromone information as well as more beta as an essential factor of heuristic information results in less exploration and more exploitation (sometimes premature convergence) in the algorithm according to the example shown in Figure 3.9. On the other hand, fewer α as well as fewer β leads to increased diversification and randomness in the search proccss. As indicated in Figure 3.9, when alpha is increased only node #3 is more probable to be chosen, so this path is bold, and the probability of choosing other paths (#1 and #2) is decreased.

On the other hand, the proper intensification or exploitation strategy makes sure that the algorithm exploits the experience of the search process and always intensifies the solution found to reduce randomness in the search process, then it can speed up convergence of the algorithm when the algorithm needs to decrease exploration. Therefore, the stronger intensification leads to local traps and premature convergence of the algorithm and provides meaningless solutions. Moreover, the weaker exploitation leads to a slowdown in the convergence of the algorithm.

To summarize, alpha and beta coefficients are important parameters in exploration and exploitation and need to be set to improve the performance of ACO algorithms for a specific problem.

3.3 Other Advanced Ant Colony Optimization

The first ACO algorithm, AS, was shown to be a viable approach for solving hard optimization problems; however, its performance is rather poor for large instance problems like TSP in comparison to other metaheuristics. To achieve better

Table 3.1 All versions of ACO.

Class	ACO Algorithm	Researched by	Date
Pheromone update strategy-based method	Ant System (AS)	[1]	1991
	Elitist Ant System (EAS)	[7]	1992
	Ant Colony System (ACS)	[2]	1996
	Rank-based Ant System (RBAS)	[8]	1997
	Max-Min Ant System (MMAS)	[9]	2000
Hybrid ACO methods	Best-Worst Ant System (BWAS)	[10]	2000
	Population-based ACO (P-ACO)	[11]	2002
	Beam-ACO (BACO)	[12]	2004
	Two-level ACO	[13]	2015
	Improved Auto-control ACO with lazy ant (IAACO)	[14]	2016

performance than AS, more studies were conducted to build an improved AS. There are several versions of advanced ACO that improve solution quality, and these new algorithms are extended by introducing a different pheromone update strategy or a hybrid ACO using other methods or operations. A brief list of extended new ACO algorithms is shown in Table 3.1.

3.3.1 Elitism Ant System

In all developments over AS, the pheromone trail updating strategy was extended to improve the algorithm's performance. A first improvement, namely AS_e (elitism strategy for AS) was developed by [7], where the global best solution (s_{gb}) is considered to update the trails and the pheromone update procedure is formulated as follows:

$$\tau_{ij}(t+1) = (1-\rho).\,\tau_{ij}(t) + \Delta\tau_{ij}^{global-best} \tag{3.5}$$

$$where \ \ \Delta\tau_{ij}^{global-best} = \begin{cases} \dfrac{e}{f(s_{gb})} & if\ ant\ k\ goes\ through\ (i,j)\ in\ this\ iteration \\ 0, & o.w. \end{cases} \tag{3.6}$$

where e is the number of elitist ants and $f(s_{gb})$ is the solution cost of the best ant that is used to update arcs after each iteration.

3.3.2 Ant Colony System

The next improvement over AS is called ant colony system (ACS), which was introduced by [2]. ACS works according to a pseudo-random proportional choice rule using a controller parameter, namely q_0. Agent #k chooses node j after node i with probability less than or equal to q_0 based on the following equation (greedy walking):

$$j = \arg\max \left\{ \left[\tau_{il}(t)\right]^{\alpha} \cdot \left[\mu_{il}\right]^{\beta} \right\} \quad if\ l \in N_i^k \tag{3.7}$$

On the other hand, agent #k chooses node j after node i with probability higher than q_0 based on an equation that pertains to the probabilistic choice rule of AS (Eq. 3.2).

There are two pheromone updating strategies in ACS, a global pheromone update and a local pheromone update. In the global update or exploitation strategy, a single ant is used, i.e. either the global-best ant so far or the iteration-best is taken into account to update as follows:

$$\tau_{ij}(t+1) = (1-\rho).\tau_{ij}(t) + \rho.\Delta\tau_{ij}^{ib/gb} \tag{3.8}$$

$$where\ \ \Delta\tau_{ij}^{ib/gb} = \begin{cases} \dfrac{Q}{f(s_{gb})} & if\ ant\ k\ goes\ through\ (i,j)\ in\ this\ iteration \\ 0, & o.w. \end{cases} \tag{3.9}$$

where ib refers to iteration-best, and gb refers to global-best ant. And in the local update or exploration strategy, ant #k decreases the pheromone when it adds a component c_{ij} to its partial solution in accordance with the following:

$$s_k \cup \{c_{ij}\} \Rightarrow \tau_{ij}(t) = (1-\gamma).\tau_{ij}(t) + \gamma.\tau_0 \tag{3.10}$$

where parameter γ controls the exploration factor and the initial pheromone (small constant value) is notated by τ_0.

3.3.3 Rank-Based AS

Rank-based AS (AS_{rank}) is another version of AS that was extended by [8]. A fixed number of better ants in the current iteration are ranked, and then the global-best ant is allowed to update trails of ranked solutions. For example, if the fixed number equals 10, it means that 10 better ants in the current iteration are ranked, and then the global-best ant is allowed to update the pheromone trails of the ranked ant's paths.

3.3.4 Max-Min AS

[9] developed the Max-Min ant system (MMAS) so that only a single ant is applied to update the pheromone trail when the iteration is passed. The updating rule of MMAS is similar to ACS but there are important requirements in MMAS that should be fulfilled:

- After the iteration, only one single ant adds the pheromone value to exploit the best solutions found during iteration or the run of the algorithm.
- An interval distance trail $[\tau_{min}, \tau_{max}]$ is considered and updated for the range of possible trails on each solution component.
- Also, pheromone trails are initialized to τ_{max} to achieve good diversification at the beginning of the algorithm. Pheromone update strategy is according to the following equation:

$$\tau_{ij}(t+1) = (1-\rho).\tau_{ij}(t) + \Delta\tau_{ij}^{best} \tag{3.11}$$

$$where\ \ \Delta\tau_{ij}^{best} = \begin{cases} \dfrac{Q}{L_{gb}} if\ ant\ k\ goes\ through\ (i,j)\ in\ this\ iteration \\ 0, \qquad\qquad\qquad\qquad\qquad\qquad o.w. \end{cases} \tag{3.12}$$

As stated, this approach applies the global-best solution found ever from the beginning of the trial (global-best ant) or the current iteration-best solution (iteration-best ant) so that both strategies are suggested. Moreover, limitations for lower and upper pheromone are written as follows:

$$\tau_{min} \leq \tau_{ij} \leq \tau_{max} \quad \forall \tau_{ij} \in T \tag{3.13}$$

$$\tau_{max} = \tau_0 \tag{3.14}$$

Ibanez comments that it would be better if the pheromone were set to prime pheromone after some iterations and that this prevents premature convergence and other solutions are found and the exploration strategy is strengthened.

3.3.4 Best-Worst AS

In another study, Cordón et al. [15] applied a new pheromone trail strategy along with evolutionary algorithm concepts to improve the performance of AS. First, they proposed the best-worst performance update rule, which is based on population-based incremental learning (PBIL). According to this strategy, edges present in global-best ant are updated as follows:

$$\tau_{ij}(t+1) = \tau_{ij}(t) + \Delta\tau_{ij}^{gb} \tag{3.15}$$

$$where \quad \Delta\tau_{ij}^{gb} = \begin{cases} \dfrac{Q}{f(s_{gb})} if\ (i,j) \in S_{gb} \\ 0 \qquad\qquad o.w. \end{cases} \tag{3.16}$$

and pheromone deposited based on a global-best solution (s_{gb}).

Then, edges present in the worst current ant are penalized in the evaporation rule:

$$\tau_{ij}(t+1) = (1-\rho).\tau_{ij}(t) \qquad \forall (i,j) \in S_{iw} \text{ and } (i,j) \notin S_{gb} \tag{3.17}$$

where s_{iw} is the worst solution in each iteration.

Finally, some random pheromone trails are mutated to make more exploration in the search method, as done in PBIL. According to this rule, each row of the pheromone matrix is mutated with probability P_m, as follows:

$$\tau'_{ij} = \begin{cases} \tau_{ij} + mutate\ (it, \tau_{th})\ if\ \ a = 0 \\ \tau_{ij} - mutate\ (it, \tau_{th})\ if\ \ a \neq 0 \end{cases} \tag{3.18}$$

where a is a random value in {0,1}, *it* is a current iteration, and τ_{th} is the mean of pheromone trail in edges that are generated by global-best.

3.3.5 Population-Based ACO

In a population-based ACO (P-ACO) proposed by [11], the population-based concepts were the inspiration, and the basic ACO is developed to a new high-quality algorithm considering the population set to solve dynamic optimization problems. First, the empty set of population, namely P, is saved such that its capacity equals the number of ants, i.e. k. In the current iteration, the best ant updates the pheromone trail matrix similar to the Max-Min AS performance, in which pheromone addition is according to the best solution ever found by the best ant; however, the evaporation or pheromone decrease is not done by ants, and then it enters the population set till the kth generation.

$$\tau_{ij}(t+1) = \tau_{ij}(t) + \Delta\tau_{ij}^{best} \tag{3.19}$$

$$where \quad \Delta\tau_{ij}^{best} = \begin{cases} \dfrac{Q}{L_{gb}} if\ ant\ k\ goes\ through\ (i,j)\ in\ this\ iteration \\ 0, \qquad\qquad o.w. \end{cases} \tag{3.20}$$

When the P set is full of the best ants obtained from all k iterations, for generation $k+1$ and m ore, the best ant of the iteration is replaced with one of the ants inside the population according to update strategies defined in the next section. Therefore, that ant is removed from the population and the best enters the population. Instea of decrasing pheromone value based on an evaporation rule, a new method is introduced, in which the removed ant decreases the pheromone matrix according to its pheromone value, i.e. the same value that it added to the pheromone matrix when it entered to the population as follows:

$$\tau_{ij}(k+1) = \tau_{ij}(k) - \Delta\tau_{ij}^{removed-ant} \tag{3.21}$$

The pseudo code of population-based ACO is presented in Figure 3.10, in which the stages of the algorithm are coded.

Algorithm 2. Population-based Ant Colony Optimization

Input: an instance P of a combinatorial problem P
Initialize Pheromone Values (τ_0)
While termination conditions not met **do**
$S_{\text{iter}} := \varnothing$, $Pset := \varnothing$, $iter := 1$
for $k=1, \ldots, m$ **do**
$s_k := \{\}$
Repeat
$c_j :=$ Probabilistic_Choice ($N(s_k), \tau$)
$s_k := s_k \cup \{c_j\}$
until s_k is a complete solution
$S_{iter} := S_{iter} \cup \{s_k\}$
Find ibs_k iteration best
$iter := iter + 1$
If $iter \leq m$
$Pset := Pset \cup \{ibs_k\}$
$\tau :=$ Pheromone Adding (τ) according to Eq. (3.19)
Else
Find rs_k the removed solution of *Pset* according to pheromone update strategies similar Eq. (3.22)
$Pset := Pset \cup \{ibs_k\}$
$Pset := Pset \setminus \{rs_k\}$
$\tau :=$ Pheromone Adding (τ)
$\tau :=$ Pheromone Subtracting (τ) according to Eq. (3.21)
End if
Return the best solution found

Figure 3.10 Pseudo code of population-based ant colony optimization.

The researchers defined several population pheromone update strategies in other research [16] that is briefed below:

- *Age of ant:* in this strategy, the oldest ant of the population (the ant that is entered to the population in the oldest iteration) leaves the population, and the newest solution is entered. This strategy is called age-based strategy or was referred to as FIFO-Queue by researchers in a previous paper [11].
- *Quality:* If the candidate solution of the current iteration is better than the worst solution inside P set, the former replaces the latter in P. Otherwise, P set does not change.
- *Elitism:* This strategy considered inserting an ant into the population set, but the removed ant from the population is not considered in this strategy, so when the best ant solution is found, it enters to population, and if the algorithm does not find the elite solution, the content of the population is not changed after generation $k+1$.
- *Probability:* This strategy focused on choosing an ant probabilistically to prevent making copies of best ant inside the population set. To choose an ant from the population, bad solutions are likely to be removed from the set according to the following equation:

$$P_i = \frac{x_i}{\sum_{g=1}^{k+1} x_g}, \qquad x_i = f(\pi_i) - \min_{g=1,..,k+1} f(\pi_g) + avg(\pi) \tag{3.22}$$

where π_i is the solution, i and $f(\pi_i)$ is an objective function of the problem corresponding to the solution from population set, and P_i is the probability of selection of an ant from the population.

3.3.6 Beam-ACO

In Beam-ACO, introduced by [12], the solution construction mechanism of ACO is hybridized with beam search (BS) to tackle an open shop scheduling problem. A greedy approach is the most straightforward algorithm and operates based on a search tree framework so that it starts from empty partial solution $s^p = \langle . \rangle$, and then it is extended by adding a new component to a partial solution at the current step from the acceptable set $N(s^p)$. It must be noted that total benefit (based on criteria) is considered in each step to construct the solution and that the partial solution with the highest benefit is selected according to the greedy strategy. As it is obvious, the greedy strategy focuses on the intensification component and many good paths may not be taken into account in this search technique.

The BS method introduced by [17] is a classical approach and is derived from the branch and bound algorithms incompletely, and so these are considered as an approximate procedure. A key idea related to BS is to search several possible ways in the graph for building the partial solution. This algorithm extends each partial solution from set *Beam* in at most k_{ext} possible ways. Then, if a new partial solution obtained is complete, it is recorded in a set that is called Beam-Complete (B_c). Assuming that new partial solution obtained is extensible, it is saved in a set that is called Beam-Extensible (B_{ext}). In other words, this set contains partial solutions: those are incomplete or are further extensible. The algorithm creates a new beam at the end of each step by selecting up to some solutions (k_{bw}, i.e. Beam-Width) from set B_{ext}. To evaluate and select each partial solution from the set, a lower bound is defined as the criterion. The minimum objective function value for any complete solution s build from s^p is calculated as a lower bound for a given partial solution s.

The pseudo code of BS is presented in Figure 3.11, in which the stages of the algorithm are coded.

Algorithm 3. Beam Search

Input: an empty partial solution $s^p = \langle\ \rangle$, beam width (k_{bw}), max number of extensions (k_{ext}) Initialize the set of the partial solutions in B ($B = \{s^p\}$) and build the empty Beam-complete set ($B_c = \varnothing$)

While B is not empty **do**

  Build an empty Beam-extensible set

  for *each partial solution* **do**

    $t = 1$

    $BS\ (s^p)$: = PreSelect ($BS(s^p)$) {optional}

    While $t \le k_{ext}$ and $BS(s^p)$ is *not empty* **do**

      Choose $c := argmax\{\mu(c) \mid c \in BS(s^p)\}c$

      $s^{p\prime} := s^p \cup \{c\}$

      $BS(s^p) = BS(s^p)\{c\}$

      If $s^{p\prime}$ is extensible **then**

        Add $\{s^{p\prime}\}$ to the Beam-extensible set

      Else

        Add $\{s^{p\prime}\}$ to the Beam-complete set

      End if

      $t = t + 1$

    End while

  End for

  Rank the partial solutions in Beam-extensible set using the lower bound $LB(.)$

  $B = \min\{k_{bw}, |B_{ext}|\}$ (this means highest ranked partial solutions from Beam-extensible set)

End while

Return a set of candidate solutions in Beam-complete

Figure 3.11 Pseudo code of beam search.

In BS techniques, the policy of choosing nodes is according to a greedy strategy. In other words, these methods are usually deterministic same as a deterministic mechanism of ACS methods (Eq. 3.6). To summarize, a BS approach builds several parallel candidate solutions and applies a *LB* to search.

As Blum outlined in his research, the BS algorithm works according to two components: (i) to weight the different possibilities of extending a partial solution by a weighting function, and (ii) to restrict the number of the partial solution at each step by *LB* value. ACO algorithms explore the search space in a probabilistic manner while, in contrast, BS explore the search space in a deterministic way. The researchers replaced the deterministic choice of a solution component in the algorithm of BS by a probabilistic transition rule in the algorithm of ACO (Eq. 3.2), and therefore the transition probability in BS is subject to the changes of the pheromone trail value, and as a consequence, the probabilistic BS will be adaptive. After that, this new approach is called Beam-ACO.

3.3.7 Two-Level ACO

In a research, [13] considered a generalization of job shop and then formulated a multi-resource flexible job shop problem (FJSP) in order to minimize makespan by applying a novel two-level ACO procedure.

Mapping cities tailor the Two-level ACO algorithm to surgical cases and thus a node tour turns to be the sequence of surgical cases. In this procedure, there are two levels: (i) a first level, in which the surgical cases are sequenced in the outer graph, and (ii) a second level, in which the required multi-resources types of every stage are allocated to surgical cases in the inner graph. Nodes inside the inner graph represent available resources of the same resource type. The resource assigned to the surgical case for each stage is determined according to the path that the ant forages in the inner graph. A mix pheromone update strategy is defined for the algorithm, and it comprises one local and two global. In the outer level, the best ant updates the trails according to global iteration-best strategy to search for the best sequence. In the inner level, the surgery-related pheromone is defined to save the information that connects the surgical case with the required resource based on global strategy, while an inner resource-related is defined to record information related to resource utilization based on local strategy. It must be noted that local updating is effective until the ant forages paths of the inner graph and it is invalid after going out of the inner. Basic pheromone updates related to AS and transition rules are by the equations from [13], which are displayed in Figure 3.12, and the following equations.

The following equations correspond to the pheromone update strategy of the surgical case sequence problem:

$$\tau_{ij}(t+1) = (1-\rho).\tau_{ij}(t) + \Delta\tau_{ij} \tag{3.23}$$

$$where \quad \Delta\tau_{ij} = \begin{cases} \dfrac{Q}{C_{max}} \; if\ ant\ k\ goes\ through\ (i,j)\ in\ current\ iteration \\ 0, \qquad\qquad\qquad\qquad\qquad\qquad\qquad o.w. \end{cases} \tag{3.24}$$

where pheromone evaporation rate is notated by ρ; $\Delta\tau_{ij}$ is the incremental pheromone on the edge (i,j), i.e. surgical case i to surgical case j; Q is an adjustable parameter; and C_{max} is the makespan of scheduling.

Algorithm 4. Two-level Ant Colony Optimization

Input: an instance SCS as a combinatorial problem P
While stopping criterion not satisfied **do**
  Position m ants on the starting node (surgical case)
  Initialize pheromone trial and parameters
  Construct an ant solution
  While stopping criterion for each ant not satisfied, $k < m$ **do**
    Initialize tabu: $= \varphi$; I
    Construct an ant solution by choosing a node i
    in outer graph based to the transition rule Eq. (3.28)
    $tabu = tabu \cup \{I_i\}$
    $I = I \backslash \{I_i\}$
    Ant enters into the inner graph for choosing resources and constructs resource set G
    Construct a resource allocation for ant solution
    for each resource type c **do**
      Construct an ant solution by choosing a node cm in inner graph
      based to the transition rule Eq. (3.29)
      Local update inner pheromone trial based on Eq. (3.27)
    End for
    Decoding & Update resource's time window
  End while
  Calculate makespan for an ant solution
  Record the best global ant (solution)
  Update pheromone on both inner and outer, globally based on Eqs. (3.23–3.26)
End while
Return the best solution found

Figure 3.12 Pseudo code of two-level ACO.

The following equations correspond to the pheromone update strategy for the resource allocation problem as part of the main problem:

$$in(\tau^{i}_{cm}(t+1)) = (1-\rho).in(\tau^{i}_{cm}(t)) + \Delta in(\tau^{i}_{tm}) \tag{3.25}$$

$$where \;\; \Delta in(\tau^{i}_{tm}) = \begin{cases} \dfrac{Q}{C_{max}} \text{ if ant k goes through surgery (i) with} \\ \qquad \text{resource graph (c,m)} \\ 0, \qquad\qquad\qquad\qquad o.w. \end{cases} \tag{3.26}$$

where $in(\tau^{i}_{cm}(t))$ denotes the pheromone for surgery i with resource m in the selection stage of cth resource type, and $\Delta in(\tau^{i}_{tm})$ is notated for the incremental pheromone on the resource m in the selection stage of cth resource as presented in Eq. (3.26).

The following equation corresponds to the local pheromone update strategy for the resource allocation problem:

$$in(\gamma^{k}_{m}) = in(\gamma^{k}_{m}) - q_0 \tag{3.27}$$

where q_0 denotes the decremented pheromone value. After updating pheromone locally, the possibility of ants moving through the same path is decreased, and hence the uneven utilization of resources can effectively be avoided.

The transition rule in the outer ant graph (surgery) is presented as the following equation. In the outer surgery graph, the choice probability $P^{k}_{ij}(t)$ of an ant means choosing node (surgical case) j from node (surgical case) i.

$$P^{k}_{ij}(t) = \frac{\left[\tau_{ij}(t)\right]^{\alpha} \cdot \left[\mu_{ij}\right]^{\beta}}{\sum_{l \in I_0}\left[\tau_{il}(t)\right]^{\alpha} \cdot \left[\mu_{il}\right]^{\beta}} \qquad if\ j \in I_0 \tag{3.28}$$

Where I_0 denotes a feasible solution inclusive of nodes to be scheduled; τ_{ij} is the outer pheromone value between cases i and j; μ_{ij} represents the heuristic value from cases i to case j, the two parameters α and β are applied to determine the relative importance of pheromone trail and the heuristic information, respectively. The next case to be chosen is according to the probability, $P^{k}_{ij}(t)$, where a roulette wheel rule is applied to choose the node.

The transition rule in the inner resource graph is denoted as $P^{ki}_{cm}(t)$, and it presents the probability of choosing resource m for the cth resource type state of surgery i by ant #k:

$$P_{cm}^{ki}(t) = \frac{\left[in(\tau_{cm}^{i}(t)) \times in(\gamma_{m}^{k})\right]^{\alpha} \cdot \left[in(\mu_{cm})\right]^{\beta}}{\sum_{g\in G}\left[in(\tau_{cg}^{i}(t)) \times in(\gamma_{g}^{k})\right]^{\alpha} \cdot \left[in(\mu_{cg})\right]^{\beta}} \quad if\ g \in G \tag{3.29}$$

where the inner surgery-related pheromone is denoted by $in(\tau_{cm}^{i}(t))$ and $in(\gamma_{m}^{k})$ denotes the inner resource-itself pheromone. The balanced utilization of resources is considered by applying the multiple of these two pheromone values. The parameter $in(\mu_{cm})$ is the heuristic information, and G is the available set of resources for selecting the next resource node.

3.3.8 Improved Auto Control ACO with Lazy Ant

In a biological study [18], it was observed that an ant moves from an active state to lazy (inactive) state in some ant colonies. Inspired by this state among ants, an advanced ACO was proposed to solve the grid scheduling problem (GSP) by Tiwari and Vidyarthi [14]. For this purpose, the Auto controlled ACO was developed by incorporating the concept of lazy ants as well as an improved auto-control mechanism to update parameters. The researchers called their novel method Improved Auto-Controlled Ant Colony Optimization (IAC-ACO), in which states of some ants were changed from active to lazy in each iteration. The researchers found that lazy ants ameliorate the intensification strategy, saving computational time.

Lazy ants help the algorithm to speed up convergence of the solution and, moreover, increase the probability of intensification of the solution near the best agent. Therefore, the computational time of the algorithm is reduced by using some lazy ants instead of active ants. To generate lazy ants, a significant portion of its path is copied from the best solution of the current generation. Moreover, at the end of the tour completion, the best ant is separated from the population. To solve the GSP, a few tasks on the different nodes/machine are exchanged. Therefore, the fitter ants in the current iteration are mutated to generate lazy ants. Until the generation of fitter lazy ants in successive iterations, the lazy ants will be alive. This is possible only if lazy ants are empowered with memory.

When the best ant of each generation in a double-layered structure is saved, the lazy ants are created. The probabilistic decision is not done for constructing the path of a lazy ant; this path is therefore free from the requirements of pheromone values and heuristics. Therefore, in comparison to the active ants, lazy ants consume less time to construct their tour. More than 80% of the route is copied from the best ants of the previous iteration for tour completion, while the rest of the task is mutually swapped among the nodes.

Once the best ant is separated from the population, a few tasks are replaced mutually on some machine, randomly based on the following equation, to generate a set of lazy ants of cardinality h. This process can be done for the k successive best ants. The fitness of lazy ants is computed as was done for the active ants. Thus, the best lazy ants are selected and recorded among the generated $k \times h$ lazy ants. The researchers assumed $k = 2$ and formulated according to the following equation to make the number of ants.

$$h = floor\left(\frac{|Current\ Population|}{2}\right) \tag{3.30}$$

Some novel and diverse researches related to the application of ACO in scheduling problems are displayed in Table 3.2.

Table 3.2 Research related to the application of ACO for scheduling problems.

Researchers	Approach	Problem
[19]	ACO	Scheduling transactions in a grid processing
[20]	ACO	Tread scheduling
[21]	ACO	Multi-objective job-shop scheduling with equal-size lot-splitting
[22]	ACO	Patient scheduling
[23]	ACO	Cyclic job shop robotic cell scheduling problem
[24]	ACO	Distributed meta-scheduling in lambda grids
[25]	Graph-based ACO	Integrated process planning and scheduling
[26]	ACO-DDVO	Irrigation scheduling
[27]	ACO	Airline crew scheduling problem
[28]	Hybrid ACO	No-wait flow shop scheduling problems
[29]	Modified ACO	Distributed job shop scheduling
[30]	Robust ACO	Continuous functions
[31]	ACO	Scheduling of agricultural contracting work
[32]	Hybrid ACO	Parallel machine scheduling in fuzzy environment
[33]	ACO	Multi-satellite control resource scheduling
[34]	Immune ACO	Routing optimization of emergency grain distribution vehicles

3.4 Introduction to Multi-Objective Ant Colony Optimization (MOACO)

In this section, we give concise information about the most recent concepts of MOACO, in which non-dominated solutions are obtained and are used to update global pheromone. These algorithms are just introduced to motivate researchers and practitioners to present new multi-objective algorithms or extend these algorithms to solve new multi-objective combinatorial optimization problems.

In the first place, we describe crowding population-based ant colony optimization (CPACO), introduced by [35] to solve multi-objective TSP. The researcher extended the traditional algorithm, i.e. population-based ant colony optimization (PACO) algorithm for solving a problem that applies the super/subpopulation scheme; however, a crowding replacement scheme is applied in the CPACO algorithm. The author emphasizes that this new scheme maintains a preset size of single population (S) and the generated solutions are used to initialize it randomly. A new population of solutions (Y) is created in every generation, and a random subset S' of S is compared with each new solution to find its closest match. The existing solution is replaced with a new solution if and only if the latter is better than the former solution.

CPACO applies different heuristic matrices with the same pheromone trail for each objective and initializes the pheromone matrix with some initial value τ_{init}. After that, the pheromone values of all solutions are updated in each generation according to their inverse of the rank as follows:

$$\Delta\tau_{ij}^{s} = \frac{1}{S_{rank}} \tag{3.31}$$

The dominance ranking method is applied to assign an integer rank for all solutions in the population. The notation of λ denotes a correction factor that is related to the heuristic information, and it is generated by CPACO for each objective function (d). Therefore, each ant can utilize a different amount of heuristic based on this factor. Transition rule probability is calculated using the following equation:

$$P_{ij}^{k} = \frac{\left[\tau_{ij}\right]^{\alpha} \cdot \prod_{d=1}^{h} \left[\mu_{ij}^{d}\right]^{\lambda_d \beta}}{\sum_{l \in N_i^k} \left[\tau_{il}\right]^{\alpha} \cdot \prod_{d=1}^{h} \left[\mu_{il}^{d}\right]^{\lambda_d \beta}} \qquad \text{if } j \in N_i^k \tag{3.32}$$

The measures that are applied to test the performance of the CPACO algorithm include dominance ranking and attainment surface comparison. To evaluate the closeness of solutions to the Pareto front and to increase the diversity of solutions, the measures above are applied. Furthermore, statistical analysis indicates that the CPACO algorithm outperforms the traditional method. The CPACO algorithm records a smaller population than the traditional algorithm. CPACO uses the sorting method and PACO applies the average-rank-weight method to rank the objectives. Consequently, the computational complexity of CPACO is lower than that of PACO.

In the next place, we present Pareto strength ant colony optimization (PSACO), which was proposed by [36] to solve the multi-objective problem. This algorithm is based on the first ant colony groups; AS and the domination concept of the strength Pareto evolutionary algorithm (SPEA-II) introduced by [37] are applied for optimizing the multi-objective TSP problem. This algorithm makes two sets of solutions; population P_t and archive A_t for each iteration t. Solutions generated by the current iteration are maintained in the set of P_t and, moreover, a fixed number of globally best non-dominated solutions are saved in the archive set. If the size of the archive exceeds the fixed value, it is truncated by the current best dominant solutions. *S(i)* in Eq. (3.33) is a strength value that shows the number of the solutions dominated in the population and archive by each individual.

$$S(i) = \left|\left\{ j \middle| j \in P_t \cup A_t,\ i > j \right\}\right| \tag{3.33}$$

where the cardinality of a set is denoted by |.| and $i>j$ represents that solution i dominates solution j. To evaluate the quality of each solution *Q(i)* as formulated in Eq. (3.33), two other partial values are calculated. The fitness of each individual is based on *S(i)* and is represented by *R(i)*. Furthermore, *D(i)* shows the density information of each individual that is computed by using the kth nearest neighbor (KNN) method.

$$Q(i) = \frac{1}{(D(i) + R(i))} \tag{3.34}$$

The pheromone update procedure of this algorithm is formulated as Eqs. (3.34) and (3.35):

$$\tau_{ij}(t+1) = (1-\rho).\tau_{ij}(t) + \sum_{k=1}^{m} \Delta\tau_{ij}^{k} \tag{3.35}$$

$$where \quad \Delta\tau_{ij}^{k} = \begin{cases} \dfrac{Q}{Q(k)} & if\ ant\ k\ goes\ through\ (i,j)\ in\ this\ iteration \\ 0, & o.w. \end{cases} \tag{3.36}$$

where *Q(k)* is the solution cost of kth ant according to the *Q(i)* calculation.

3.5 Keywords Analysis for Application of ACO in Scheduling

After a brief review of ACO history, the scientometric analysis was applied using the keywords “Ant colony optimization” and “Scheduling.” To this end, the titles “Ant colony optimization” and “Scheduling” were searched in SCOPUS database, which returned approximately 1746 scientific articles between 1970 and early 2018. The search was done among keywords in “article title,” “author keywords,” and “abstract.” Figure 3.13 shows a cognitive map where the number of documents is equal to the size of the node on the mentioned term, and links among disciplines are presented by a line, and its density is relative to the level of which two areas were being used in one paper. Shadings are used to show the cluster of each item to which it belongs.

The significant keywords (five most significant and five least significant keywords) and their number of occurrences have been presented in Table 3.3. The goal of this analysis (keyword search) is to analyze the terms as regards accuracy. This analysis mainly uses brainstorming to find the crucial keywords that still have a high or low number of searches.

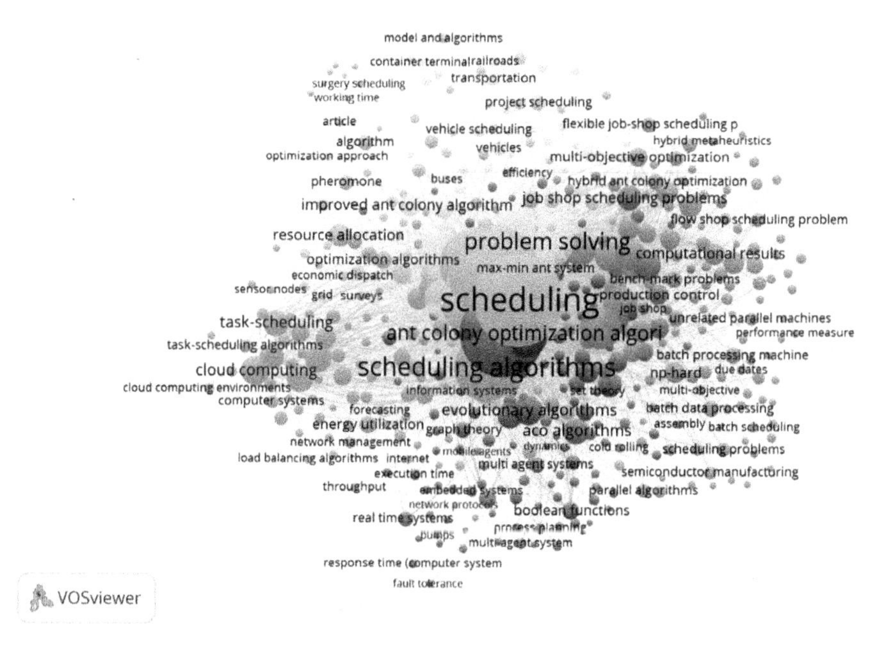

Figure 3.13 Cognitive map (co-occurrences for keyword search).

Table 3.3 The significant keywords in the application of ACO in scheduling.

No.	Keyword	Occurrences
1	Scheduling	915
2	Ant colony optimization	897
3	Optimization	810
4	algorithms	625
5	Artificial intelligence	562
6	Surgery scheduling	6
7	Hospital management	4
8	Human	4
9	Working time	5
10	Cellular manufacturing	8

3.6 Application of Bi-Objective Ant Colony Optimization in Healthcare Scheduling

3.6.1 Problem Statement

In this section, we give a brief description of surgical case processing from input to output in operating theaters (Figure 3.14). Firstly, the patient is transported from either the ward as an inpatient or ASU as an outpatient to PHU. While the patient is being held in PHU, the nurse checks their documents and prepares him/her for surgery.

The patient occupies both a nurse and a PHU bed. Then he/she is moved to an operating room where an anesthetist manages the anesthesia process, and a specific surgeon performs a surgical procedure on the case. During this stage, other resources such as a nurse, OR, anesthetist, medical technicians, scrubs, and surgeon are allocated to the surgical case. At the end of the surgical process, anesthesia is reversed by anesthetist, and then the patient is transported to PACU, where he/she recovers from the residual effects of anesthesia under the care of a PACU nurse. For the third stage, a nurse and a PACU bed are assigned to the patient.

In addition to patient flow in Figure 3.14, the sequence of the surgical case is presented. As shown, surgical cases ($SC_1,...,SC_j$) are sequenced and, moreover, efficient available resources are allocated to them in order to minimize makespan (C_{max}) and minimize the number of the non-scheduled surgical cases within the interval between the end of the time window (EW) and the start of the time window (SW). Also, multi-resources assigned to the jth surgical case for each stage are presented by blue boxes over the SC_j.

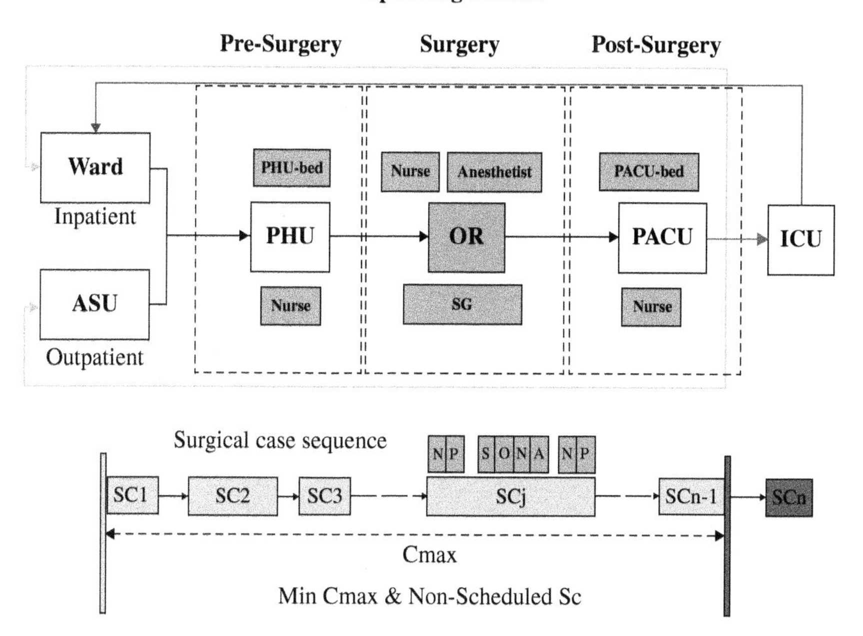

Figure 3.14 Patient processing in operating theater and surgical case scheduling in operating theater.

The various structures of the shop are taken into account to model and solve the SCS problem. For instance, [38] formulated a two-stage hybrid flow shop model in order to minimize total overtime in operating rooms, and [39] modeled the SCS problem as a four-stage flow shop under open scheduling policy. In other studies [13, 40, 41], the similarities between the operating room scheduling environment and job shop scheduling were observed. An FJS is introduced by Pinedo [42] as a generalization of the job shop with the incorporation of the parallel machine environments. Each order has its route to follow through the shop. Even though the flow shop may be modeled for an operating theater in the case of identical surgical procedures, in the real world each surgical case is processed by a special surgeon and therefore the job shop environment is suitable for this problem. Pham and Klinkert [40] developed novel multi-mode blocking job shop scheduling to model the SCS problem in order to minimize makespan. However, [13] considered the generalization of the job shop and then formulated a multi-resource FJSP in order to minimize makespan. They assumed that the operating sequence of three stages must be followed completely and in sequence, so this assumption makes a constraint that follows the rules of a no-wait flow shop.

Pinedo defined the no-wait requirement as a phenomenon that may occur in flow shops with zero intermediate storage. In the no-wait situation, orders are not permitted to wait between two successive machines. By using this constraint in FJSP, the starting time of the first stage in PHU for the surgical case has to be delayed to ensure that the case can go through the FJS without having to wait for any resource. Therefore, the surgical cases are pulled down the line by resources that have become idle. On the other hand, [13] assumed that three general stages are essential for all cases in FJS. We can observe flexibility in all stages because of the diversification in resources of each stage.

3.6.2 Introduction to Pareto Enveloped Selection Ant System (PESAS)

In this section, we suggest metaheuristic approaches in order to tackle the combinatorial nature of the bi-objective surgical case scheduling (BOSCS) problem. In many studies in the field of operating room scheduling problems, some heuristic or metaheuristic procedures such as NSGA-II from genetic algorithm family [43, 44], tabu search [45, 46], column generation based heuristic [47], Monte-Carlo along with genetic algorithm [41], and ACO [13, 48] were developed to achieve near-optimal solutions.

As noted, [13] proposed an ACO algorithm with a two-level hierarchical graph (outer and inner graph) to solve SCS. They took into account outer and inner graphs in order to integrate sequencing surgical cases and to allocate resources simultaneously. Therefore, we extend a two-level ACO algorithm to a bi-objective two-level ACO for solving the optimization problem. To the best of our knowledge, in the area of BOSCS in literature, no research is available that employs any versions of the multi-objective ACO algorithm. Therefore, we apply pheromone updating of MMAS along with non-dominated concepts of a Pareto envelope-based selection algorithm (PESA-II) in order to construct a new algorithm from multi-objective ACO groups for solving the problem. Then we compare our proposed solution procedure with Pareto strength ant colony optimization (PSACO), which was articulated in the fourth section, and hence the efficiency of the method is determined according to the several metrics of multi-objective algorithms. The traditional algorithm is based on the proposed MOACO by [36], and our proposed algorithm called Pareto envelope-based selection ant system (PESAS), which was developed by applying rules of PESA-II along with the pheromone updating strategy of the Xiang algorithm. Therefore, we contribute to new knowledge with the extension of Xiang's model so that we develop a single objective ACO algorithm to the bi-objective algorithm. In the proposed algorithm, two-level ACO is extended to the bi-objective ant colony optimization approach by using the rules of pheromone update strategy of MMAS along with the domination concept of PESA-II [49].

In the PESAS algorithm, two agent sets include the population of ants (Ant_t) and the repository (Rep_t) is made in each iteration. Solutions produced by the current iteration are recorded in the set of Ant_t and then a fixed number of globally non-dominated solutions are saved in repository set, and this fixed number is based on a threshold of the repository (Rep_{th}). In the current iteration, the solutions of the repository are updated by generated non-dominated agents. Since an agent solution from the repository must be selected for pheromone updating, a grid is created according to the evolutionary concepts in PESA-II that solution selecting is region-based not individual-based. In region-based selection, each cell space of the grid with less population is more probable to be selected to improve the diversity of the algorithm. After creation of grid, it is rechecked that the number of repository solutions must be less than Rep_{th}; otherwise, the excessive population of the repository is eliminated according to the selection probability of cell space that cells with more population are more probable for selection. For example, as indicated in Figure 3.15, light gray (near the numbers) are non-dominated solutions of the repository and dark gray (further to the right) show dominated ant populations. For example, cells #1 and #3 are selected with more probability for updating pheromone, and the population of cell #2 is more likely to be eliminated if the total number in the repository exceeds the threshold.

The selection probability rule of the cells for pheromone updating is based on Roulette Wheel Selection (RWS) and is formulated as follows:

$$P_i = \frac{e^{-\beta n_i}}{\sum_{j=1}^{n} e^{-\beta n_j}} \quad if \quad n_i \leq n_j \Leftrightarrow P_i \geq P_j \tag{3.37}$$

where β is the selection pressure of the cell for pheromone updating, n denotes the number of the cells in repository space, n_i is the population of ith cell, and P_i is the probability selection of ith cell. It should be noted that if the population of the selected cell is more than 1, one is selected randomly according to a uniform distribution.

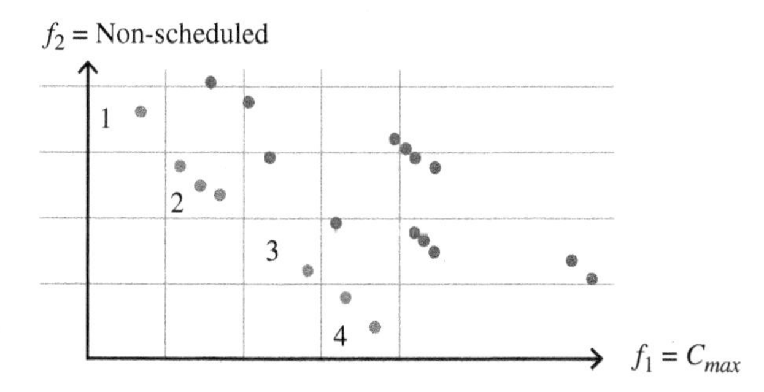

Figure 3.15 Sample of bi-objective space of ant solutions (f_1,f_2) with grid.

Besides, the selection probability rule of the cells for eliminating is based on RWS and is formulated as follows:

$$q_i = \frac{e^{\gamma n_i}}{\sum_{j=1}^{n} e^{\gamma n_j}} \quad if \quad n_i \geq n_j \Leftrightarrow q_i \geq q_j \tag{3.38}$$

where γ is the selection pressure of the cell for eliminating, n denotes the number of cells in repository space, n_i is a population of ith cell, and q_i is the probability selection of ith cell. It must be noted that if the population of the selected cell is more than 1, one is selected randomly according to a uniform distribution.

To update the trail pheromone, we follow the updating equations of MMAS proposed by [9]:

$$\tau_{ij}(t+1) = (1-\rho).\tau_{ij}(t) + \Delta\tau_{ij}^{gb} \tag{3.39}$$

$$where \quad \Delta\tau_{ij}^{gb} = \begin{cases} \frac{Q}{Q(nd)} & if\ (i,j) \in Q(nd) \\ 0 & o.w. \end{cases} \quad and \quad Q(i) = \frac{1}{S(i)} \tag{3.40}$$

and pheromone deposited based on a selected non-dominated solution by using region-based selection (*nd*). Also, *S(i)* is called strength value and indicates the number of the solutions dominated in current population and repository by each, as used in the PSACO. In this approach, the ant is selected according to Eq. (3.30) for updating.

3.6.3 Results and Discussion

3.6.3.1 Illustrative Examples

To evaluate the proposed approaches, we took three test surgery cases. These cases are classified into small, medium, and large, which differed as regards surgery duration, the number of the surgery cases, and the allocated resources. Case categories and their specifications are shown in Table 3.4. As can be seen, the three problems of each case are different in size of surgeries (column 3), size of resources (column 4–9), and surgery type structure (column 10). The algorithms were coded in Matlab language and ran on an Intel Core (TM) Duo CPU T2450, 2.00 GHz computer with 1 GB of RAM.

3.6.3.2 Performance Evaluation Metrics

In this section, some main performance metrics are applied in order to evaluate the quality and diversity of the obtained non-dominated solutions in the Pareto

Table 3.4 Test cases and structure.

Cases	Problem	Surgical case	PHU bed	Nurse	Surgeons	ORs	PACU bed	Anesthesia	Surgery type (S:M:L:EL:S)
Case1	1	8	1	5	5	2	2	5	2:4:1:1:0
	2	10	2	8	6	4	4	6	2:6:1:1:0
	3	10	2	8	6	4	4	6	2:5:2:1:0
Case2	1	15	3	10	6	4	3	8	3:9:2:1:0
	2	20	3	15	10	5	4	8	4:12:3:1:0
	3	20	3	15	10	5	4	8	4:10:3:3:0
Case3	1	30	4	19	10	6	5	9	7:16:3:2:2
	2	30	4	22	12	6	5	11	5:15:3:4:3
	3	30	5	22	12	6	6	12	3:15:3:4:5

repository set of the PESAS algorithm. There are three critical metrics in the literature of multiple objective problems [50], which we describe as follows:

- *Quality metric* (QM). This metric was proposed by [51] and takes into account the number of Pareto solutions obtained by each algorithm. In other words, non-dominated solutions of all algorithms are compared and then the algorithm with a higher number of final Pareto solutions has more quality. It is clear that the larger the *QM*, the better the obtained solution set.

$$Q = N(S^{P_1 \cup P_2}) = \left| S^{P_1 \cup P_2} - \left\{ x \in S^{P_1 \cup P_2}\ \exists y \in S^{P_1 \cup P_2} : x\, dom\, y \right\} \right| \tag{3.41}$$

where, $S^{P_1 \cup P_2}$ denotes the final Pareto solution set of both algorithms and N is the number of the final obtained Pareto solutions that contains N_{p1} and N_{p2}.

- *Spacing metric* (SM).This metric was employed by [52] in order to assess the uniformity of the spread of the solutions in the final Pareto solution set obtained by each algorithm, and it is computed as follows:

$$S = \frac{\sum_{i=1}^{N-1} \left| d_i - \bar{d} \right|}{(N-1)\bar{d}} \tag{3.42}$$

where, d_i denotes the Euclidean distance between the two solutions of the Pareto set, $\bar{d}$ denotes the average value of all Euclidean distances, and N is the number of final Pareto solutions found. It is obvious that the smaller the *SM*, the better spacing of the Pareto set and the best value for this metric is equal to 0.

- *Diversity metric* (DM). This metric was used by [53] and was employed to evaluate the spread of the solutions in the final Pareto set of each algorithm. This metric is formulated as follows:

$$D = \sqrt{\sum_{i=1}^{n} \max\left(\left\| x_i - y_i \right\|;\ \vec{x}, \vec{y} \in F\right)} \tag{3.43}$$

where F denotes the set of found Pareto, $\vec{x}$ and $\vec{y}$ denotes two solution vectors of Pareto frontier, and n represents the dimension of the solution space or, in other words, the number of the objective functions. It is clear that the larger the *DM*, the better the generated Pareto set.

3.6.3.3 Comparison between the Algorithm's Performance on all Considered Instances

The proposed algorithm was run 10 times for each instance tested in order to compare with other traditional algorithms.

Table 3.5 indicates the average performance of our proposed algorithm and comparison to PSACO as an important method for solving all instances. The first and second columns display considered test problems and approaches, respectively. The next three columns represent the average, best, and worst of the QM for solutions obtained. The average, best, and worst of the diversity metric for solutions are given from the sixth to eighth columns of the table. Also, the following three columns present the results of SM. [54] introduced a two-time index to compare the computational time of algorithms, namely $A_1(s)$ the average computational times for obtaining one solution and $A_2(s)$ the average computational times for all solutions found. The last two columns of the table give results of $A_1(s)$ and $A_2(s)$ after 10 runs. These computational time criteria are formulated as follows:

$$A_2(s) = \frac{\sum_{i=1}^{w} t_i}{w} \tag{3.44}$$

$$A_1(s) = \frac{A_2(s)}{\sum_{i=1}^{w} \left| S_k^i \right| / w} \tag{3.45}$$

Where S_k^i denotes the number of the solutions found in the ith run, t_i represents the computational times for the ith run, and w is running time.

As the table shows, PESAS outperforms PSACO for solving all instances from small to large size in three metrics (quality, diversity, and spacing), with the exception of the diversity metric for instance #1 and the best SM obtained for instances #6 and #9, where PSACO yields better metrics. Nevertheless, the optimal

Table 3.5 Comparison of the performance of the algorithms on all considered test case problems.

Problems	Algorithms	QM			DM			SM				
		Average	Best	Worst	Average	Best	Worst	Average	Best	Worst	$A_1(s)$[a]	$A_2(s)$[a]
Instance1	PESAS	**3.9**	**5**	**3**	139.066	156	**128.03**	**0.7228**	**0.268**	**0.99**	**4.9443**	19.283
	PSACO	0.1	1	0	**145.814**	**163.05**	120.04	1.059	0.55	1.68	186.87	**18.687**
Instance2	PESAS	**6.4**	**7**	**5**	**167.404**	**179.1**	**159.1**	**0.568**	**0.4**	**0.8**	**4.1542**	**26.587**
	PSACO	0.3	1	0	141.613	159.11	132	0.993	0.46	1.43	89.34	26.802
Instance3	PESAS	**7.8**	**9**	**7**	**182.619**	**191.16**	**170.14**	**0.631**	**0.47**	**0.8**	**3.3124**	**25.837**
	PSACO	0.1	1	0	167.535	188.2	122.14	0.831	0.58	1.16	265.5	26.55
Instance4	PESAS	**9.5**	**10**	**8**	**177.79**	**196.2**	**162.25**	**0.497**	**0.29**	**0.76**	**3.9363**	37.395
	PSACO	0	0	0	172.48	196.16	155.16	0.846	0.43	1.12	infinite	**31.218**
Instance5	PESAS	**11.4**	**12**	**10**	**189.777**	**206.4**	**171**	**0.466**	**0.34**	**0.59**	**7.9447**	90.57
	PSACO	0.5	1	0	184.47	205	167.3	0.787	0.5	1.12	158.34	**79.17**
Instance6	PESAS	**7.8**	**9**	**5**	**185.9873**	**212.8**	**169.9**	**0.4239**	**0.12**	**0.7**	**13.871**	108.2
	PSACO	0.6	3	0	165.67	188	144.3	0.636	0.42	1.09	161.583	**96.95**
Instance7	PESAS	**11.5**	**17**	**8**	**290.404**	**307.59**	**254.33**	**0.5019**	**0.29**	**0.67**	**17.0143**	195.665
	PSACO	1.6	7	0	259.611	291.4	220.5	0.675	0.48	0.89	116.461	**186.339**
Instance8	PESAS	**9.3**	**13**	**3**	**324.888**	344.37	**290.5**	**0.577**	**0.34**	**0.82**	**21.5433**	200.353
	PSACO	2.9	7	0	289.327	**345**	246.4	0.663	0.37	0.91	64.8520	**188.071**
Instance9	PESAS	**7.1**	**14**	**3**	**320.423**	**376.6**	**299.6**	0.528	0.32	**0.82**	**29.7419**	**211.168**
	PSACO	2.9	7	0	302.033	355	263.4	**0.523**	**0.22**	0.9	72.8351	211.222

[a] Time unit: second.

non-dominated solutions found by PESAS are better than those of PSACO for all test cases. Since our proposed algorithm uses only a diverse dominant agent to update the pheromone trail, it seems that the results of PESAS outperform the results of PSACO, which uses all dominant ants to update the pheromone. It seems that having diverse dominant ants increases the diversification strategy of the algorithm to find a better Pareto front, and the results indicate this assumption. On the other hand, the average of the computational time (A_2) of PSACO is less than that of PESAS, and the reasons for this extra time relate to the existence of the grid-based selection procedure in our proposed algorithm.

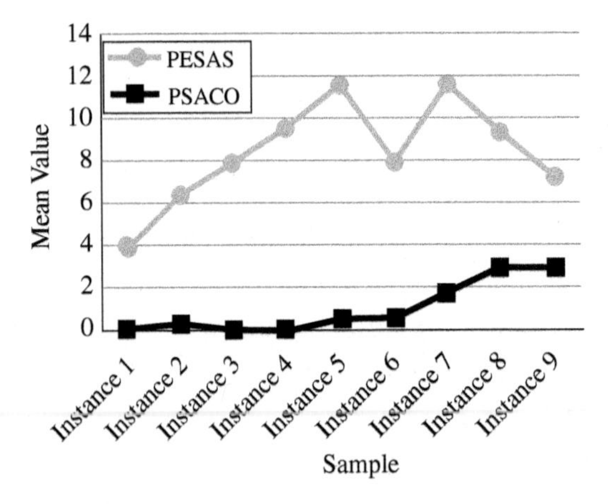

Figure 3.16 Interaction between the algorithms and test samples for the quality measure.

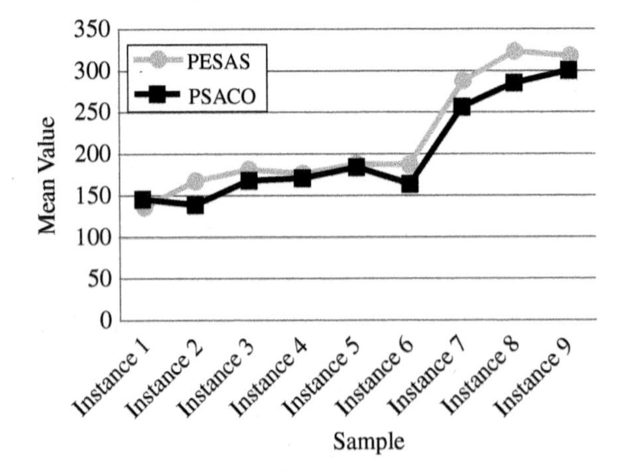

Figure 3.17 Interaction between the algorithms and test samples for the diversity measure.

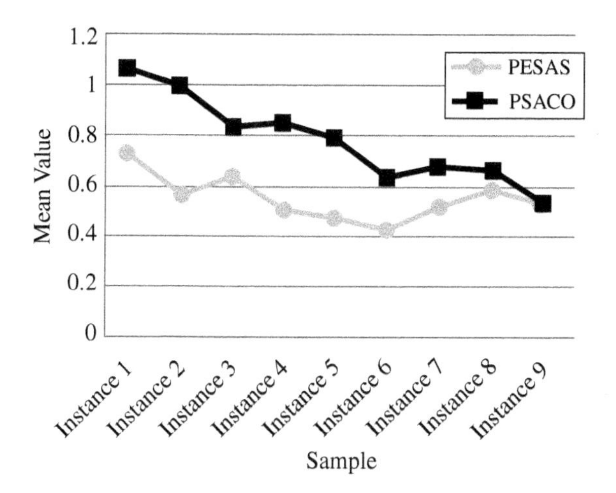

Figure 3.18 Interaction between the algorithms and test samples for the spacing measure.

However, our proposed algorithm requires less effort for finding an optimal solution in comparison with PSACO based on $A_1(s)$ measurement.

Figure 3.16 shows that the proposed PESAS algorithm obtains more Pareto solutions than the other algorithm for each instance and therefore our method outperforms PSACO. Figure 3.17 indicates that the proposed PESAS algorithm obtains more diverse solutions than the other algorithm for each instance except instance 1. Thus our method outperforms PSACO. Figure 3.18 indicates that the proposed PESAS algorithm obtains solutions with less space metric than another algorithm for each instance except instance #6 and #9 in best SM obtained. Therefore, our method outperforms PSACO.

References

1 Dorigo, M., Maniezzo, V., and Colorni, A. (1991). Positive feedback as a search strategy (Tech. Rep. No. 91-016). *Politecnico di Milano, Milan, Italy.*

2 Dorigo, M. and Gambardella, L.M. (1997). Ant colony system: a cooperative learning approach to the traveling salesman problem. *IEEE Trans. Evol. Comput.* 1: 53–66.

3 Deneubourg, J.-L., Aron, S., Goss, S., and Pasteels, J.M. (1990). The self-organizing exploratory pattern of the argentine ant. *J. Insect Behav.* 3: 159–168.

4 Blum, C. and Roli, A. (2003). Metaheuristics in combinatorial optimization: overview and conceptual comparison. *ACM Comput. Surveys (CSUR)* 35: 268–308.

5 Dorigo, M. and Blum, C. (2005). Ant colony optimization theory: a survey. *Theor. Comput. Sci.* 344: 243–278.

6 Yang, X.-S. (2010). *Nature-Inspired Metaheuristic Algorithms*. Luniver Press.

7 Dorigo, M. (1992). Optimization, learning and natural algorithms. PhD thesis. Politecnico di Milano.

8 Bullnheimer, B., Hartl, R.F., and Strauss, C. (1997). A new rank based version of the Ant System. A computational study. *Centr. Eur. J Operat. Res.* 7: 25–38.

9 Stützle, T. and Hoos, H.H. (2000). MAX–MIN ant system. *Future Gener. Comput. Syst.* 16: 889–914.

10 Cordon, O., de Viana, I.F., Herrera, F., and Moreno, L. (2000). A new ACO model integrating evolutionary computation concepts: The best-worst Ant System. Proc Ants'2000. 22–29.

11 Guntsch, M. and Middendorf, M. (2002b). A population based approach for ACO. Workshops on Applications of Evolutionary Computation 72–81. Springer.

12 Blum, C. (2005). Beam-ACO—hybridizing ant colony optimization with beam search: an application to open shop scheduling. *Comput. Oper. Res.* 32: 1565–1591.

13 Xiang, W., Yin, J., and Lim, G. (2015). An ant colony optimization approach for solving an operating room surgery scheduling problem. *Comput. Ind. Eng.* 85: 335–345.

14 Tiwari, P.K. and Vidyarthi, D.P. (2016). Improved auto control ant colony optimization using lazy ant approach for grid scheduling problem. *Future Gener. Comput. Syst.* 60: 78–89.

15 Cordón, O., de Viana, I.F., and Herrera, F. (2002). Analysis of the best-worst ant system and its variants on the QAP. International Workshop on Ant Algorithms, 228–234. Springer.

16 Guntsch, M. and Middendorf, M. (2002). Applying population based ACO to dynamic optimization problems. International Workshop on Ant Algorithms, 111–122. Springer.

17 Ow, P.S. and Morton, T.E. (1988). Filtered beam search in scheduling. *Int. J. Prod. Res.* 26: 35–62.

18 Gordon, D.M., Goodwin, B.C., and Trainor, L.E.H. (1992). A parallel distributed model of the behavior of ant colonies. *Theoret. Biol.* 156 (3): 293–307.

19 Mahato, D.P., Singh, R.S., Tripathi, A.K., and Maurya, A.K. (2017). On scheduling transactions in a grid processing system considering load through ant colony optimization. *Appl. Soft Comput.* 61: 875–891.

20 Anjaria, K. and Mishra, A. (2017). Thread scheduling using ant colony optimization: an intelligent scheduling approach towards minimal information leakage. *Karbala Int. J. Modern Sci.* 3: 241–258.

21 Huang, R.-H. and Yu, T.-H. (2017). An effective ant colony optimization algorithm for multi-objective job-shop scheduling with equal-size lot-splitting. *Appl. Soft Comput.* 57: 642–656.

22 Obiniyi, A. (2015). Multi-agent based patient scheduling using ant colony optimization. *Afr. J. Comput. ICT* 8: 91–96.

23 Elmi, A. and Topaloglu, S. (2017). Cyclic job shop robotic cell scheduling problem: ant colony optimization. *Comput. Ind. Eng.* 111: 417–432.

24 Pavani, G.S. and Tinini, R.I. (2016). Distributed meta-scheduling in lambda grids by means of ant colony optimization. *Future Gener. Comput. Syst.* 63: 15–24.

25 Wang, J., Fan, X., Zhang, C., and Wan, S. (2014). A graph-based ant colony optimization approach for integrated process planning and scheduling. *Chin. J. Chem. Eng.* 22: 748–753.

26 Nguyen, D.C.H., Ascough, J.C. II, Maier, H.R. et al. (2017). Optimization of irrigation scheduling using ant colony algorithms and an advanced cropping system model. *Environ. Model. Software* 97: 32–45.

27 Deng, G.-F. and Lin, W.-T. (2011). Ant colony optimization-based algorithm for airline crew scheduling problem. *Expert Syst. Appl.* 38: 5787–5793.

28 Engin, O. and Güçlü, A. (2018). A new hybrid ant colony optimization algorithm for solving the no-wait flow shop scheduling problems. *Appl. Soft Comput.* 72: 166–176.

29 Chaouch, I., Driss, O.B., and Ghedira, K. (2017). A modified ant colony optimization algorithm for the distributed job shop scheduling problem. *Procedia Comput. Sci.* 112: 296–305.

30 Chen, Z., Zhou, S., and Luo, J. (2017). A robust ant colony optimization for continuous functions. *Expert Syst. Appl.* 81: 309–320.

31 Alaiso, S., Backman, J., and Visala, A. (2013). Ant colony optimization for scheduling of agricultural contracting work. *IFAC Proc. Vol.* 46: 133–137.

32 Liao, T.W. and Su, P. (2017). Parallel machine scheduling in fuzzy environment with hybrid ant colony optimization including a comparison of fuzzy number ranking methods in consideration of spread of fuzziness. *Appl. Soft Comput.* 56: 65–81.

33 Zhang, Z., Zhang, N., and Feng, Z. (2014). Multi-satellite control resource scheduling based on ant colony optimization. *Expert Syst. Appl.* 41: 2816–2823.

34 Zhang, Q. and Xiong, S. (2018). Routing optimization of emergency grain distribution vehicles using the immune ant colony optimization algorithm. *Appl. Soft Comput.* 71: 917–925.

35 Angus, D. (2007). Crowding population-based ant colony optimisation for the multi-objective travelling salesman problem. IEEE Symposium on Computational Intelligence in Multicriteria Decision Making, 333–340.

36 Thantulage, G.I. (2009). Ant colony optimization based simulation of 3D automatic hose/pipe routing. PhD thesis. Brunel University School of Engineering and Design.

37 Zitzler, E., Laumanns, M., and Thiele, L. (2001). SPEA2: Improving the Strength Pareto Evolutionary Algorithm. TIK-report 103.

38 Guinet, A. and Chaabane, S. (2003). Operating theatre planning. *Int J. Prod. Econ.* 85: 69–81.

39 Augusto, V., Xie, X., and Perdomo, V. (2010). Operating theatre scheduling with patient recovery in both operating rooms and recovery beds. *Comput. Ind. Eng.* 58: 231–238.

40 Pham, D.-N. and Klinkert, A. (2008). Surgical case scheduling as a generalized job shop scheduling problem. *Eur. J. Oper. Res.* 185: 1011–1025.

41 Lee, S. and Yih, Y. (2014). Reducing patient-flow delays in surgical suites through determining start-times of surgical cases. *Eur. J. Oper. Res.* 238: 620–629.

42 Pinedo, M.L. (2008). *Scheduling: Theory, Algorithms, and Systems*. New York: Springer.

43 Marques, I. and Captivo, M.E. (2015). Bicriteria elective surgery scheduling using an evolutionary algorithm. *Oper. Res. Health Care* 7: 14–26.

44 Marques, I., Captivo, M.E., and Pato, M.V. (2014). Scheduling elective surgeries in a Portuguese hospital using a genetic heuristic. *Oper. Res. Health Care* 3: 59–72.

45 Lamiri, M., Grimaud, F., and Xie, X. (2009). Optimization methods for a stochastic surgery planning problem. *Int J. Prod. Econ.* 120: 400–410.

46 Saremi, A., Jula, P., Elmekkawy, T., and Wang, G.G. (2013). Appointment scheduling of outpatient surgical services in a multistage operating room department. *Int J. Prod. Econ.* 141: 646–658.

47 Fei, H., Meskens, N., and Chu, C. (2010). A planning and scheduling problem for an operating theatre using an open scheduling strategy. *Comput. Ind. Eng.* 58: 221–230.

48 Behmanesh, R., Zandieh, M., and Hadji Molana, S.M. (2019). The surgical case scheduling problem with fuzzy duration time: an ant system algorithm. *Sci. Iran.* 26 (3): 1824–1841.

49 Corne, D.W., Jerram, N.R., Knowles, J.D., and Oates, M.J. (2001). PESA-II: region-based selection in evolutionary multiobjective optimization. Proceedings of the 3rd Annual Conference on Genetic and Evolutionary Computation, 283–290. Morgan Kaufmann Publishers Inc.

50 Noori-Darvish, S., Mahdavi, I., and Mahdavi-Amiri, N. (2012). A bi-objective possibilistic programming model for open shop scheduling problems with sequence-dependent setup times, fuzzy processing times, and fuzzy due dates. *Appl. Soft Comput.* 12: 1399–1416.

51 Schaffer, J.D. (1985). Multiple objective optimization with vector evaluated genetic algorithms. Proceedings of the First International Conference on Genetic Algorithms and Their Applications, Lawrence Erlbaum Associates. Inc., Publishers.

52 Svinivas, N. (1995). Multiobjective optimization using nondominated sorting in genetic algorithms. *IEEE Trans. Evol. Comput.* 2: 221–248.

53 Zitzler, E. (1999). *Evolutionary Algorithms for Multiobjective Optimization: Methods and Applications*. Citeseer.

54 Li, J.-Q., Pan, Q.-K., and Tasgetiren, M.F. (2014). A discrete artificial bee colony algorithm for the multi-objective flexible job-shop scheduling problem with maintenance activities. *App. Math. Model.* 38: 1111–1132.

4

Task Scheduling in Heterogeneous Computing Systems Using Swarm Intelligence

S. Sarathambekai and K. Umamaheswari

Department of Information Technology, PSG College of Technology, Coimbatore, Tamilnadu, India

4.1 Introduction

A distributed system is a collection of multiple heterogeneous processing elements that communicate with one another to solve a problem. The Task Scheduling (TS) problem is one of the most challenging issues in distributed environments, and is known to be a Nondeterministic Polynomial time (NP) hard problem [1]. To solve NP-hard problems, heuristics/metaheuristic methods have been used instead of traditional optimization methods in order to get a near-optimal solution within a finite duration. DPSO is a well-known swarm technique. Each particle in a DPSO algorithm moves in N-dimensional problem space with a velocity. The velocity dynamically changes based on its own experience and the experience of neighboring particles. Suppose the neighborhood of a particle is the entire swarm, then the best position in the neighborhood is called the Global best (Gbest) particle; otherwise called the Local best (Lbest) particle [2].

The effect of neighborhood topologies such as star, ring, and Von Neumann (VN) on PSO have been investigated and it has been demonstrated that the different neighborhood topologies significantly improved the performance of the PSO [3–5]. The neighborhood topologies such as fully connected, ring, and VN for unimodal and multimodal functions were evaluated by Pimpale [6] and stated that the Lbest topologies performed less well in unimodal but provided better results for multimodal optimization.

Ni and Deng [7] suggested an improved random topology which was based on graph theory. Here, the neighboring particles are randomly selected for each

Evolutionary Computation in Scheduling, First Edition. Edited by Amir H. Gandomi, Ali Emrouznejad, Mo M. Jamshidi, Kalyanmoy Deb, and Iman Rahimi.

particle and Dijkstra's algorithm is applied if there are any unconnected particles in the randomly generated topology. The author illustrated the improvement of random topology-based PSO when compared to other topologies such as fully connected, ring- and star-based PSO.

Reyes Medina et al. [8] studied and analyzed the tree-based communication topology. It is constructed as a binary tree where the root node is randomly selected and the remaining particles are distributed in the tree branches. Once the tree is constructed, the neighboring particles will not be changed during the entire execution of the algorithm.

Sarathambekai and Umamaheswari [9] explained the significance of dynamic topology and proposed a Binary Heap Tree (BHT)-based dynamic topological structure for TS problem. The author analyzed the performance of BHT with various topologies. The paper [9] provides detailed steps to build BHT and very minimal information for constructing other neighborhood structures in the literature. This chapter explains how to make various static topological models such as mesh, star, ring, VN, and binary tree for the TS problem in distributed systems. The concept of BHT was taken from the previously published paper [9]. Therefore, this chapter is an extended version of [9].

The algorithm presented in this chapter considers the scheduling of independent tasks with minimizing tri-objectives [9] such as makespan, mean flow time and reliability cost using DPSO with various neighborhood topologies.

The remainder of the chapter is organized as follows: Section 4.2 describes the problem formulation. Section 4.3 explains swarm intelligence (SI) in the TS problem. The dynamic topology is presented in Section 4.4. Section 4.5 presents the performance metrics of the system. Implementation details are reported in Section 4.6. Section 4.7 describes the real-time scenario for SI in the scheduling problem. Finally, Section 4.8 concludes the chapter.

4.2 Problem Formulation

4.2.1 Task Model

A distributed system consists of a number of heterogeneous processors connected with a mesh topology. Let $T = \{T_1, T_2..., T_n\}$ denote the n tasks that are independent of each other to be scheduled on m processors $P = \{P_1, P_2..., P_m\}$. Expected Time to Compute (ETC) matrix is a n × m matrix which is used to represent expected execution times of tasks on processors. One row of the ETC matrix represents estimated execution time for a specified task on each processor. Similarly, one column of the ETC matrix consists of the estimated execution time of a specified processor for each task.

The TS problem is formulated based on the following assumptions:

- ETC values are known in advance and all tasks are available at zero time.
- Processors are always available and preemption is not allowed.
- Each processor can process only one task at a time.
- A task cannot be processed on more than one processor at a time.
- Each processor uses the First-Come, First-Served (FCFS) method for performing the received tasks.

4.2.2 Scheduler Model

A static scheduler model in distributed systems is shown in Figure 4.1. Let $T = \{T_1, T_2..., T_n\}$ denote the n tasks that are independent of each other to be scheduled on m processors $P = \{P_1, P_2..., P_m\}$. The two queues such as a Task Queue (TQ) and a Processor Queue (PQ) are managed by the scheduler. Initially the particles are in TQ. The particle is encoded using a permutation-based method. In the permutation vector, the position of a task represents the sequence of the task scheduled and the corresponding value indicates a task number. A sample particle with size 6 (Task number: 1, 2, 3, 4, 5, and 6) is shown in Figure 4.2.

Each processor has a PQ which maintains the tasks to be executed on that processor. The scheduling algorithm in the central scheduler is started working with the TQ. Based on the workload of each processor in the distributed systems, the

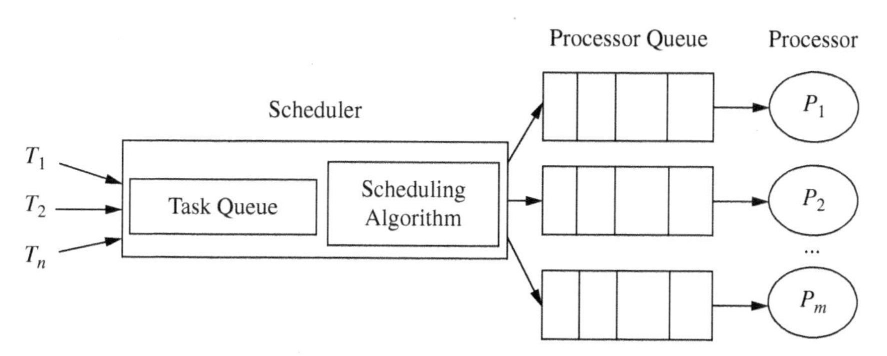

Figure 4.1 Scheduler model for distributed systems.

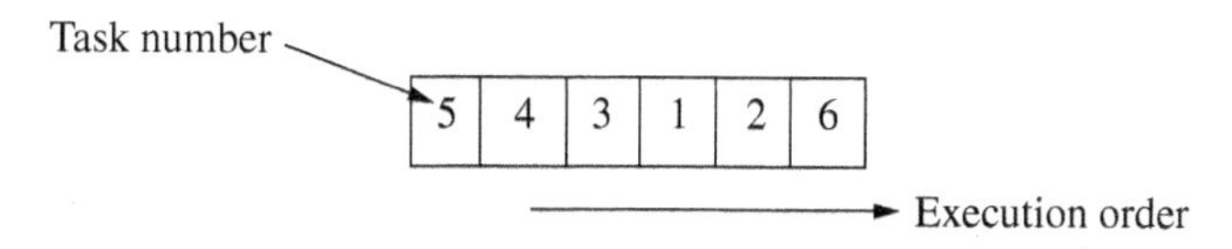

Figure 4.2 A particle in permutation based format.

scheduler distributes each task in the particle from the TQ to the individual PQ. After completing the placement of tasks in the particle from the TQ to PQ, the processors (P_1, P_2... P_m) will start executing the tasks in their own PQ using the FCFS method.

4.3 SI in the TS Problem

The SI algorithms are inspired by the behavior of certain social living beings (ants, birds, fishes, etc.). The two main notable characteristics of SI-based algorithms are self-organization and decentralized control [9, 10]. Particle swarm optimization (PSO) is a prominent SI algorithm for addressing the meta-tasks scheduling problem in distributed systems.

4.3.1 Discrete Particle Swarm Optimization (DPSO)

PSO is an optimization algorithm based on population, where the system is initialized with a population of random particles. The population of the possible particles in PSO is known as a swarm and each particle moves in the D-dimensional problem space with a certain velocity. The velocity is dynamically changed based on the flying knowledge of its own (Pbest) and the knowledge of the swarm (Gbest). The velocity of a particle is controlled by three components: inertial momentum, cognitive, and social. The inertial component simulates the inertial behavior of the bird to fly in the previous direction, the cognitive component models the memory of the bird about its previous best position, and the social component models the memory of the bird about the best position among the particles. The velocity and position of the particles in PSO are updated using Eqs. (4.1) and (4.2) respectively.

$$V_i^{(t+1)}(j) = WV_i^t(j) + C_1 r_1 (Pb_i^t(j) - p_i^t(j)) + C_2 r_2 (Gb^t(j) - p_i^t(j)) \tag{4.1}$$

$$p_i^{(t+1)}(j) = p_i^t(j) + V_i^{(t+1)}(j) \tag{4.2}$$

Where $i = 1, 2, 3...N$; $j = 1, 2, 3...n$; N denotes the swarm size and n is the size of particle.

- W is to the inertia weight used to control the impact of the previous history of velocities V_i^t on the current velocity of a given particle.
- $V_i^{(t+1)}(j)$ is the jth element of the velocity vector of the ith particle in $(t+1)^{th}$ iteration, which determines the direction in which a particle needs to move.
- $p_i^t(j)$ is the jth element of the ith particle in the tth iteration.

- r_1 is the random value in range[0,1] sampled from a uniform distribution.
- C_1 and C_2 are positive constants, called acceleration coefficients, which control the influence of Personal best (Pb) and Global best (Gb) on the search process.

PSO is different from other evolutionary techniques in that it does not require filtering operations such as crossover and mutation. In PSO, the members of the entire swarm are preserved through the search procedure, so that information is socially shared among particles to direct the search toward the finest position in the search space. In addition to being easily implemented, it is also computationally inexpensive, because its memory and CPU speed necessities are significantly lower [11]. Due to the simplicity in concept, easy implementation, and quicker convergence [11], this chapter presents an SI based on a PSO variant. The PSO cannot be used directly in the TS problem because their positions happen to be continuous values. Therefore, discretization methods were introduced to transform a real (continuous) solution vector into a discrete solution vector to specifically address the discrete optimization problem [11]. The pseudo code of the PSO algorithm for TS problem with swarm size (N), particle size (n), and number of processors (m) is presented in Figure 4.3.

Sarathambekai and Umamaheswari [12] presented a DPSO algorithm in which the particles can update their positions in a discrete domain directly. Therefore, the mapping techniques (Sigmoid Function, Smallest Position Value) are not required for transforming continuous positions of particles into discrete values, thus saving a considerable amount of computational time. The neighborhood

Pseudo code 4.1: PSO for Task scheduling problem
Input: N, n, m; Output: Global best

begin
    Initialize the swarm randomly.
    Initialize each particle position and velocity.
    Using Discretization method to map the particle's position from continuous space into discrete space
    Evaluate each particle and find the Personal best and the Global best.
repeat
    Update velocity of each particle using Eq. (4.1).
    Update position of each particle using Eq. (4.2).
    Using Discretization method to map the particle's position from continuous space into discrete space
    Evaluate fitness value of each new particle.
    Update Personal best and the Global best for each new particle.
until stopping condition is true.

Figure 4.3 PSO with discretization method for task scheduling problem.

topology of DPSO significantly improves the performance of the algorithm, because it determines the rate at which the information transmits through the swarm. The following section presents the neighborhood communication of the DPSO algorithm.

4.3.2 DPSO with Neighborhood Communication

4.3.2.1 Neighborhood Model Based on Structure

The neighborhood topological structure is broadly categorized into two models:

- Global best model
- Local best model.

4.3.2.2 Global Best Model (Gbest Model)

The Gbest model is a fully connected topology model or mesh topology model where all the particles are connected with each other. This topological structure has $N(N-1)/2$ communication links to connect N particles in the swarm. Figure 4.4 shows the Gbest model of DPSO with a swarm size of 6 (six particles).

In this model, all the particles will be attracted by the best particle found in the whole swarm. The sample fitness values (more information regarding the procedure to calculate fitness value is presented in [13]) of the swarm size of 6 for the Gbest model is given in Figure 4.5. It is a minimization problem. Therefore, the particle with the least fitness value will be selected as Gbest particle. The neighborhood structure of Gbest model is fixed for all the iterations. However, only the Gbest particle varies from generation to generation based on the fitness value.

Since all the particles are connected, the broadcasting of the best position and fitness information is very fast. This fast broadcasting might result in a premature convergence (local optima) problem and also prevent further exploration (diversity) of the search space.

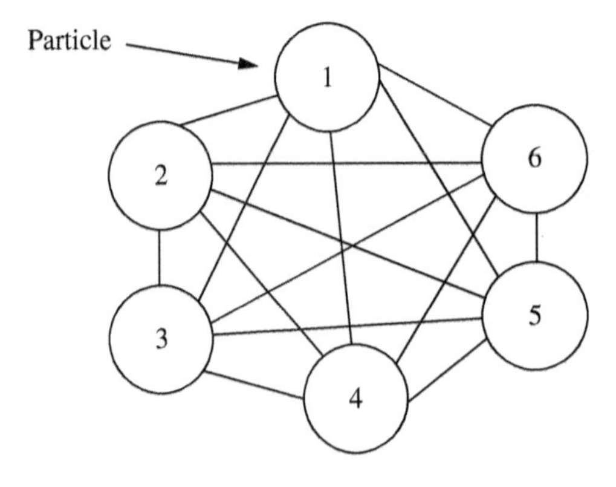

Figure 4.4 A neighborhood structure for Gbest model with swarm size 6.

Particle	Sample Fitness Value
1	2.11
2	1.14
3	4.03
4	2.55
5	3.67
6	5.04

Iteration 1

Gbest particle: 2 (minimum fitness value)

The particles (1, 3, 4, 5, 6) move towards the Gbest particle: 2.

Particle	Sample Fitness Value
1	3.11
2	2.14
3	3.03
4	1.05
5	4.67
6	3.04

Iteration 2

Gbest particle: 4 (minimum fitness value)

The particles (1, 2, 3, 5, 6) move towards the Gbest particle: 4.

Figure 4.5 An illustrative example of Gbest model with swarm size of 6.

4.3.2.3 Local Best Model (Lbest Model)

The Lbest model is constructed with only a few neighbors. It takes time to propagate the information to other particles in the swarm. This slow propagation will allow the particles to explore more areas in the search space, thus reducing the possibility of premature convergence.

The Lbest model does not support elitism, which means that the best particle is not retained in every iteration. The best particle is selected based on the fitness value of the particle in the current swarm within the current iteration.

4.3.2.3.1 Star Topology In the star-based Lbest model, all particles in the swarm are connected to the central particle, known as a hub, which is randomly selected from the swarm. The entire swarm moves toward the hub and the hub directs its flight toward the best particle of the neighborhood. The graphical representation of star-based Lbest model is given in Figure 4.6. In this figure, particle 3 acts as a hub which is selected by random.

Figure 4.7 demonstrates finding Lbest particles of the swarm size of 8 with the sample fitness values. The neighborhood structure of the star-based Lbest model is fixed for all iterations. However, only the best particle varies from generation to generation based on the fitness value.

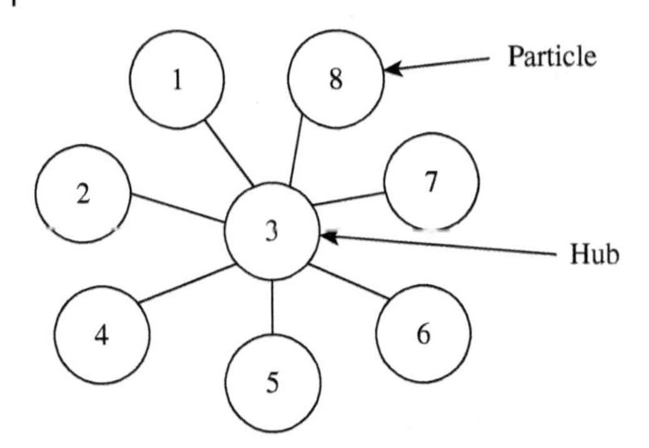

Figure 4.6 Star-based Lbest model with swarm size of 8.

Particle	Sample Fitness Value
1	2.11
2	1.14
3	4.03
4	2.55
5	3.67
6	5.04
7	3.75
8	2.50

Iteration 1

Hub: Particle 3 (randomly selected)

The particles (1, 2, 4, 5, 6, 7, 8) move towards the hub (Particle 3).

The particle 3 (hub) moves towards the best particle in the swarm that is particle 2 which is selected based on its fitness.

Particle	Sample Fitness Value
1	3.11
2	2.14
3	3.03
4	1.75
5	4.67
6	3.04
7	2.75
8	3.04

Iteration 2

Hub: Particle 3 (randomly selected)

The particles (1, 2, 4, 5, 6, 7, 8) move towards the hub (Particle 3).

The particle 3 (hub) moves towards the best particle in the swarm that is particle 4 (selected based on the fitness).

Figure 4.7 Example for finding Lbest particle in star model.

4.3.2.3.2 Ring Topology A swarm is a collection of particles where the particles are connected to each other in such a way that they form a closed loop. In this loop, each particle has two immediate adjacent neighboring particles. Also, each particle is affected only by its immediate adjacent neighbors in the topological swarm. Figure 4.8 presents the neighborhood structure of the swarm based on the ring model.

The Lbest particle of a particle in the swarm is calculated based on the ring model. In Figure 4.8, the neighboring particles (two adjacent particles) of particle 1 are Particles 2 and 8. The best particle (minimum fitness value) among these two adjacent particles will be the Lbest particle of particle 1. In this example Particle 2 is the Lbest particle of particle 1.

Figure 4.9 depicts the sample fitness value, neighboring particles and the corresponding Lbest particles of the swarm with size 8 based on the ring model given in the previous Figure 4.8.

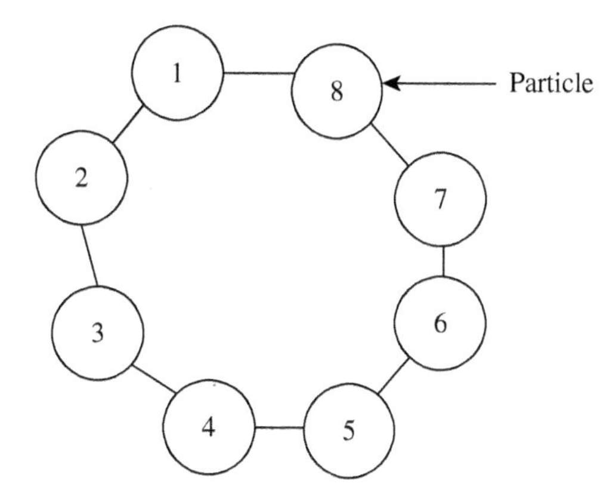

Figure 4.8 Ring model with swarm size of 8.

P. No.	Sample Fitness Value
1	2.11
2	1.14
3	4.03
4	2.55
5	3.67
6	5.04
7	3.75
8	2.50

P. No.	Neigh. Particles
1	2,8
2	1,3
3	2,4
4	3,5
5	4,6
6	5,7
7	6,8
8	7,1

P. No.	Lbest Particles
1	2
2	1
3	2
4	5
5	4
6	5
7	8
8	1

*P. No. - Particle Number; Neigh. Particle - Neighboring particle

Figure 4.9 Illustration in finding Lbest particle of the ring model.

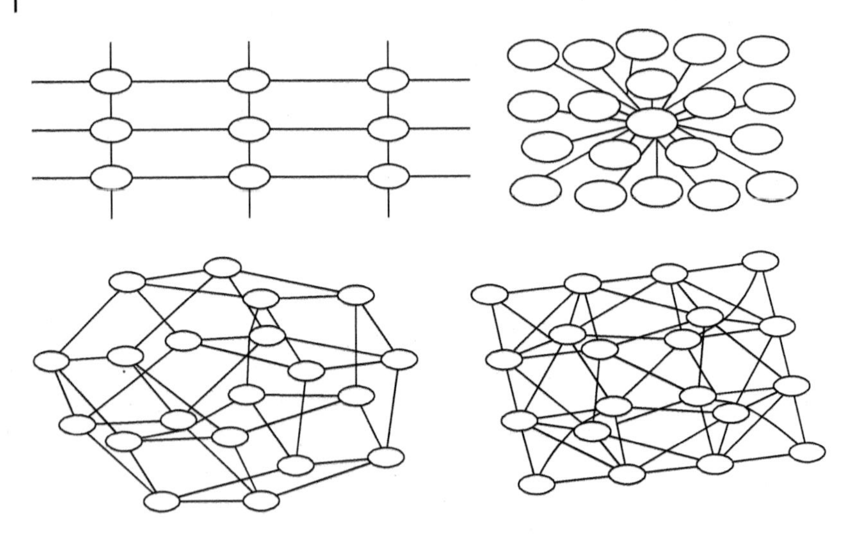

Figure 4.10 Sample structure for the VN model.

4.3.2.3.3 VN Topology In VN topology, particles are connected using a two-dimensional rectangular lattice. The velocity of each particle is affected only by the best performance of its four neighboring particles (above, below, right, and left particles). The graphical structure of VN model is given in Figure 4.10.

The procedure to find the Lbest particles of the swarm based on the VN model [14] is given in Figure 4.11.

Figure 4.12 is an example for finding the neighboring particles and the Lbest particles in the swarm with size 8 based on the VN model. The particles are arranged in R rows (R = 2) and C columns (C = 4) and the neighboring particles are calculated using Eqs. (4.3)–(4.6). The best particle (minimum fitness value) among these four neighboring particles would be the Lbest particle of the particles in the swarm.

4.3.2.3.4 Tree Topology The tree topology [8] is constructed as a binary tree. It has a single root particle (node). This particle is selected by random from the swarm. The remaining particles are equally (balanced) distributed in the tree branches. While the root particle is affected by the best performance of their children in the second level, the second level (or) next level nodes are affected by the best performance of their children and its parent. The leaf nodes are affected only by the best performance of the parent. Figure 4.13 shows the sample overview structure of the tree topology with a swarm size of 8.

Figure 4.14 demonstrates the neighboring particles and their corresponding Lbest particles of the tree based neighborhood structure. From Figure 4.14,

Procedure 4.2: Find Lbest particles in the VN model	
Input: Swarm with size *N*, Fitness values, VN model; **Output**: Lbest particles	
Step 1: Arrange the *N* particles in *R* rows and *C* columns, that is $N = R * C$	
Step 2: For each ith particle in the swarm, $i \in \{1,2, \dots, N\}$	
//Four steps used to find four adjacent particles of ith particle in the swarm:	
a) Above neighbor:	
$\text{Neighbor}_i(1) = (i - C) \bmod N$, if $\text{Neighbor}_i(1) == 0$, $\text{Neighbor}_i(1) = N$	(4. 3)
b) Left neighbor:	
$\text{Neighbor}_i(2) = (i - 1)$, if $(i - 1) \bmod C == 0$, $\text{Neighbor}_i(2) = i - 1 + C$	(4.4)
c) Right neighbor:	
$\text{Neighbor}_i(3) = (i + 1)$, if $i \bmod N == O$, $\text{Neighbor}_i(3) = i + 1 - C$	(4.5)
d) Below neighbor:	
$\text{Neighbor}_i(4) = (i + C) \bmod N$, if $\text{Neighbor}_i(4) == 0$, $\text{Neighbor}_i(4) = N$	(4.6)
Step 3 : Stop	

**Note:* The negative neighborhood number in mod calculation is determined by subtracting the left operand from the right operand. e.g.: $(-5) \bmod 20 = 15$

Figure 4.11 Procedure for finding Lbest particles in VN model.

Particle No.	Sample Fitness Value
1	2.11
2	1.14
3	4.03
4	2.55
5	3.67
6	5.04
7	3.75
8	2.50

Particle No.	Neighboring Particles (a, 1, b, r) & Lbest Particle	
1	Neigh.:5,4,5,2;	Lbest:2
2	Neigh.:6,1,6,3;	Lbest:1
3	Neigh.:7,2,7,4;	Lbest:2
4	Neigh.:8,3,8,5;	Lbest:8
5	Neigh.:1,4,1,6;	Lbest:1
6	Neigh.:2,5,2,7;	Lbest:2
7	Neigh.:3,6,3,8;	Lbest:8
8	Neigh.:4,7,4,5;	Lbest:4

*a - above; 1 - left; b - below; r - right; Neigh. - Neighboring Particles.

Figure 4.12 Example of the VN model with swarm size 8.

it is evident that the neighboring particles of a particle 2 are 1 (Parent node), 4 (Left child), and 5 (Right child). Since particle 1 has a minimum fitness value among these neighboring particles, this would be the Lbest particle of the particle 2.

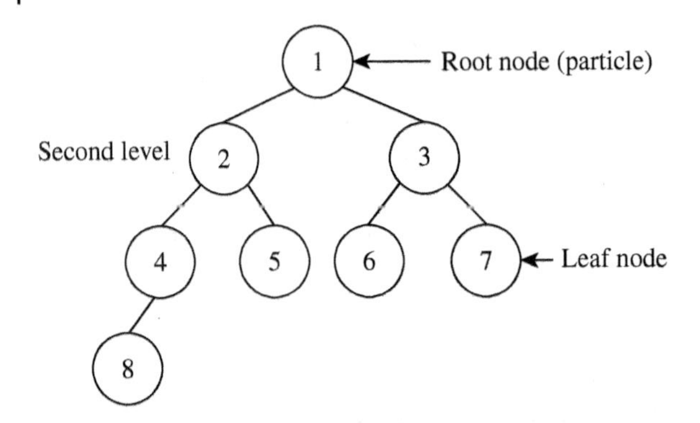

Figure 4.13 Sample neighborhood structure based on tree topology.

Particle No.	Sample Fitness Value
1	2.11
2	1.14
3	4.03
4	2.55
5	3.67
6	5.04
7	3.75
8	2.50

Particle No.	Neighboring Particles & Lbest Particle	
1	Neigh.2,3;	Lbest:2
2	Neigh.1,4,5;	Lbest:1
3	Neigh.1,6,7;	Lbest:1
4	Neigh.2,8;	Lbest:2
5	Neigh.:2;	Lbest:2
6	Neigh.3;	Lbest:3
7	Neigh.:3;	Lbest:3
8	Neigh.:4;	Lbest:4

Figure 4.14 Illustration of tree topology with swarm size 8.

4.3.2.4 Neighborhood Topology Based on Communicating Particles

Based on the changes in the neighboring particles, the neighborhood topology is broadly categorized into two types as follows:

4.3.2.4.1 Static Topology The neighboring particles of the swarm are computed only at the beginning of the execution, i.e. during the swarm initialization. After constructing the neighborhood structure, the neighboring particles are fixed throughout the execution of all processes.

4.3.2.4.2 Dynamic Topology The neighboring particles of the swarm are computed in each iteration based on the fitness value of the particles. Therefore, the neighboring particles vary from iteration to iteration. The Gbest model and all the Lbest models (star, ring, VN, and tree) are static topological models which would decrease the diversity of the swarm. Therefore, the static topology-based PSO might get stuck in local optima. To reduce the possibilities of premature convergence and improve the diversity of the swarm, the dynamic topology-based Lbest model is presented in this chapter.

The dynamic Lbest model presented is BHT. The BHT is constructed based on the fitness value. Therefore, the neighboring particles vary from generation to generation. This dynamic nature increases exploration capability because it permits particles every time to search solutions in the new areas of the search space.

The following section presents the way to incorporate the dynamic BHT into DPSO to improve the performance of the algorithm.

4.4 Dynamic Topology: Binary Heap DPSO (BHDPSO) in Scheduling Problem

The DPSO algorithm applies a heap tree model (instead of Gbest model) for communication among the neighboring particles in the swarm to reduce the local optima problem and improve the diversity.

4.4.1 Procedure to Build BHT

The binary heap is a heap data structure which is created using a binary tree. The binary heap satisfies the heap ordering property. The ordering can be one of two types:

- *Min-heap property:* The value of each node is greater than or equal to the value of its parent, with the minimum value element at the root.
- *Max-heap property:* the value of each node is less than or equal to the value of its parent, with the maximum value element at the root.

The objectives (makespan, flow time, and reliability cost) of the TS problem are a minimization problem and hence the BHDPSO algorithm searches the minimum fitness value (minimization of weighted sum of makespan, mean flow time, and reliability cost) in the swarm. Therefore, the BHT is constructed based on a min-heap property.

In BHT, the fitness value of a node (particle) is less than or equal to the fitness values of its children. The particle with the least fitness value is always at the heap's root. Each node has a single parent node in the min-heap tree and this parent node acts as the Lbest particle of the node in BHT.

The procedure and an illustrative example of constructing the BHT of the swarm are given below:

Step 1: Place the fitness value of the particles in the swarm in a single dimensional array. Figure 4.15 shows the sample fitness value of swarm with size 7 and the corresponding initial tree.

Step 2: Use min-heap property to heapify the initial tree. This heapify function rearranges the tree such that the parent is always lesser than its children. The heapify function would finally provide BHT. Figure 4.16 gives the step-by-step procedure to construct BHT from the initial heap tree Tree 1 which is shown in Figure 4.15.

In min-heap BHT, the parent node is the Lbest particle of a node in the swarm. Since each node has a single parent, there is no comparison among the neighboring particles (parent and children) in the min-heap. In the case of the binary tree, the Lbest particle of a node is the best value among the parent and their children. Figure 4.17 is an example for finding the Lbest particles (Parent node is the Lbest

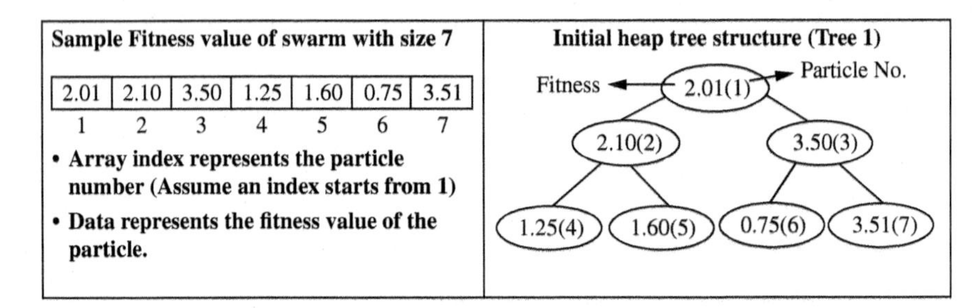

Figure 4.15 Construction of initial heap tree.

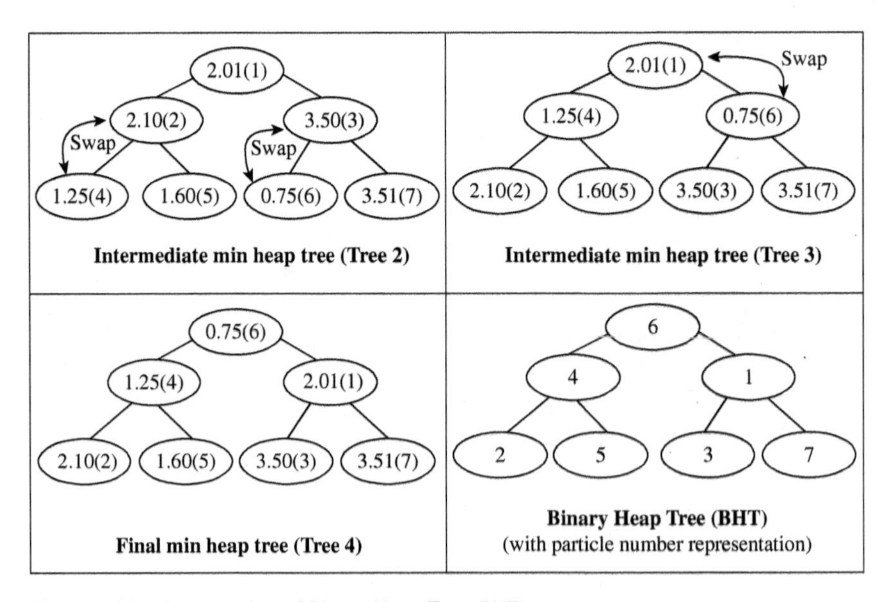

Figure 4.16 Construction of Binary Heap Tree (BHT).

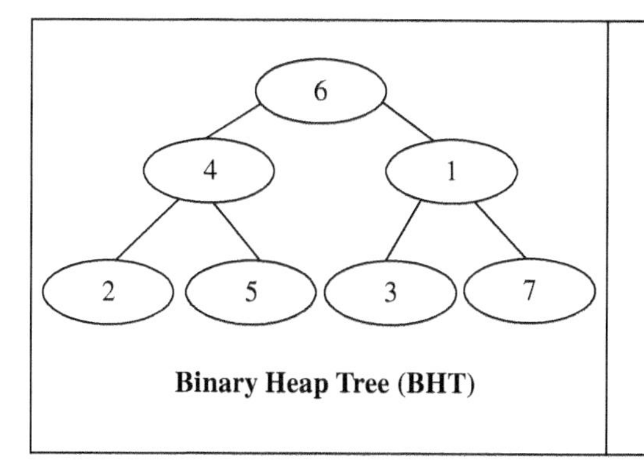

Particle No.	Lbest particle
1	6
2	4
3	1
4	6
5	4
6	6
7	1

Lbest Particles

Figure 4.17 Finding Lbest particles in BHT model.

Pseudo code 4.3: BHDPSO

Input: Swarm size (N), Particle size (n), Number of processors (m)
Output: Root node in BHT (BHT)

Begin
Initialize the swarm using intelligent techniques.
Set count=max//max is the number of iterations to be performed
for Iteration$_i$ varies from 1 to count
  for each Particle$_j$ in the swarm //j varies from 1 to N
    Evaluate the particles (calculate fitness value)
    Update Pbest particles.
  end for
  Construct BHT according to the fitness value of the particles.
  Find Lbest particles from BHT
  Update velocity and position using equation (4.7) and (4.2) // Instead of Gbest use Lbest

$$V_i^{(t+1)}(j) = WV_i^t(j) + C_1 r_1 (Pb_i^t(j) - p_i^t(j)) + C_2 (1 - r_1)(Lb^t(j) - p_i^t(j)) \quad (4.7)$$

Iteration$_i$=Iteration$_i$+1
end for
Return root particle in BHT
End

Figure 4.18 Pseudo code for BHDPSO.

particle of a node) of the swarm from the generated BHT given in Figure 4.16. The DPSO-BHT algorithm is presented in Figure 4.18.

4.5 Evaluation Metrics

DPSO with various neighborhood structures is assessed by the following metrics [15]:

- *Fitness function:* This is used to test the quality of the solution. Since it is a minimization problem, a better solution has a lower fitness value. The three objectives such as makespan, mean flow time and reliability cost are involved in the calculation of fitness function, which is given in Eq. (4.8), where the weights W_1, W_2, and W_3 indicate the importance of the objectives in the meta-TS problem.

$$\text{Fitness} = W_1\text{Makespan} + W_2\text{Mean Flow Time} + W_3\text{Reliability Cost} \tag{4.8}$$

where

- Makespan (MS) is used to calculate the throughput of the systems, assuming $C_{i,j}$ (i ε{1, 2, ..., n}, j ε {1, 2, ..., m}) is the execution time for performing i^{th} task in j^{th} processor and W_j(j ε {1, 2, ..., m}) is the previous workload of P_j. This is estimated using Eq. (4.9):

$$\text{MS} = max\left\{\sum\nolimits_{ij} C_{i,j} + W_j\right\} j\,\varepsilon(1,2,3\ldots,m) \tag{4.9}$$

- Mean Flow Time (MFT) measures the quality of service of the distributed system. Assume k to be the total number of tasks assigned to processor P_i and F_{ji}, the finishing time of task T_j on a processor P_i, (iε{1, 2, ..., m}, jε {1, 2, ..., n}, the calculation of mean flow time is given in Eqs. (4.10) and (4.11).

$$\text{MFT} = \frac{\sum_{i=1}^{m} \text{M\_Flow}_i}{m} \tag{4.10}$$

$$\text{M\_Flow}_i = \frac{\sum_{j=1}^{k} F_{ji}}{k_i} \tag{4.11}$$

- Reliability Cost (RC) is the indicator of how reliable a given system is when a group of tasks are assigned to it. It is indirectly proportional to reliability. It is the summation of processor reliability and link reliability. In processor reliability, processor failures are assumed to be independent and it follows a Poisson process with a constant failure rate. The RC is defined in Eq. (4.12), where $X(T_i) = j$ indicates that task T_i is allocated to P_j and λj is the failure rate of processor P_j.

$$\text{RC} = \sum\nolimits_{j=1}^{m} \sum\nolimits_{X(T_i)=j} {}_{j}C_{ij}(T_i) \tag{4.12}$$

- *Relative Percentage Deviation (RPD):* The average RPD is calculated using Eq. (4.13), where P is the average result of the DPSO variant and AC_i is the average result provided by another DPSO variant.

$$\text{RPD} = (\text{AC}_i - \text{P})/\text{P} * 100 \tag{4.13}$$

- *Resource Utilization (RU):* This is the performance criterion for the scheduler to perform scheduling with balancing the load. The RU is the average of the processor's utilization, which is calculated using Eq. (4.14). The processor's utilization is defined as the percentage of time that processor P_j is busy during the scheduling time. The processor's utilization PU_j for the processor P_j is calculated using Eq. (4.15), where Aval(P_j) is the processor's availability time. The processor availability time is the time when the processor P_j completes the execution of all the assigned tasks.

$$\text{RU} = \frac{\sum_{j=1}^{m} \text{PU}_j}{m} \tag{4.14}$$

$$\text{PU}_j = \frac{\text{Aval}(\text{P}_j)}{\text{Makespan}} \quad \text{for } j = 1 \ldots m \tag{4.15}$$

- *Hypothesis Test:* This is an essential procedure in statistics. The Wilcoxon signed-rank test [16] is a non-parametric statistical hypothesis test which is used to statistically compare the performance of the proposed algorithms with the existing algorithms. The hypothesis test needs two inputs: the statistical results of the proposed algorithm (A) and that of the existing algorithm (B). Initially, it sets the value of two hypothesis parameters such as the null hypothesis (no difference between the proposed algorithm and the existing algorithm) and the alternate hypothesis (proposed algorithm is better than the existing algorithm).

  The procedures followed to obtain the test statistic have been represented in Figure 4.19 (Wilcoxon signed-rank test). Based on the test statistic, this hypothesis test accepts or rejects the null hypothesis. If "alternate hypothesis" is accepted and "null hypothesis" is rejected, then it can be concluded that "Proposed algorithm is statistically better than the existing algorithm"; otherwise it would indicate that "There is no difference between the proposed algorithm and the existing algorithm."

4.6 Simulation Results

The DPSO with various neighborhood algorithms was coded in Java and executed on an Intel Xeon processor.

4.6.1 Benchmark Dataset

The simulation results were attained using a set of benchmark ETC instances (www.fing.edu.uy/inco/grupos/cecal/hpc/HCSP/HCSP_inst.htm) for the distributed heterogeneous systems. The ETC instances in the dataset consist of 512 tasks and 16 processors.

Algorithm 4.4 : Wilcoxon Signed Rank Test

Input: Statistical results of A and B; **Output:** Accept H_1 or Reject H_1

Step 1: Formulate the null hypothesis and alternate hypothesis.

Null hypothesis (**H_0**): $H_0 : \mu_B = \mu_A$ (The mean difference is zero)

Alternate hypothesis (**H_1**): $H_1 : \mu_B > \mu_A$ (The mean difference is positive)

Step 2: Calculate the mean difference between B and A using equation (4.16)

$$\text{Diff} = \mu_B - \mu_A \tag{4.17}$$

Step 3: Find the absolute value of "Diff" using equation (7.12)

$$\text{Abs} = |\text{Diff}| \tag{4.18}$$

Step 4: Assign the rank to the absolute difference in ascending order.

Step 5: Set W+ as the sum of the positive ranks and W– as the absolute value of the sum of the negative ranks.

(*Note:* Here, the proposed algorithm expects to see more higher and positive ranks (W+ is much larger than W–)).

Step 6: Set the wilcoxon test statistic W = W +

Step 7: Find the upper critical value W_α for upper tailed one sided test from the wilcoxon signed-rank test table with the significance level $\alpha = 0.05$

Step 8: Accept H_1 and reject H_0 if $W \geq W_\alpha$, else accept H_0 and reject H_1

Figure 4.19 Procedure for Wilcoxon signed-rank test.

All instances consisting of 512 tasks and 16 processors are classified into 12 different types of ETC matrices according to the following metrics.

The instances are labeled as a_bb_cc as follows:

- a shows the type of consistency; c – consistent, i – inconsistent, and s – semi-consistent
- bb indicates the heterogeneity of the tasks; hi – high and lo – low
- cc represents the heterogeneity of the machines; hi – high and lo – low

The algorithms are stochastic-based algorithms. Each independent run of the same algorithm on a particular problem instance may yield a different result. To make a better comparison of the algorithms, each experiment was repeated 10 times with different random seeds and the average of the results was tabulated and presented below.

4.6.2 Parameter Setup

- *Swarm size and iteration count*: The DPSO algorithm was tested in different swarm sizes 32, 50, 70, and 100 with different iterations such as 1000, 500, and 250. From the experimental results, the swarm size 50 and the iteration 500 provided better results than others. Therefore, the swarm size and the maximum numbers of iteration of DPSO with various topologies were set to 50 and 500 respectively.
- *Weights in the fitness Eq. (4.8)*: The influence of MT, MFT, and RC in fitness is parameterized by the weights, where the weights w_1, w_2, and w_3 are chosen such that $\Sigma w_i = 1$, where $i = 1$ to 3. The MS and MFT are equally important performance criteria in independent tasks scheduling problems and hence the weights w_1 and w_2 are set as equal. If the scheduler schedules dependent tasks, then the RC is an important criterion because it would consider both the link reliability and processor reliability. In the presented work, the scheduler schedules only the independent tasks. Therefore, the RC is a less important criterion when compared to MS and MFT because it considers only the processor reliability. Therefore, an experimental analysis was performed with different weights such as *Experiment 1*: $w_1 = 0.35$, $w_2 = 0.35$, and $w_3 = 0.3$; *Experiment 2:* $w_1 = 0.4$, $w_2 = 0.4$, and $w_3 = 0.2$ and *Experiment 3*: $w_1 = 0.45$, $w_2 = 0.45$, and $w_3 = 0.1$. Based on the obtained results, Experiment 2 provided better objective values when compared to others. Therefore, the weights w_1, w_2, and w_3 were set to 0.4, 0.4, and 0.2 for the MS, MFT, and RC respectively.
- *Inertia weight, cognitive, and social coefficients in Eq. (4.1)*: DPSO uses hamming distance-based inertia weight [12], which ensures that the particles move only toward the leader particle in the swarm. Here, cognitive and social coefficients $C_1 = C_2 = 1$.
- The failure rate for each processor was uniformly distributed in the range from 0.95×10^{-6} /h to 1.05×10^{-6}/h [15].

4.6.3 Performance of DPSO with Star and Binary Tree Topologies

4.6.3.1 DPSO-Star Hub vs. DPSO-Star Best

In Star Lbest model, the entire particles move toward the central node called the hub (central particle). This hub particle travels toward the best particle among its neighboring particles. Therefore, the hub particle in the star model might not be the best particle in the swarm. The variations in MS, MFT, and RC between the hub particle and the best particle found in the star model are presented in Figures 4.20 to 4.22.

Looking at Figures 4.20 to 4.22, the objective values of the hub particle are not the same as the best particle found in the star model in the majority of the ETC instances. Therefore, the star model takes time to find the optimal solution in the swarm.

4.6.3.2 DPSO-Binary Tree Root vs. DPSO-Binary Tree Best

A tree has a root node, and the root node in the binary tree-based Lbest model travels toward the best particle among its children. All the remaining particles except the leaf particles move toward the best particle among the children and its parent. The leaf particle flies toward its parent particle in the tree model. Since the root particle is selected in a random manner in binary tree model, it might not be the best particle in the swarm. The variations in MS, MFT, and RC between the root particle and the best particle found in binary tree model are presented in Figures 4.23 to 4.25.

It can be observed in Figures 4.23 to 4.25 that, due to the random selection of root particle, there is a deviation in the objective values of root particle and the best particle found in the binary tree model in the majority of the ETC instances. Hence, the binary tree model takes time to find the optimal solution in the swarm.

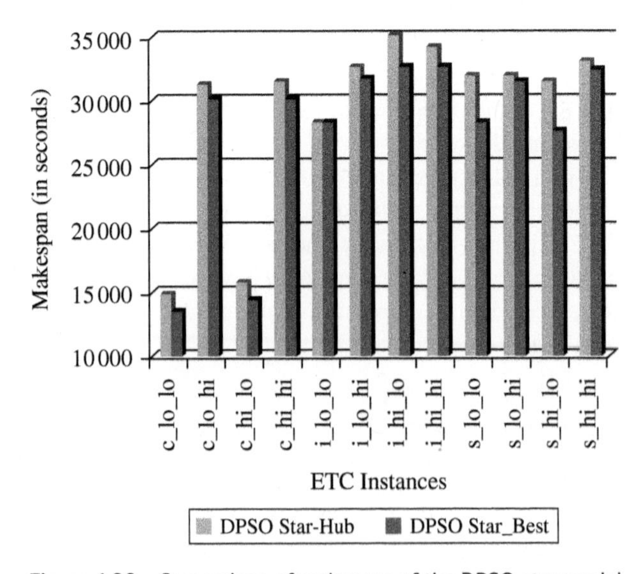

Figure 4.20 Comparison of makespan of the DPSO-star model.

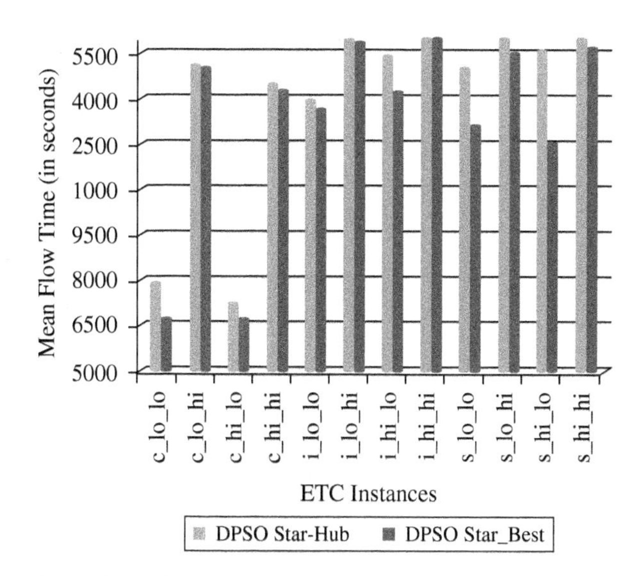

Figure 4.21 Mean flow time comparison of the DPSO-star model.

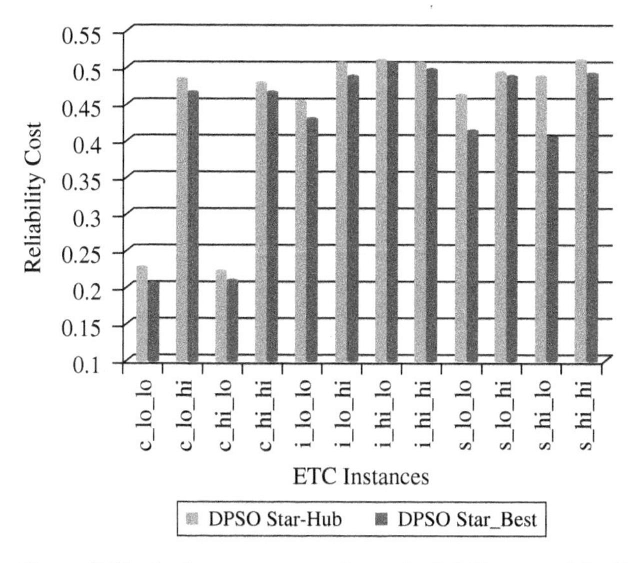

Figure 4.22 Performance comparison of reliability cost of the DPSO-star model.

4.6.4 Performance of DPSO with Static and Dynamic Topologies

The objective values such as MS, MFT, and RC obtained for DPSO with different topologies are presented in Tables 4.1 to 4.3. The optimal values obtained by the topology are indicated in bold.

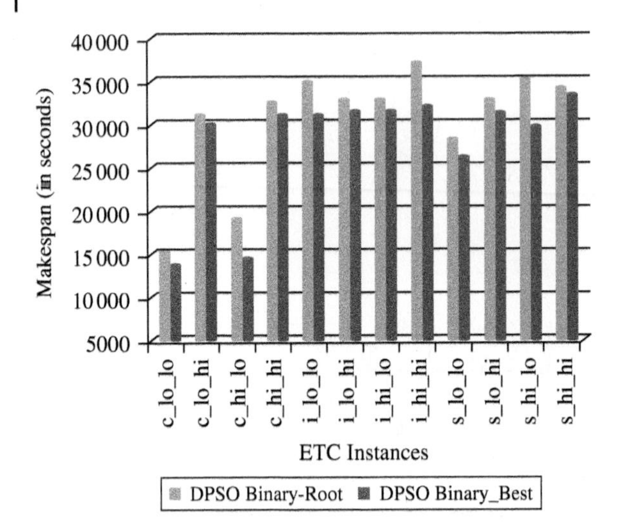

Figure 4.23 Comparison of MS of the DPSO-binary tree model.

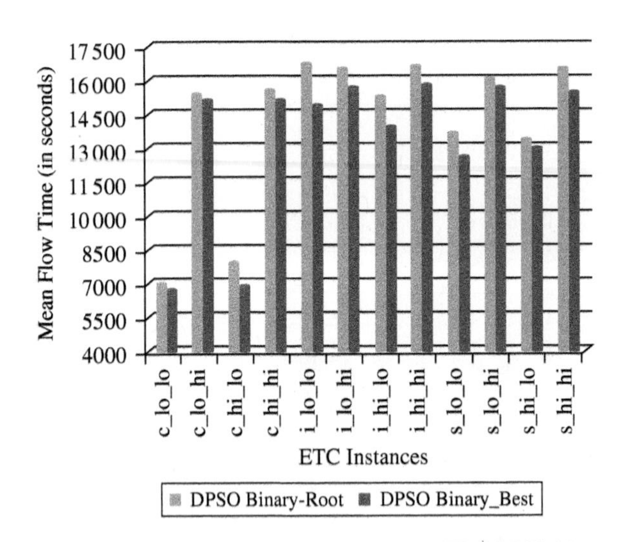

Figure 4.24 Performance comparison of MFT of DPSO-binary tree model.

It can be seen from Tables 4.1 to 4.3 that the dynamic BHT provided better results in most of the ETC instances compared to static topologies. The average percentage of performance improvements of dynamic BHT compared with other neighborhood topologies of star, ring, VN, and binary tree are:

- Makespan improvement: 18.57, 19.65, 19.66, and 19.80%
- Mean flow time improvement: 20.94, 22.40, 20.36, and 21.84%
- RC improvement: 23.73, 24.31, 23.12, and 22.62%.

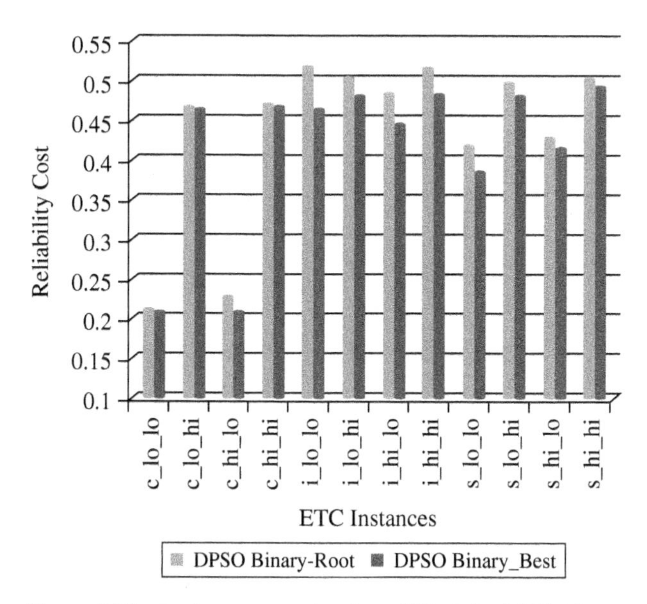

Figure 4.25 Performance evaluation of RC of the DPSO-binary tree model.

4.6.5 Resource Utilization

The RU of DPSO with different neighborhood topologies (mesh, star, ring, VN, binary, and BHT) were calculated using Eq. (4.9) and the values are plotted in Figure 4.26.

The average RU of DPSO with different topologies and the rank of the algorithms (Rank 1: better performance and Rank 6: less performance) based on RU are given in Table 4.4.

From Table 4.4, it is inferred that the BHT and mesh topologies utilized the resources efficiently when compared to others. The mesh topology is a Gbest model in which the neighborhood structure is static but the Gbest particle varies from generation to generation. Therefore, it is a partial dynamic model. For this reason, DPSO-Mesh was able to give better performance improvements in terms of objectives and also RU when compared to other neighborhood topologies of star, ring, VN, and binary tree.

4.6.6 Performance Comparison of DPSO-Mesh (A) and DPSO-BHT (B) with Respect to Statistical Hypothesis Tests

The detailed procedure to perform the Wilcoxon signed-rank hypothesis test is presented in Figure 4.19 and Section 4.5. It is essential that the following two parameters are set:

- Null hypothesis ($\boldsymbol{H_0}$) is set to "No difference between two algorithms (A and B)."
- Alternate hypothesis ($\boldsymbol{H_1}$) is set to "The algorithm B performs better than the algorithm A."

Table 4.1 DPSO with various neighborhood topologies in terms of makespan.

	Makespan (in seconds)					
ETC Instances	DPSO Mesh	DPSO Star	DPSO Ring	DPSO VN	DPSO Binary	DPSO BHT
c_lo_lo	13430.51	13587.05	13821.98	13412.69	13873.76	**13178.05**
c_lo_hi	30005.21	30060.92	30083.88	30093.39	30083.88	**27724.31**
c_hi_lo	14398.72	14469.74	14688.51	14279.78	14435.31	**13059.35**
c_hi_hi	29918.33	30232.79	30785.23	30615.19	31085.66	**20220.51**
i_lo_lo	29087.81	28247.95	31777.78	31609.91	31140.68	**27581.99**
i_lo_hi	31907.81	31700.67	31703.48	31728.13	31599.78	**30292.47**
i_hi_lo	28275.43	32644.67	29052.91	29934.94	31555.68	**25617.19**
i_hi_hi	31858.32	32548.96	32764.42	32353.15	31967.48	**30074.69**
s_lo_lo	25877.51	28264.13	29286.99	27781.06	26080.77	**25275.83**
s_lo_hi	**30049.22**	31538.61	31708.65	31833.03	31477.51	30963.36
s_hi_lo	16397.11	27662.89	27975.38	30029.11	29562.93	**13289.39**
s_hi_hi	32183.11	32479.21	32863.28	32230.03	33444.91	**30964.36**

Table 4.2 Comparison of mean flow time of DPSO with topologies.

	Mean flow time (in seconds)					
ETC Instances	DPSO Mesh	DPSO Star	DPSO Ring	DPSO VN	DPSO Binary	DPSO BHT
c_lo_lo	6795.09	6700.01	7076.19	6968.98	6735.25	**6449.38**
c_lo_hi	14969.91	15036.42	15180.06	15204.68	15180.06	**13028.21**
c_hi_lo	6326.37	6665.44	6807.18	6736.48	6836.01	**6205.41**
c_hi_hi	14452.01	14315.52	15112.77	14815.96	15076.84	**13075.56**
i_lo_lo	14253.11	13659.44	14488.95	14941.87	14763.66	**13127.56**
i_lo_hi	15623.91	15935.13	15895.71	16145.93	15702.87	**14039.73**
i_hi_lo	**12177.22**	14244.05	13408.55	13821.86	13874.12	12531.76
i_hi_hi	15423.32	16071.95	15101.33	15299.44	15764.61	**14235.47**
s_lo_lo	11066.11	13095.37	13103.59	12833.23	12592.85	**10008.87**
s_lo_hi	15252.03	15571.79	15833.19	15652.13	15660.92	**13023.25**
s_hi_lo	**6144.45**	12591.14	13097.73	11610.51	12978.23	6285.19
s_hi_hi	14975.21	15705.12	15827.37	15601.52	15370.46	**13016.89**

Table 4.3 Reliability cost comparison of DPSO with different topologies.

ETC Instances	Reliability cost					
	DPSO Mesh	DPSO Star	DPSO Ring	DPSO VN	DPSO Binary	DPSO BHT
c_lo_lo	0.202652	0.203091	0.206381	0.205393	0.203133	**0.190886**
c_lo_hi	**0.461131**	0.464439	0.462283	0.461447	0.462283	0.462184
c_hi_lo	**0.191642**	0.209442	0.209304	0.210796	0.207226	0.192498
c_hi_hi	**0.463532**	0.465259	0.466188	0.468874	0.465751	0.466072
i_lo_lo	0.359362	0.429998	0.465904	0.466519	0.462627	**0.300782**
i_lo_hi	0.493252	0.486712	0.487086	0.493064	0.479145	**0.440281**
i_hi_lo	0.412341	0.501868	0.436484	0.448574	0.445437	**0.400137**
i_hi_hi	0.483491	0.493428	0.487181	0.486681	0.482759	**0.440129**
s_lo_lo	**0.281172**	0.412149	0.414496	0.416664	0.384713	0.282976
s_lo_hi	0.475981	0.484875	0.483854	0.479761	0.479151	**0.440449**
s_hi_lo	**0.203571**	0.399459	0.419634	0.391051	0.414065	0.204667
s_hi_hi	0.391861	0.490659	0.499232	0.479878	0.492024	**0.370017**

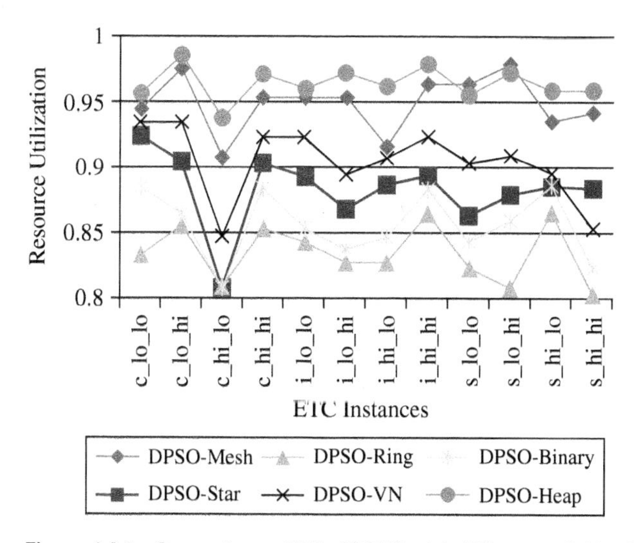

Figure 4.26 Comparison of RU of DPSO with different neighborhood topologies.

Table 4.4 Average RU of DPSO with various neighborhood topologies.

S. No.	Neighborhood Topology	Average RU (%)	Rank
1	Mesh	95.91	2
2	Star	88.26	3
3	Ring	83.42	6
4	VN	90.41	4
5	Binary	85.59	5
6	BHT	96.41	1

Table 4.5 presents the statistical analyses of the fitness value of DPSO-Mesh and DPSO-BHT algorithms with 512×16 ETC instances.

From Table 4.5, the obtained Wilcoxon test statistic W is 73, which is greater than the upper critical value obtained from the Wilcoxon signed-rank test table [16]. Hence, $\mathbf{H_0}$ stands rejected and $\mathbf{H_1}$ accepted. Thus, there is evidence that at the 5% level of significance that the DPSO-BHT presented performed better than DPSO-Mesh.

4.6.7 Performance Evaluation of DPSO with a Large Dataset

The efficiency of BHT and Mesh were tested with a large ETC dimensions (n×m matrix; where n is the number of tasks; m is the number of processors) such as 1024×32, 4096×128, and 8192×256 and the results are presented in Table 4.6.

From Table 4.6, it was inferred that the fitness improvement of DPSO-BHT in terms of 1024×32, 4096×128, and 8192×256 by 5.58%, 2.58%, and 3.01% compared with DPSO-Mesh across all ETC instances respectively.

From the simulation results and the hypothesis test, the DPSO-BHT has shown significant improvement in the TS problem when compared to other neighborhood topologies in the literature.

4.7 Real-Time Application – Smart Traffic System

Traffic lights are used as road safety equipment, but sometimes traffic lights cause people to waste time. At a junction, sometimes the red light is on but there is no traffic, and the road users must wait until the red light changes to a green light. Because of traffic jams, emergency vehicles such as police cars, ambulances, and

Table 4.5 Performance comparison of fitness value of DPSO-Mesh with DPSO-BHT with respect to hypothesis test.

ETC Instance	DPSO-Mesh (A)	DPSO-BHT (B)	Diff. (B-A)	A. Diff.	O. Diff.	R.	S.R.
c_lo_lo	8090.28	7971.01	119.26	119.26	113.03	1	1
c_lo_hi	17990.14	17877.11	113.03	113.03	119.26	2	2
c_hi_lo	8290.07	7905.94	384.13	384.13	121.48	3	3
c_hi_hi	17748.22	13838.52	3909.71	3909.71	243.56	4	4
i_lo_lo	17336.43	16295.88	1040.55	1040.55	326.14	5	−5
i_lo_hi	19012.78	18652.97	359.81	359.81	359.81	6	6
i_hi_lo	16181.14	16059.66	121.48	121.48	384.13	7	7
i_hi_hi	18912.75	18284.15	628.59	628.59	502.83	8	8
s_lo_lo	14777.51	14533.93	243.56	243.56	628.59	9	9
s_lo_hi	18120.59	18446.73	−326.14	326.14	666.78	10	10
s_hi_lo	9016.66	8349.87	666.78	666.78	1040.55	11	11
s_hi_hi	18863.41	18360.57	502.83	502.83	3909.71	12	12
						W+	73
						W−	5
				Wilcoxon test statistic W			73

Note: *Diff. – Difference; A. Diff. – Absolute Difference; O. Diff. – Ordered Difference; R. – Rank; S.R. – Signed Rank.

the fire brigade can get stuck at the traffic lights, because users are waiting for the green light to go on. An efficient scheduling of the cycle program is required to solve these types of issue in traffic signals. The cycle program is tied to the time spans that a set of traffic lights (in a junction) keeps their color states. In other words, the traffic signals are controlled based on the cycle program. The optimal cycle programs have to organize traffic lights in neighboring intersections so as to improve the global flow of vehicles. In order to optimize the traffic cycle program, the traffic controller uses a BHDPSO algorithm.

In this system, a video sensor is fixed at the traffic signal to take a video of the traffic. Then an image processing technique is used to find the number of vehicles present at the traffic signal. The number of vehicles count is given as the input for the SI (BHDPSO) to find the optimized traffic cycle program. This optimized cycle program would improve the movement of automobiles in vehicular traffic [17]. The sample traffic structure is given in Figure 4.27. In this structure, the controller will use BHDPSO to optimize the green signal timings.

Table 4.6 Comparison of fitness value of BHDPSO and DPSO-Mesh with various ETC dimensions.

	1024×32			4096×128			8192×256		
ETC Instances	**BHDPSO**	**DPSO-Mesh**	**RPD**	**BHDPSO**	**DPSO-Mesh**	**RPD**	**BHDPSO**	**DPSO-Mesh**	**RPD**
c_lo_lo	**230.22**	239.22	3.91	186.03	**185.46**	−0.31	**178.97**	179.23	0.14
c_lo_hi	**2182.51**	2258.57	3.48	**1858.45**	1898.55	2.16	1727.92	**1718.51**	−0.54
c_hi_lo	**2 260 501**	2 391 563.51	5.79	**1 874 492.21**	1 893 136.91	0.99	**1 771 084.91**	1 778 612.51	0.42
c_hi_hi	**2.30E+07**	2.40E+07	4.65	**1.89E+07**	1.95E+07	3.28	**1.73E+07**	1.78E+07	2.83
i_lo_lo	**432.49**	467.47	8.08	**465.85**	481.17	3.28	**467.13**	476.57	2.02
i_lo_hi	**4365.46**	4666.17	6.88	**4584.41**	4663.46	1.72	**4694.04**	4838.33	3.07
i_hi_lo	**4 362 632**	4 679 823.51	7.27	**4 653 622.51**	4 731 489.51	1.67	**4 662 291.51**	4 804 507.51	3.05
i_hi_hi	**4.44E+07**	4.73E+07	6.52	**4.67E+07**	4.77E+07	2.25	**4.71E+07**	4.81E+07	2.01
s_lo_lo	**419.47**	445.13	6.11	**426.69**	444.87	4.26	**75.63**	80.73	6.74
s_lo_hi	**4083.69**	4102.65	0.46	**4229.45**	4426.59	4.66	**760.88**	781.62	2.72
s_hi_lo	**4 208 786**	4 477 223.51	6.37	**4 245 952**	4 472 696	5.34	**774 728.21**	828 071.31	6.88
s_hi_hi	**4.17E+07**	4.49E+07	7.45	**4.22E+07**	4.28E+07	1.64	**7 565 201.51**	8 071 759.51	6.69
Average			**5.58**			**2.58**			**3.01**

RPD – Relative Percentage Deviation (Refer Eq. (4.13) in Section 4.5).

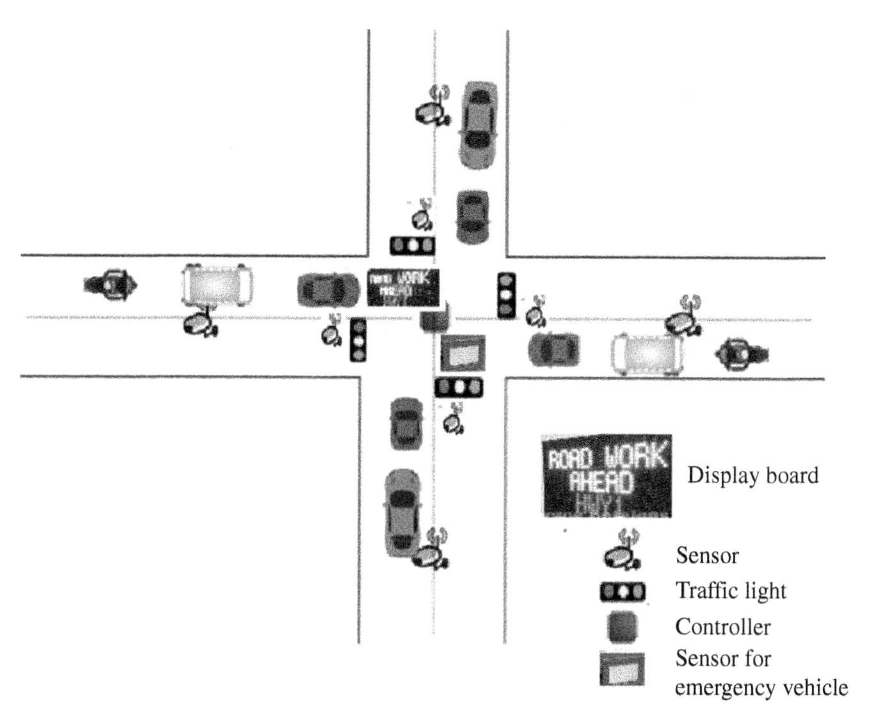

Figure 4.27 DPSO-BHT in traffic cycle scheduling.

4.8 Conclusion

DPSO is a newly developed popultion-based metaheuristic technique for discrete optimization problems. This chapter presents a problem of scheduling independent tasks in heterogeneous multiprocessor systems using a novel DPSO metaheuristic algorithm which minimizes the tri-objectives such as makespan, mean flow time, and RC. Extensive simulation results have been conducted to analyze the efficiency of the DPSO with various neighborhood topologies. The DPSO-BHT performs significantly better when compared with other topological structures in the TS problem. The efficiency of the algorithm was also tested in a traffic control system. This algorithm provided better performance in a smart traffic system when compared to the existing classical PSO.

In this chapter, the algorithm was analyzed using static data in which all the tasks to be scheduled are known in advance. However, often this information is not available in real time and tasks must be dynamically scheduled as they appear on the system. Here, the algorithm considered only meta-tasks. Many real-world scheduling problems require that certain tasks are completed before some other tasks can run. Therefore, the algorithm can be extended to schedule tasks with inter-task precedence constraints. This type of problem is referred to as Directed Acyclic Graphs (DAGs), in which the nodes represent tasks and the arcs represent communication between the tasks.

References

1 Cristina, M. (2011). Evolutionary computation in scheduling. PhD thesis. University of Babeş-Bolyai.

2 Engelbrecht, A. (2013). Particle swarm optimization: Global best or local best. 1st BRICS Countries Congress on Computational Intelligence. Piscataway, NJ: IEEE Computer Society. DOI 10.1109/BRICS-CCI.&.CBIC.2013.2.

3 Toscano-Pulido, G., Reyes-Medina, A.J., and Ramirez-Torres, J.G. (2011). A statistical study of the effects of neighborhood topologies in particle swarm optimization. In: *Computational Intelligence* (eds. K. Madani, A.D. Correia, A. Rosa and J. Felipe), 179–192. Springer.

4 Reyes Medina, A.J., Pulido, G.T., and Ramirez Torres, J.G. (2013). A comparison study of PSO neighborhoods. In: *Advances in Intelligent Systems and Computing*, 251–265. Springer.

5 Dallaki, H., Lari, K.B., Hamzeh, A. et al. (2015). Scaling up the hybrid particle swarm optimization algorithm for nominal data-sets. *Intell. Data Anal.* 19 (4): 825–844.

6 Pimpale, R.A. (2012). Guaranteed coverage particle swarm optimization using neighborhood topologies. *Int. J. Comput. Sci. Eng. Technol.* 3 (12): 597–604.

7 Ni, Q. and Deng, J. (2013). A new logistic dynamic particle swarm optimization algorithm based on random topology. *Hindawi – Sci. World J.* 2013 (2): 1–8.

8 Reyes Medina, A.J., Toscano Pulido, G., and Ramirez Torres, J.G. (2013). A comparison study of PSO neighborhoods. Proceeding of the International Joint Conference on Computational Intelligence, Funchal, Madeira, Portugal, 175: 251–265.

9 Sarathambekai, S. and Umamaheswari, K. (2017). Task scheduling in distributed systems using heap intelligent discrete particle swarm optimization. *Int. J. Comput. Intell.* 33 (4): 737–770.

10 Krause, J., Cordeiro, J., Parpinelli, R.S., and Lopes, H.S. (2013). A survey of swarm algorithms applied to discrete optimization problems. In: Swarm Intelligence and Bio-Inspired Computation, (eds. X.-S. Yang, Z. Cui, R. Xiao et al.). DOI: https://doi.org/10.1016/B978-0-12-405163-8.00007-7: 169–191.

11 Uysal, O. and Bulkan, S. (2008). Comparison of genetic algorithm and particle swarm optimization for bicriteria permutation flowshop scheduling problem. *Int. J. Comput. Intell. Res.* 4 (2): 59–175.

12 Sarathambekai, S. and Umamaheswari, K. (2017). Task scheduling using hamming particle swarm optimization in distributed systems. *J. Comput. Inf.* 36: 1001–1021.

13 Sarathambekai, S. and Umamaheswari, K. (2018). Multi-objective optimization techniques for task scheduling problem in distributed systems. *Comput. J.*, https://doi.org/10.1093/comjnl/bxx059 61 (2): 248–263.

14 Zou, W., Zhu, Y., Chen, H., and Shen, H. (2012). Artificial Bee Colony Algorithm based on Von Neumann topology structure. Proceeding of third International Conference on Computer & Electrical Engineering, 53 (2): 46–53.

15 Vairam, T., Sarathambekai, S., and Umamaheswari, K. (2018). Multiprocessor task scheduling problem using hybrid discrete particle swarm optimization. *Sādhanā* 43 (206): 1–13.

16 David, M.L. (2012). E-study guide for statistics for managers using MS Excel: Chapter 12.8 – Wilcoxon signed-ranks test: Nonparametric analysis for two related populations. Cram101 Textbook Reviews.

17 Sarathambekai, S., Vairam, T., and Umamaheswari, K. (2016). Swarm intelligence technique for automating traffic cycle program. *Natl. J. Technol.* 12 (4): 25–34.

5

Computationally Efficient Scheduling Schemes for Multiple Antenna Systems Using Evolutionary Algorithms and Swarm Optimization

Prabina Pattanayak[1] and Preetam Kumar[2]

[1] *Department of Electronics and Communication Engineering, National Institute of Technology Silchar, Assam, Silchar, India*
[2] *Department of Electrical Engineering, Indian Institute of Technology Patna, Bihar, Patna, India*

5.1 Introduction and Problem Statement Formulation

In this section, different problem statements/scenarios of multi-antenna wireless communication systems are discussed, where GA and PSO have been implemented to provide computationally efficient scheduling schemes. Also, an introduction to these problem statements/scenarios is presented for understanding the background.

Single-antenna systems offer less system capacity as they can transmit data to only one user instantaneously. The system capacity enhancement of wireless systems and lower delay wireless packet data systems is achieved by multi-user multiple-input multiple-output (MU-MIMO) systems by transmitting data to several users, as many as the number of antennas at the base station (BS) without requiring additional bandwidth or transmit power. This simultaneous data transfer by BS to a number of users is proposed in a dirty paper coding (DPC) scheme. Hence, BS always searches for a subset of users who are the best in channel conditions. To accomplish this activity, BS needs full channel state information (CSI) of every user in the reverse channel according to DPC. DPC is an example of an extensive search scheme (ESS) which is hard to realize for a higher number of users and BS antennas due to the constraint in hardware implementation.

MIMO systems achieve higher system sum-rate than single antenna systems by spatial multiplexing, where multiple parallel data streams can be transmitted

Evolutionary Computation in Scheduling, First Edition. Edited by Amir H. Gandomi, Ali Emrouznejad, Mo M. Jamshidi, Kalyanmoy Deb, and Iman Rahimi.

from the transmitter to the receiver simultaneously. Multi-user diversity (MUD) helps to achieve further gain in system sum-rate for MU-MIMO systems [1, 2]. This is because of independent and uncorrelated fading channels between the BS and various user equipment (UE) present at different geographical locations. In the broadcast channel (BC), the BS sends the data message to multiple UEs simultaneously according to DPC [3]. The number of simultaneously served UEs by the BS is equal to the number of antennas at the BS [4, 5]. Because of this, it becomes the duty of the BS to select/schedule M_T number of UEs among K UEs with the best channel parameters for getting service from the BS, where M_T is the number of antennas present at the BS. A variety of MU-MIMO scheduling algorithms [6–10] have been proposed in literature, which take advantage of MUD.

Good cross-layer scheduling schemes make both the physical layer and the multiple access control layer work in tandem. This kind of scheduling algorithm results in optimal transmit power allocation to the best channel conditioned UEs for reception of data streams, which result in the maximized system sum-rate. Some of the important cross-layer scheduling schemes for MU-MIMO BC have been formulated and elaborated in [11, 12]. More often, these scheduling problems are treated as optimization scenarios, where either the utility function is maximized or the cost function is minimized. Authors in [13] have been trying to maximize the system sum-rate of the closed-loop MU-MIMO system by scheduling users and antenna simultaneously. In this chapter, the optimization of system sum-rate of various multi-antenna communication systems has been the prime goal to be achieved with the least feedback overhead from the users to the BS.

The system model which has been considered by [13] is discussed for formulating the problem statement. A cellular system has been considered where a BS provides service to K number of UEs. M_T and N_R number of antennas have been installed at the BS and each UE respectively. As DPC is an interference pre-cancelation technique, the encoding is done in a certain sequence at the BS and the decoding at the UEs is exactly in the opposite sequence. The encoding done by the BS for the k UE is to precancel the interference observed by the $(k-1)$ UEs encoded earlier. The hardware implementation of this process is quite exhaustive and complex. Moreover, the feedback overhead required for DPC implementation is also the highest. So, as replacement for DPC, suboptimal user and antenna scheduling algorithms with limited feedback and less complexity are proposed in the literature [14–23]. Authors in [13] have assumed the system sum-rate to be the utility function. To achieve the best system sum-rate, the BS should select/schedule the best M_T number of receive antennas among KN_R number of total receive antennas. The total number of ordered selections that can be possible for DPC has been expressed as [11]:

$$N_{OrderedUsers} = \sum_{k=1}^{M_T} (k!) \binom{K}{k}. \tag{5.1}$$

It will be very difficult to perform this ESS DPC for a practical MU-MIMO BC. The total number of searches required for DPC as computed by 5.1 will be very high, which will not be feasible for real-time communications where the scheduling interval is a few milliseconds. To fill this gap, the authors in [13] have implemented an evolutionary algorithm (Binary Genetic Algorithm [BGA]) to search the set of best users/antennas, which will result in the maximum system sum-rate well within the span of the scheduling interval. This is explained in Section 5.2. The authors in [11, 12] have used GA for cross-layer scheduling of MU-MIMO systems with single and multiple carriers, respectively. Elitism and adaptive mutation (AM) has been proposed to be a preferred mutation process for GA as discussed by [11, 24]. The authors of [13] have been motivated by the aforementioned works to develop a combined user and antenna scheduling (CUAS) scheme for single carrier MU-MIMO broadcast systems using BGA with elitism and AM. Moreover, they considered random beamforming [19, 23]. In [13], it has been shown that using BGA with elitism and AM for CUAS achieves the same system sum-rate as that of ESS with much less computation overhead. Moreover, CUAS using BGA with elitism and AM attains higher system sum-rate than some of the classical sub-optimal limited feedback scheduling schemes [23] for MU-MIMO downlink systems.

The importance of CSI at the transmitter (CSIT) has been discussed widely and exhaustively in the literature for multi-antenna systems [17–19, 23, 25–27]. The CSIT helps the transmitter for beamforming/precoding to transmit the message symbols to the appropriate antenna at UE from each of the antennas at the transmitter. This practice helps in achieving the maximum system sum-rate for the broadcasting scenario of MU-MIMO. However, it has also been discussed that to send the CSI from each UE to the transmitter, an uplink channel is utilized. This reduces the spectrum efficiency of the data communications by consuming the probable useful bandwidth for carrying the CSI. Moreover, this phase of communication introduces delay into the system, which should be minimized. Therefore, limited feedback scheduling schemes have been proposed and discussed [16, 17, 19, 23, 28–35], where the main motive of the research is to reduce the feedback data volume from UE to the BS. Signal-to-interference-plus-noise ratio (SINR) has been considered as a suitable form of CSI to be fed back from UE to the BS. The authors in [23] discussed the quantization of SINR with one-bit and a fixed quantization threshold, which will reduce the feedback overhead. It has been shown that this technique of one-bit SINR quantization fails to exploit the concept of MUD. Furthermore, this scheme attains less system sum-rate compared to scheduling schemes with the full feedback of CSI.

Therefore, to mitigate the limitations pertaining to one-bit quantization process, the authors in [17] studied the effect of multi-bit SINR quantization to achieve system sum-rate close to that of full feedback scheduling scheme with not so much feedback overhead as the full feedback of CSI from UEs to the BS. For multi-bit quantization, the selection of optimal quantization thresholds is very important to get the benefit of using the multi-bit quantization process. Therefore, optimal quantization threshold selection is necessary. The system sum-rate of the MU-MIMO BC is a function of scheduled users' SINR value, number of transmit antennas (M_T), best receive antenna of the scheduled users for corresponding transmit antenna, and number of users (K). Moreover, it has been shown that the system sum-rate is a function of number of users (K) and quantization threshold (Ξ) for a given set of M_T, N_R, and the received signal-to-noise ratio (SNR) (Ξ), i.e. $C_{sum-rate} = f(K, \Psi)$, where $C_{sum-rate}$ is the achievable system sum-rate. For finding the optimal quantization threshold, a solution to $\frac{\partial f(K,\Psi)}{\partial \Psi} = 0$ has to be obtained. However, it has also been discussed in [23] that a closed-form solution to $\frac{\partial f(K,\Psi)}{\partial \Psi} = 0$ is not tractable. Due to this multi-variable relation, deriving closed-form solution of the optimal quantization thresholds (Ξ) is very much more complex for MU-MIMO systems. Moreover, the range over which the optimal quantization thresholds (Ξ) can acquire value is very wide. For this reason, the exhaustive search over this wide range of values is required for efficient user/antenna scheduling process for multi-antenna systems. To address these difficulties, GA has been used to select the optimal quantization thresholds (Ξ) for obtaining the highest system sum-rate. From a system implementation point of view, the optimal quantization thresholds have been expressed in terms of system SNR value and number of antennas at the transmitter.

It has been shown in [17] that quantization with four bits along with optimum quantization thresholds is sufficient to achieve system throughput close to that of full feedback scheduling [23]. The authors in [17] have discussed two user/antenna scheduling schemes based on four-bit quantization of SINR. In the first scheme, all UEs send the four quantized bits representing the highest SINR among all $M_T N_R$ SINR values and the corresponding transmit antenna index to the BS for scheduling purposes. However, this scheme has the limitation of assigning a particular receive antenna of a UE to multiple transmit antennas and keeping some of the antennas at the BS idle by not sending any message signal from them. Therefore, the second scheme, named the optimistic scheme, has been proposed where these two aforementioned limitations are addressed properly. The methodology of GA has been adopted to find the optimal quantization threshold values for both of these scheduling schemes. Furthermore, for attaining the least feedback overhead of CSI a four-bit quantization process has been proposed in the literature. This scheme of four-bit quantization is also successful in providing a resulting system throughput very close to the optimum values. It is well known that the quantization threshold plays an absolutely vital role in the multi-bit quantization process. However, selection of the optimum quantization thresholds is also cumbersome. Hence GA is proposed to find the optimum quantization threshold values for MU-MIMO systems. This process is explained in detail in Section 5.3.

The advantages of the MU-MIMO system have been discussed earlier in this chapter. The requirement for wide-band communication channels increases due to the huge increase of users' data traffic. These wide-band communication channels are more frequency-selective in nature [36, 37]. To overcome the limitations of these wide-band frequency selective channels, the orthogonal-frequency-division-multiplexing (OFDM) technique has been used widely. OFDM converts the wide-band frequency-selective channel into multiple narrow-band frequency-flat fading sub-channels, which can be modulated independently [38]. This robustness of OFDM makes it a suitable candidate for the underlying technology of enhanced data rate communication framework. Therefore, MIMO and OFDM are integrated together (MIMO-OFDM) to be the backbone of current and future wireless technologies [39].

The limitations associated with feedback data from UEs to the BS have been discussed in detail previously in this section. The feedback load burden increases with number of users, number of antennas at the transmitter, and number of sub-carriers for the MU MIMO-OFDM system. The MU MIMO-OFDM system has a detrimental effect on the system by increasing the feedback overload by many fold. Design and development of limited feedback user/antenna scheduling schemes becomes very important for MU MIMO-OFDM systems. The computational complexity increases further for MIMO-OFDM systems as MIMO-OFDM systems have multiple carriers. The scheduling process explained for MU-MIMO systems needs to be implemented for all the sub-carriers present in a MU MIMO-OFDM system. Quantization of real-time SINR values has also been shown to be a promising technique for reducing the feedback burden for MU MIMO-OFDM systems [25]. The authors in [25] have shown that quantization of SINR with four-bits is sufficient to have achievable system sum-rate close to the optimal with the least feedback overhead. As discussed previously in this chapter, finding the optimal quantization threshold values for MU MIMO-OFDM systems is absolutely vital from a system design point of view. Moreover, the range of the possible values for the quantization threshold is very wide for MIMO-OFDM systems. Therefore, to facilitate the process of finding optimal quantization threshold values the technique of GA has been used to find the optimal values swiftly [25]. This is explained in detail in Section 5.4.

The quantization process helps in achieving reduction of feedback overhead. However, in the MU MIMO-OFDM system, clustering of adjacent sub-carriers has to be carried out to attain a further reduction of feedback load. Therefore, each UE finds the highest SINR value corresponding to each transmit antenna and adds them for a cluster of sub-carriers. Then each UE sends the four-bit quantized bits representing the added value to the BS, which employs both the techniques of clustering and quantization with four-bits. The importance of selecting proper quantization threshold has been discussed earlier in this section. Therefore, to find optimal quantization thresholds for a different MU MIMO-OFDM system is absolutely crucial to attain reduction of the feedback overhead. Moreover, it has been shown that the system sum-rate is a function of the number of users (K) and quantization threshold (Ψ) for a given set of M_T, N_R, the received SNR (Ξ), and cluster size (L_C), i.e. $C_{sum-rate} = f(K, \Psi)$, where $C_{sum-rate}$ is the achievable system sum-rate. To find the optimal quantization threshold, a solution to $\frac{\partial f(K,\Psi)}{\partial \Psi} = 0$ has to be obtained. However, it has also been discussed in [23] that a closed-form solution to $\frac{\partial f(K,\Psi)}{\partial \Psi} = 0$ is not tractable. Moreover, the search space for the quantization thresholds for MU MIMO-OFDM is large. Therefore, authors in [25] have used BGA for finding the optimal four-bit quantization threshold values, which is discussed in Section 5.4.

Also, the real-time joint selection of a transmit and receive antenna pair in MIMO systems involves high computational complexity. This complexity increases with the number of transmit and receive antennas. Hence, Binary Particle Swarm Optimization (BPSO) is used for a low complex solution to this problem. Moreover, it is shown that convergence rate is reduced by applying cyclically shifted initial population. MIMO systems have offered a lot of benefits but also incur higher hardware cost by using a multiple radio frequency (RF) chain, which includes power amplifier, analog-to-digital converters, etc. This hardware cost can be brought down by using a subset of these antennas, which have good channel conditions [40]. Therefore, users in [41] have studied joint transmit and receive antenna selection (JTRAS) for single-user MIMO systems. The exhaustive search scheme for this is computationally inefficient. The system parameters used in this chapter till now hold true for this subsection as well. The MIMO system has a transmitter with M_T transmit antennas and a receiver with N_R receive antennas. It has been further assumed that the transmitter has M_t and the receiver has N_r number of RF chains available to them. Here, the transmitter has been assumed to have no information about the channel and the receiver has the CSI. The receiver performs the selection of both N_r and M_t number of receive and transmit antennas. The MIMO system considered here is closed-loop MIMO channel, where the receiver sends the information about the selected transmit antennas to the transmitter.

The number of possible ways of joint selection of transmit and receive antenna is given as

$$N_{ESS} = \binom{M_T}{M_t} \times \binom{N_R}{N_r}. \tag{5.2}$$

This value grows with the number of transmit and receive antennas. The channel capacity associated with each of these combinations has been expressed as:

$$C_{sum-rate} = \log_2 \left| I_{N_r} + \Xi H_i H_i^* \right| \tag{5.3}$$

where I_a is the $a \times a$ identity matrix, H_i is the channel between the receive and transmit antenna associated with ith combination and $(A)^*$ is the complex conjugate transpose of a matrix (A). The utility function or the cost function for this problem has been assumed to be 5.3. However, evaluation of 5.3 for N_{ESS} number of combinations of transmit and receive antennas could not be accomplished during a packet time which is of the order of a few milliseconds. To overcome this, a low computationally burdened joint transmit and receive antenna selection scheme employing BPSO has been suggested and discussed in [41]. This process is being elaborately presented in Section 5.5.1.

Further, the usage of BPSO will be shown for a MU-MIMO system, where the joint user scheduling and receive antenna selection (JUSRAS) will be performed by BPSO as proposed and described in [42]. The computational complexity of JUSRAS increases with number of users in the system K and number of receive antennas at each UE (N_R). As a consequence, BPSO has been used for JUSRAS as explained in Section 5.5.2. Moreover, authors in [43] have proposed a hybrid discrete particle swarm optimization (DPSO) algorithm with Levy flight for scheduling MIMO radar tasks. An efficient scheduling scheme using GA has been proposed by the authors in [44] for massive MIMO systems. In this paper, it has been shown that the achievable system throughput by this scheduling scheme using GA is almost the same as that of an exhaustive search scheduler with extensively less implementation complexity.

5.2 CUAS Scheme Using GA

A closed loop MU-MIMO broadcasting system has been considered in [13], where each UE sends its CSI in the form of SINR to the BS for scheduling/selection of best channel conditioned UEs for data transmission. The instantaneous SINR at the n_rth antenna of the kth UE considering the data transmitted from m_t transmit antenna as the required signal is expressed as

$$SINR_{m_t,n_r}^{(k)} = \frac{\left| h_k(n_r, m_t) \right|^2}{\dfrac{M_T}{\Xi_k} + \sum_{m'=1, m' \neq m_t}^{M_T} \left| h_k(n_r, m') \right|^2} \tag{5.4}$$

where $h_k(n_r, m_t)$ is the channel element between the n_r antenna of the kth UE and the m_t antenna of the BS, Ξ_k is the received SNR of the kth UE. This is independent and identically distributed (i.i.d) complex Gaussian channel gain coefficient with zero mean and unit variance, i.e. $h_k(n_r, m_t) \sim \mathcal{CN}(0,1)$. The total transmit power by all the antennas at the BS is assumed to be unity. The achievable sum-rate of the MU-MIMO system is expressed as

$$C_{sum-rate} = \sum_{m_t=1}^{M_T} \log_2 (1 + \max_{1 \le k \le K, 1 \le n_r \le N_T} SINR_{m_t,n_r}^{(k)}) \tag{5.5}$$

The possible number of ways of scheduling M_T number of users from all K users according to DPC is given in Eq. (5.1), which is very high for a higher number of users. To have a subset of this large set, the authors in [13] removed some of the possible user combinations by considering such constraints as:

a) User combinations having same M_T users will be evaluated only once.
b) No user should be served by more than one transmit antenna.
c) User combinations having duplicate users should be removed.

By considering these constraints, the total number of user combinations evaluated by the CUAS is given by

$$N_{UniqueUserSequence} = \binom{K}{M_T}. \tag{5.6}$$

This number also grows very fast with a higher number of transmit antennas and users. Because of this, the number of evaluation of the utility function is very high. This much computation cannot be done within the time frame of current high-speed communication systems. Therefore, BGA with elitism and AM has been used for the CUAS by the authors of [13] to reduce the computational complexity, so as to complete the scheduling process well within the stipulated time frame. Moreover, to facilitate this process, high-end digital signal processing (DSP) units are used at the BS.

5.2.1 GA Methodology Followed for CUAS

Evolutionary algorithms were considered by Holland [45] in 1975, which mimic the biological system's evolution process. GA delivers solutions close to the optimal ones swiftly with less computational complexity. Solutions to non-convex optimization utility functions can also be delivered by GA. BGA has been considered for solving the discussed user/antenna scheduling scenario for the MU-MIMO broadcast network due to the features of GA highlighted above. In this section, BGA has been chosen to be used. Therefore, the chromosomes will have a string of combination of bit "0" and bit "1."

The set of constraints below have been considered by the authors in [13] for obtaining near-optimal scheduling solution incurring less computational complexity:

a) Each antenna at BS should send independent messages to unique UEs. Therefore, no rows in the population should contain a duplicate UE's index.
b) Unique population should be considered, i.e. all the UE indices in every row should be unique.
c) The UE index has been constrained to acquire any value ranging from 1 to K. It should not go beyond this interval.

During the process of population initialization, these three constraints need to be checked. Initialization of the population should be continuously repeated till none of these three constraints violate. In each generation, the fitness value of each chromosome has to be evaluated and then sorted in descending order. The probability of crossover has been assumed to be unity. After getting the children from the crossover process, a process of mutation has been carried out on the newly formed children by flipping any gene randomly. Adaptive mutation has been considered by the authors of [13], where the probability of mutation was continuously evaluated rather than keeping it constant. According to AM, the mutation probability changes according to the mean and standard deviation of the present population, according to the relation mentioned below:

$$P_m = \frac{1}{\beta_1 + \beta_2 \sigma_P / \mu_P} \tag{5.7}$$

where σ_P, μ_P are the standard deviation and mean of the present population's fitness value, $\beta_1 = 1.2$ and $\beta_2 = 10$ (adopted from [11]). For comparison purposes, the authors assumed P_m to be 0.1 when simple mutation is considered. The aforementioned three constraints need to be satisfied by the children formed after mutation. If any constraints get violated, then randomly some genes are flipped till no constraints are violated. Moreover, elitism has also been adopted in the BGA, where the best chromosomes of the present generation are passed on to the next generation. Better children get produced by this process. The crossover and mutation process has been depicted in Figure 5.1. This BGA process is explained in Figure 5.2.

5.2.2 Simulation Results and Discussion

Different performances of the discussed BGA have been demonstrated to emphasize the benefits obtained as compared to various other schemes. The number of antennas at each UE is assumed to be two. The number of generations assumed for BGA is 25. In Figure 5.3, the system sum-rate obtained by this BGA process with AM and elitism has been compared with a limited feedback scheduling scheme

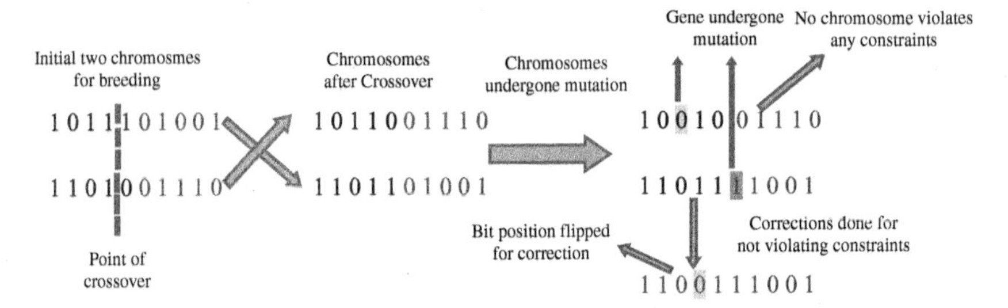

Figure 5.1 The crossover and mutation process of BGA. The system parameters considered are $K = 25, M_T = 2$.

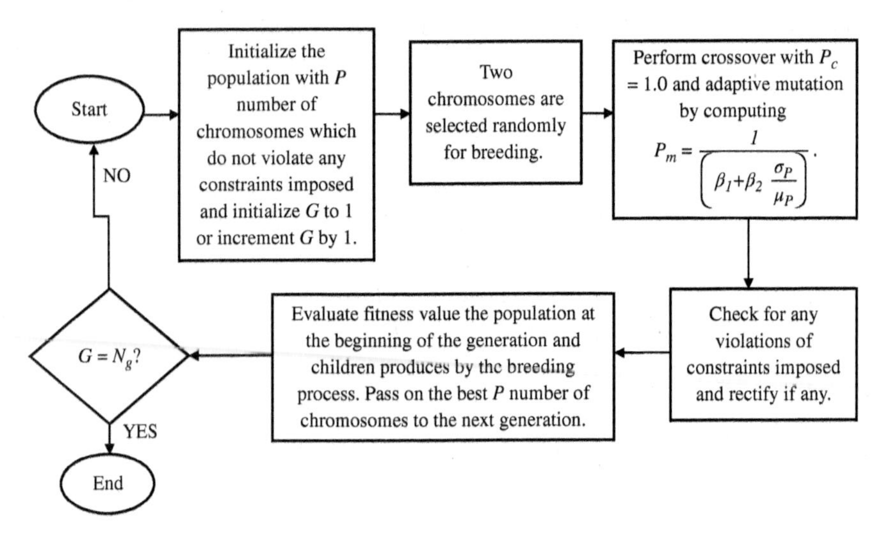

Figure 5.2 The process of BGA used for sections 5.2, 5.3, and 5.4.

of [23], ESS, and BGA process with normal mutation (with P_m to be 0.1). Each point in these figures is the averaged value of 100 independent simulations. From these figures, it has been observed that BGA with elitism and AM succeed in achieving system sum-rate close to that of ESS (DPC scheme). It has also been shown that the BGA process discussed with AM and elitism attains a higher system sum-rate than that of the limited feedback scheduling scheme discussed in [23].

The authors of [13] have also discussed a parameter called the percentage deviation from optimal (PDFO) as defined below:

$$PDFO(\theta) = \left(\frac{(C_{sum}^{ESS}) - C_{sum}^{\theta}}{C_{sum}^{ESA}} \right) \times 100, \tag{5.8}$$

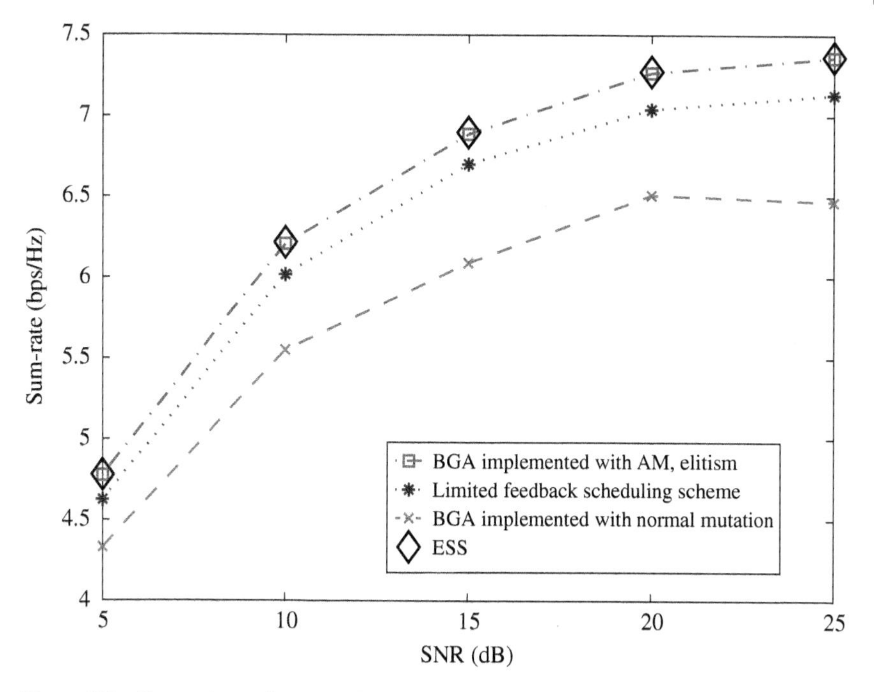

Figure 5.3 Comparison of system throughput obtained by different methods. The MU-MIMO system parameters considered are $M_T = 6, N_R = 3, K = 20$, and $P = 20$.

where θ can be the BGA process with and without AM and elitism, limited feedback scheduling scheme of [23]. This parameter gives an idea about the extent by which the system sum-rate attained by the methods other than ESS (DPC scheme) differ from the system sum-rate obtained by ESS (DPC scheme). In Figure 5.4, the PDFO for the above scenarios is presented. Each point in these figures is the averaged value of 100 independent simulations. The number of generations considered is 25 for all these figures. In [13], the authors have also shown the generation-wise performance of the proposed BGA with and without elitism and AM. The system sum-rate obtained by the proposed BGA with and without elitism and AM has been plotted with respect to number of generations in Figure 5.5. It has been observed that the system sum-rate attained at a particular generation is always more than that of the previous generation. This shows the correctness of the BGA implementation. The mutation probability (P_m) for the normal mutation is assumed to be 0.1. Here also, each point in these figures is the averaged value of 100 independent simulations.

The stochastic behavior of the proposed BGA has been shown in [13]. The empirical distribution for the histogram of the attained sum-rate by the proposed BGA with and without AM and elitism is showcased in Figure 5.6.

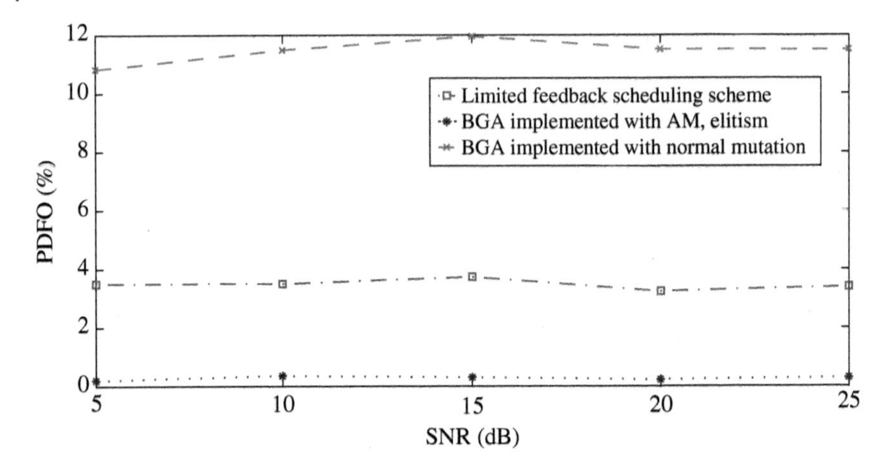

Figure 5.4 The comparison of PDFO obtained by different schemes. The MU-MIMO system parameters considered are $M_T = 8, N_R = 3, K = 25$, and $P = 25$.

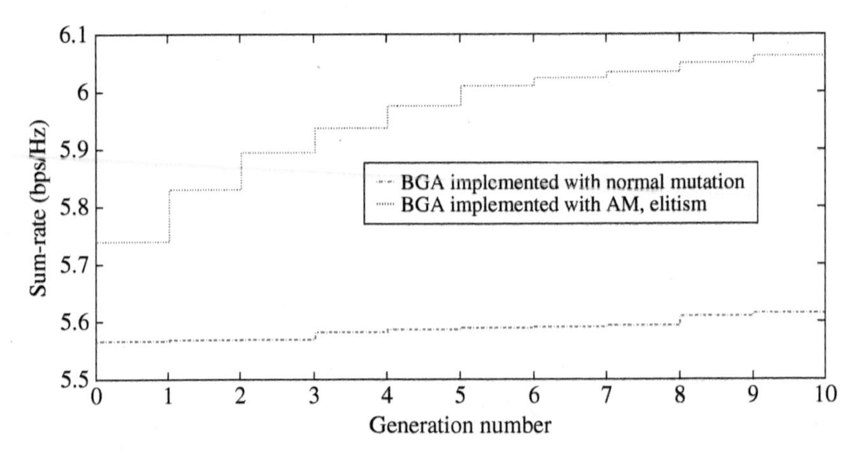

Figure 5.5 Generation wise performance of variants of BGA. The MU-MIMO system parameters considered are $M_T = 8, N_R = 3, K = 25, P = 25$, and system SNR is 10 dB.

Each point in these figures is the averaged value of 1000 independent simulations. The system sum-rate (expressed in bps/Hz) is presented in the horizontal axis and the total number of incidences during 1000 independent simulations are presented in the vertical axis of these histograms. It has been observed that the interval of sum-rate achieved by the BGA with elitism and AM is higher than the BGA without elitism and AM. Moreover, the occurrences of higher system sum-rate are also more for the BGA with elitism and AM than the BGA without elitism and AM. These are the explanations for achieving higher system sum-rate by the proposed BGA.

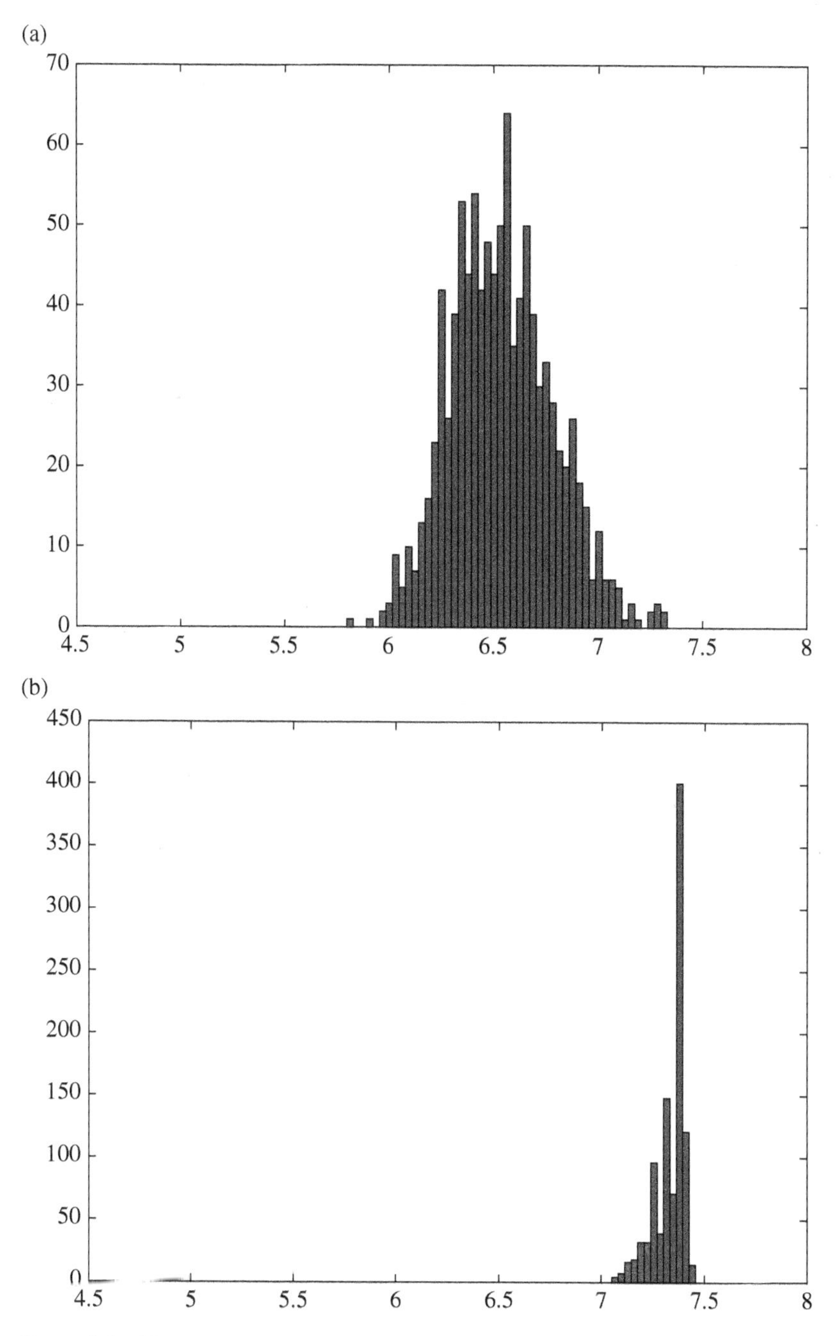

Figure 5.6 The histogram of the system sum-rate obtained by BGA with or without elitism and AM. The MU-MIMO system parameters considered are $M_T = 8$, $N_R = 3$, $K = 30$, $P = 25$, $N_g = 10$, and the system SNR is 20 dB. a) Without elitism and AM. b) With elitism and AM.

5.2.3 Computational Complexity for CUAS Scheme

It is essential to demonstrate the computational complexity analysis of the proposed BGA mechanism in [13]. The authors in [13], have represented the computational complexity in terms of the number of complex multiplications and additions (CMAs) as shown in [41, 42]. There can be a total $M_T N_R$ number of SINRs per UE. For calculation of each SINR, $2M_T$ CMAs are required as per Eq. (5.4). Therefore, altogether $2M_T^2 N_R$ CMAs will be computed for Eq. (5.5). The total number of CMA computations required for ESS (DPC) is:

$$N_{CMA}^{ESS(DPC)} = \left[2M_T^3 N_R \binom{K}{M_T} \right] \tag{5.9}$$

The total number of CMA computations required for the BGA with elitism and AM is:

$$N_{CMA}^{BGA)} = \left[2M_T^3 N_R \times \left(P + 2 \times \left[P \times P_c \right] \right) \times N_g \right] \tag{5.10}$$

where P is the population size, P_c is the crossover probability, and N_g is the number of generations required for the BGA. In [13], the authors show the time complexity of ESS (DPC) and the proposed BGA technique by considering high computational DSP processors like the multi-core ARM and DSP processor 66AK2Ex series of Texas Instruments (this processor can perform up to 44.8 Giga multiply-accumulate per second) for different MU-MIMO configurations. It has also been shown that the implementation of the BGA with elitism and AM is very much possible within the time duration of packet data communications, which is in the range of a few milliseconds [11, 12, 46].

In Table 5.1, the time complexity required by ESS (DPC) and the BGA scheme with AM and elitism has been shown for some MU-MIMO scenarios. Here, the

Table 5.1 Time complexity comparison.

System Parameters				
	ESS (DPC)		BGA	
$\{M_T, N_R, K, P, N_g\}$	CMAs	Time (ms)	CMAs	Time (ms)
5, 3, 25, 25, 10	39 847 500	0.88945	562 500	0.012556
7, 3, 30, 30, 15	4.1897e+09	93.5	2 778 300	0.062016
8, 2, 20, 20, 20	257 986 560	5.8	2 457 600	0.054857

number of CMA computations are tabulated along with the estimated execution time, taking into account the Texas Instruments DSP processor series 66AK2Ex for computation.

5.3 Selection of Optimum Quantization Thresholds for MU-MIMO Systems Using GA

The importance of selecting the optimal quantization thresholds for multi-bit quantization process has been discussed in Section 5.2. Here, SINR quantization with four bits is considered for MU-MIMO systems. The four bits representing a floating point SINR value are defined as below:

$$four\ quantized\ bits = \begin{cases} 4-\text{bit representation of "0", if } 0 \leq SINR < \Psi_1 \\ 4-\text{bit binary representation of j,} \\ \text{if } \Psi_j \leq SINR < \Psi_{j+1}, 2 \leq j \leq 14 \\ 4-\text{bit representation of "1", if } SINR > \Psi_{15} \end{cases}$$

where Ψ is the highest quantization threshold. For different quantization thresholds (from Ψ_1 to Ψ_15), it has been assumed that

a) $\Psi_0 = 0$
b) $\Psi_{15} = \Psi$
c) $\Psi_{16} = \infty$
d) the separation between two successive quantization levels is given by $\frac{\Psi}{15}$.

5.3.1 GA Methodology Followed for CUAS

BGA has been used for finding the optimal values of the four-bit quantization thresholds for SINR. For this case, the parameters which are considered while designing the BGA methodology are given as: (a) M_T, (b) K, (c) N_R, (d) the system SNR Ξ, (e) Ψ_{15}, and (f) Δ, which will be the fixed gap between two probable values of Ψ_{15}. To have minimum number of utility function executions, each chromosome of the population should satisfy the following constraints:

- No chromosomes will be allowed to have the binary coding of decimal zero(0).
- No chromosomes will be allowed to have the binary coding of decimal value > $\lceil(\Psi_{15}/\Delta)\rceil$.

After initialization of the population, two chromosomes are selected randomly to take part in a breeding process to give rise to two children. Like the previous BGA technique discussed in Section 5.2, the one-point crossover probability assumed is unity and AM with the adaptive probability as Eq. (5.7) has been used in this section.

The mutation probability has been calculated dynamically, but taking the mean and standard deviation of the current population. After the crossover and mutation process, the two constraints mentioned above are checked for each child. If any of these two constraints are violated, some of the genes have to be toggled ("0" to "1" and vice versa) till all the constraints are satisfied. The best chromosomes of the current generation are passed on to the next generation. The parameters used are the population size $P = 10$, $\Delta = 0.1$, $\Psi_{15} = 20 \times \Xi$, and number of generations $N_g = 10$. This BGA process is explained in Figure 5.2.

5.3.2 Simulation Results and Discussion

Here, a heterogeneous MU-MIMO broadcasting system has been considered for finding the optimal quantization threshold values. In this system, it has been assumed that all the users present in a cellular area of a BS do not enjoy the same received SNR. This is due to different path losses and the shadowing effect at various positions with respect to the transmitter. In this model, for simplicity it has been assumed that all the users K have been divided into Z number of user groups. Each user group has same number of users, i.e. K/Z.

The instantaneous SINR for this system has been given as below:

$$SINR_{m_t,n_r}^{(k,z)} = \frac{\left|h_k(n_r,m_t)\right|^2}{\frac{M_T}{\Xi_k^z} + \sum_{m'=1,m'\neq m_t}^{M_T} \left|h_k(n_r,m')\right|^2} \tag{5.11}$$

where Ξ_k^z represents the average received SNR of user k belonging to the user group z. The achievable average sum-rate of the system can be computed as:

$$C_{sum-rate} = IE\left[\sum_{m_t=1}^{M_T} \log_2\left(1 + SINR_{m_t,n_r^*}^{(k^*,z)}\right)\right], \tag{5.12}$$

where IE[.] is the expectation operator and $SINR_{m_t,n_r^*}^{(k^*,z)}$ is the SINR value achieved by the scheduled user k^* among the users who have sent the highest decimal value of the four-bit SINR quantized feedback corresponding to the m_t antenna at the BS. This scheduled user is assumed to the present in the zth user group according to the received SNR value. The optimum quantization threshold values required by the four-bit quantized feedback scheduling scheme and four-bit quantized optimistic scheduling scheme for various MU-MIMO systems have been computed by the process discussed above using BGA. These optimum quantization threshold values are tabulated in Table 5.2.

It has been observed from Figures 5.7 and 5.8 that the system sum-rate achieved by these two schemes employing four-bit SINR quantization is close to that of the full feedback scheduling scheme of [23] and the limited feedback scheduling

Table 5.2 Optimal quantization thresholds required by 4-bit quantized and 4-bit optimistic quantized feedback scheduling for various MIMO systems for different system SNRs [17].

Received SNR Ξ	Number of antennas at the BS (M_T)	Ψ_{15} for four-bit quantized feedback scheduling scheme as a multiple of received SNR	Ψ_{15} for four-bit quantized optimistic feedback scheduling scheme as a multiple of received SNR
5 dB	2	3	2.5
	3	2	2
	4	1	1
	5	1	1
10 dB	2	2	2
	3	1	0.9
	4	0.6	0.6
	5	0.5	0.4

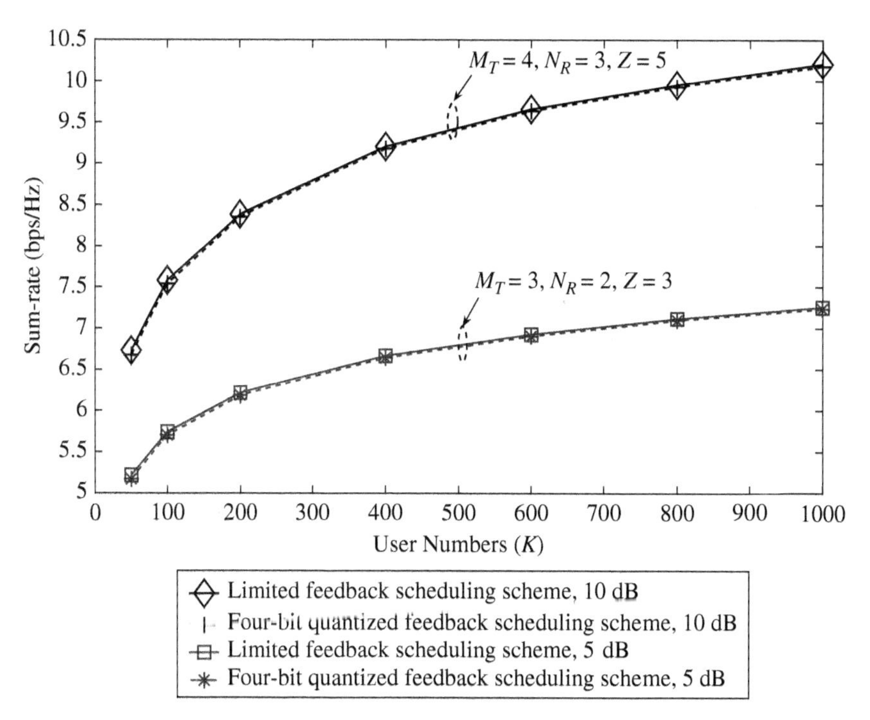

Figure 5.7 Achievable system sum-rate comparison between full feedback scheduling scheme [23] and four-bit quantized scheduling scheme with optimum quantization threshold values as per Table 5.2.

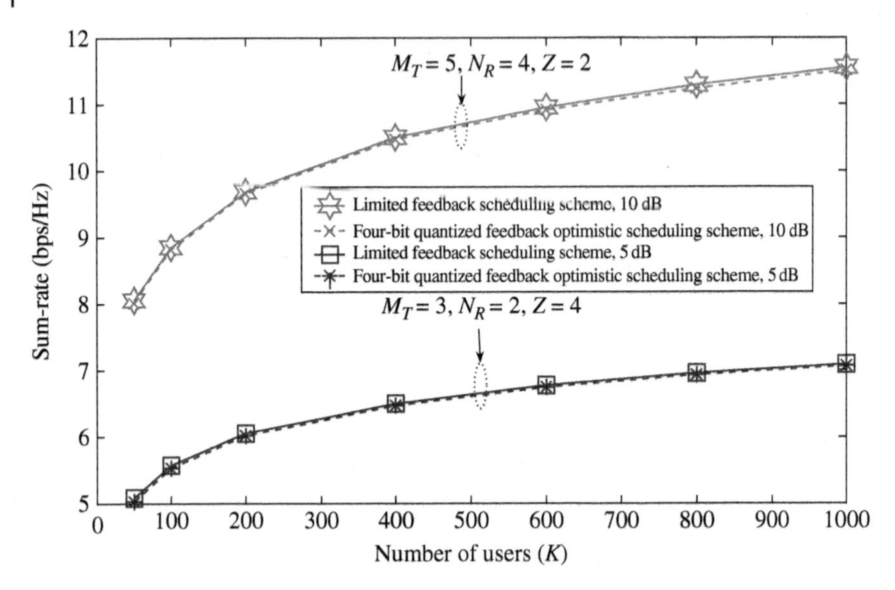

Figure 5.8 Achievable system sum-rate comparison between limited feedback scheduling scheme [18] and four-bit quantized optimistic scheduling scheme with optimum quantization threshold values according to Table 5.2.

scheme of [18]. Moreover, the feedback overhead of these two scheduling schemes employing four-bit SINR quantization is less than that of full feedback scheduling scheme of [23] and the limited feedback scheduling scheme of [18].

5.4 Selection of Optimum Quantization Thresholds for MU MIMO-OFDM Systems Using GA

The previous system parameters used in Sections 5.2 and 5.3 are valid with the introduction of some more system parameters like number of sub-carriers (L) and cluster size (L_C). The channel matrix between user k over sub-carrier l can be formulated as [36, 47]:

$$\mathbf{H}_{(k,l)} = \sum_{a=1}^{\mathrm{A}} \sigma_a \mathbf{W}_{(k,a)} \exp\left(-j2\pi \frac{l}{\mathrm{L}} a\right), 1 \leq l \leq \mathrm{L}, 1 \leq k \leq \mathrm{K}, \tag{5.13}$$

where the exponential function is denoted by exp(.). The power delay profile (PDP) of the MIMO-OFDM broadcast network is expressed as [47]:

$$\sigma_a^2 = \frac{1-\exp(-1/\mathrm{D}_{\exp})}{1-\exp(-\mathrm{A}/\mathrm{D}_{\exp})} \exp\left(-a/\mathrm{D}_{\exp}\right), 1 \leq a \leq \mathrm{A}. \tag{5.14}$$

The values of A and D_{exp} have been assumed to be 8 and 2 respectively as [36, 47]. These PDPs are normalized as

$$\sum_{p=1}^{P} \sigma_p^2 = 1. \tag{5.15}$$

The instantaneous SINR of user k over lth sub-carrier is given as:

$$SINR_{(k,l)}^{(m_t,n_r)} \triangleq \frac{\left|h_{(k,l)}^{(m_t,n_r)}\right|^2}{\left(\frac{M_T}{\Xi}\right) + \sum_{m'=1,m'\neq m_t}^{M_T} \left|h_{(k,l)}^{(m',n_r)}\right|^2}, \tag{5.16}$$

The achievable system sum-rate of this broadcast system can be computed as:

$$C_{sum-rate} = \left(\frac{1}{L}\right) \sum_{l=1}^{L} C_{sum-rate}^{(l)}, \tag{5.17}$$

where $C_{sum-rate}^{(l)}$ is the sum-rate of subcarrier l, which is expressed as

$$C_{sum-rate}^{(l)} = IE\left[\left\{\sum_{m_r=1}^{M_T} log_2\left(1 + \max_{1\leq k\leq K} SINR_{(k,l)}^{(m_t,n_r)}\right)\right\}\right], 1 \leq n \leq N_R, \tag{5.18}$$

5.4.1 GA Methodology Followed to Find the Optimal Ψ for MU MIMO-OFDM

The BGA process is described below. A small change in quantization threshold will have greater impact on the achievable system sum-rate due to the process of clustering of adjacent sub-carriers. To address this, the process of elitism and AM has been followed with a very narrow increment in the successive probable quantization threshold values. The utility function is Eq. (5.17). The minimum and maximum probable values for the quantization threshold are 0.1 and 100 respectively. Moreover, two successive probable quantization threshold values are separated by 0.1. The number of binary genes in each chromosome has been assumed to be $\log_2\left[\frac{100}{0.1}\right]$. The probability of crossover (P_c) has also been fixed at 0.8. The mutation probability for AM has been computed as per Eq. (5.7). The probability of breeding of any chromosome j from a population of size P_g has been computed as:

$$\text{Probability}_{Breeding}^{j} = \frac{C_{sum-rate}^{j}}{\sum_{j=1}^{P_g} C_{sum-rate}^{j}} \tag{5.19}$$

The process of BGA needs to be followed for Ng number of generations.

The constraints mentioned below need to be checked for every chromosome of population for each generation:

a) No chromosome should be allowed with all its genes as bit "0."
b) No chromosome should be allowed to have a decimal value greater than 1000.
c) No two chromosomes of any population should have duplicate values.

For any generation g, the fitness value of each chromosome (including the parents and children resulted from crossover and mutation) of the population is computed and they are sorted in the descending order. Then, the first P_g number of chromosomes are allowed to move to the next generation, according to elitism. This BGA process is explained in Figure 5.2.

5.4.2 Simulation Results and Discussion

By following these steps, the optimal quantization threshold values for the four-bit quantization process have been computed and presented in Table 5.3. These values are valid for $L = 64, 128, 256$. The GA parameters used for the data presented in Table 5.3 are $P_g = 10$ and $N_g = 3$.

It has been shown in Figure 5.9 that the four-bit quantized feedback scheduling scheme with optimal quantization thresholds obtains a system sum-rate equal to that of limited feedback scheduling scheme of [18] with less feedback overhead. Moreover, in Figure 5.10, it has been shown that the four-bit quantized feedback scheduling scheme with optimal quantization thresholds according to Table 5.3 achieves higher system sum-rate than the center-subcarrier based scheduling scheme of [37].

5.4.3 Achievement of Reduction in Computational Complexity by BGA

It has been shown in [25] that reduction in computational complexity has been achieved by implementing the BGA discussed above for MU MIMO-OFDM broadcasting network. The total number of times the utility function gets executed for

Table 5.3 The optimized quantization threshold values Ψ_{15} for $L_C = 16, 32$ and system SNR $\Xi = 5$ dB and 10 dB.

	$\Xi = 5$ dB		$\Xi = 10$ dB	
$M_T = N_R$	$L_C = 16$	$L_C = 32$	$L_C = 16$	$L_C = 32$
2	45	75	35	60
3	25	65	20	30
4	20	35	10	20
5	15	25	10	15

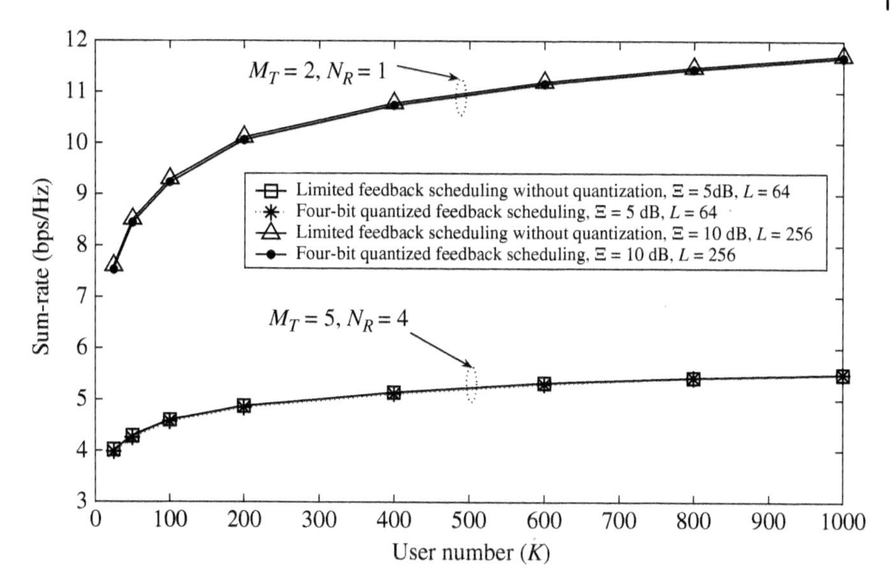

Figure 5.9 Comparison of sum-rate between the four-bit quantized scheduling scheme with optimum quantization thresholds according to Table 5.3 and the limited feedback scheduling scheme of [18] for various values of L, Ξ, and $L_C = 16$.

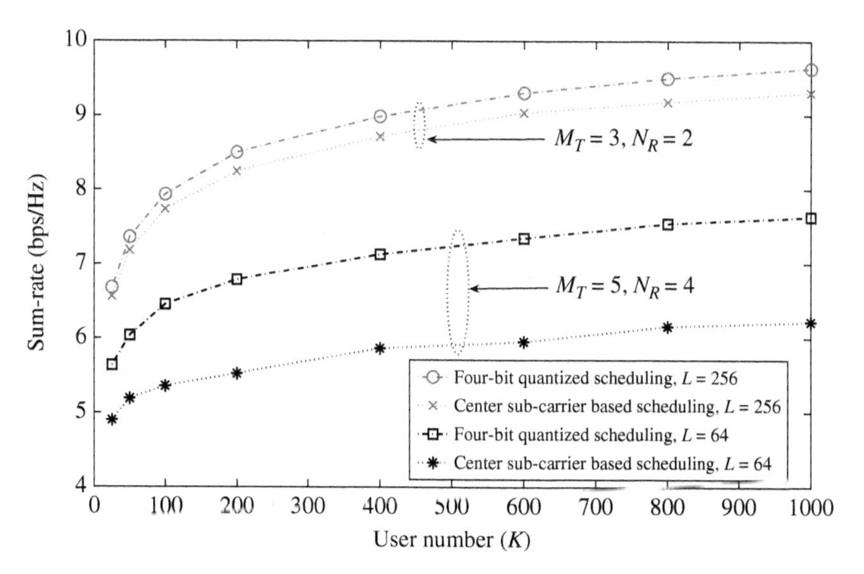

Figure 5.10 Comparison of sum-rate between the four-bit quantized scheduling scheme with optimum quantization thresholds according to Table 5.3 and the center sub-carrier-based scheduling scheme of [37] for various values of L, $\Xi = 10$ dB, and $L_C = 32$.

the ESS process and the BGA process are given by N_{ESS} and N_{BGA} respectively. $N_{ESS} = 100/0.1 = 1000$, which includes all the possible options for quantization threshold as 0.1, 0.2, ..., 100. The expression for N_{BGA} has been given as the expression below:

$$N_{BGA} = \{P_g + (2 \times P_g \times P_c)\} \times N_g \tag{5.20}$$

By using $P_g = 10$, $P_c = 0.8$, and $N_g = 3$, N_{BGA} is computed to be 78. Therefore, the reduction in computation complexity achieved by the BGA is $\left\{\frac{1000-78}{1000}\right\} \times 100\% = 92.2\%$.

5.5 Scheduling for MIMO Systems Using PSO

In recent years, PSO has also been used for these kinds of user/antenna scheduling schemes [41, 42]. Authors in [41, 42] have used the BPSO technique for joint transmit and receive antenna selection for single-user MIMO systems and joint user and receive antenna selection for multi-user MIMO systems respectively. It has been shown in these papers that low computational complexity solutions to scheduling problems have been achieved by using BPSO. Moreover, the authors of [41, 42] have presented that using a particular technique of BPSO which uses cyclically shifted initial population achieves an earlier convergence than the conventional BPSO method.

5.5.1 JTRAS in Single-User MIMO Systems Using BPSO

In this section, the use of BPSO will be shown for JTRAS in single-user MIMO systems. The social behavior of fish schooling and bird flocking has given rise to PSO. PSO has been proved to be a robust optimization approach which can find the global optimum very swiftly [48]. PSO is an population-based evolutionary algorithm, which consists of particles as individuals and the collection of particles is called a swarm or population. In this process, particles move in a N-dimensional space, where N is the number of variables present in the cost function. Particles change their position to find the optimal solution based on their own and other particles' experience. In BPSO [49], the position of each particle is represented as a string of binary bits of "0" and "1."

The variables and notations used in [41] are as below:

- The search space consisting of all possible solutions, which is denoted as the combinations presented in N_{ESS}.
- The fitness or cost function which is denoted as Eq. (5.3).
- The size of population (or swarm) (P).

- Number of iterations (N_g).
- Dimension of the particle position (i.e. the number of variables in the cost function).
- Two parameters c_1 and c_2 are used for social and cognitive control respectively over the movement of the particle. It has been assumed that $c_1 = c_2 = 2$.
- In conventional BPSO, the initial population is assumed to be selected randomly. However, in this paper, authors have used an initial feed as the initial population to reduce the convergence time. Then some cyclic shift of this inital feed is also considered for some of the particles in the initial swarm.
- A cyclically shifted initial swarm also does not allow biasing to the initial population.

5.5.1.1 Simulation Results and Discussion

The 10% outage capacity for a MIMO system has been presented in Figure 5.11 by considering the system parameters as those of [41]. For this figure, the system parameter values are $M_T = 6$, $N_R = 30$, $M_t = 3$, $N_r = 3$, $N_g = 25$, $P = 25$, $c_1 = 2$, $c_2 = 2$. The GA used is priority-based GA, as discussed in [50].

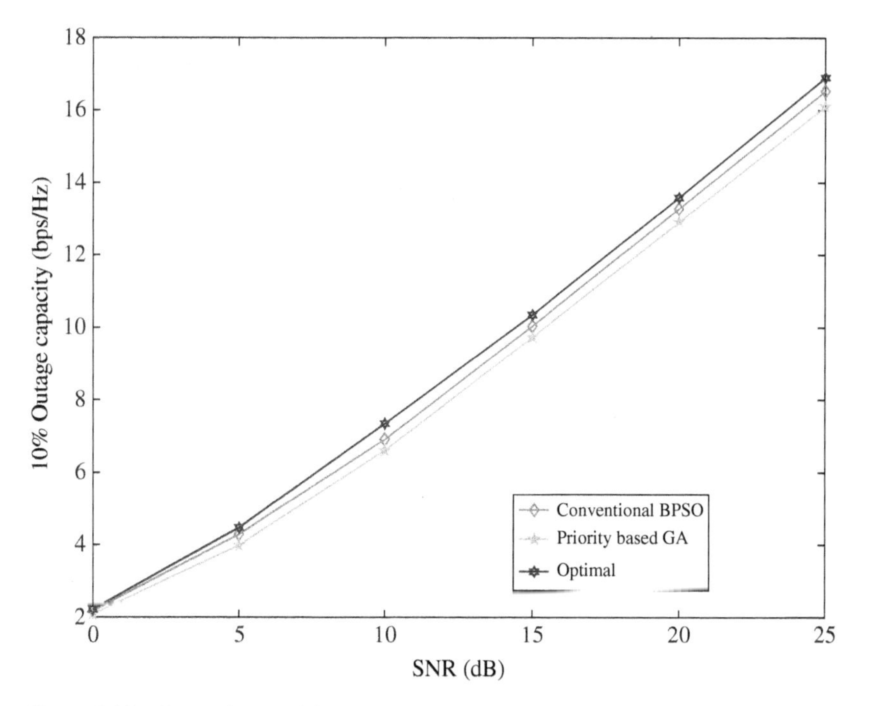

Figure 5.11 Comparison of 10% outage capacity achieved by GA [50], conventional BPSO, and the optimal scheme.

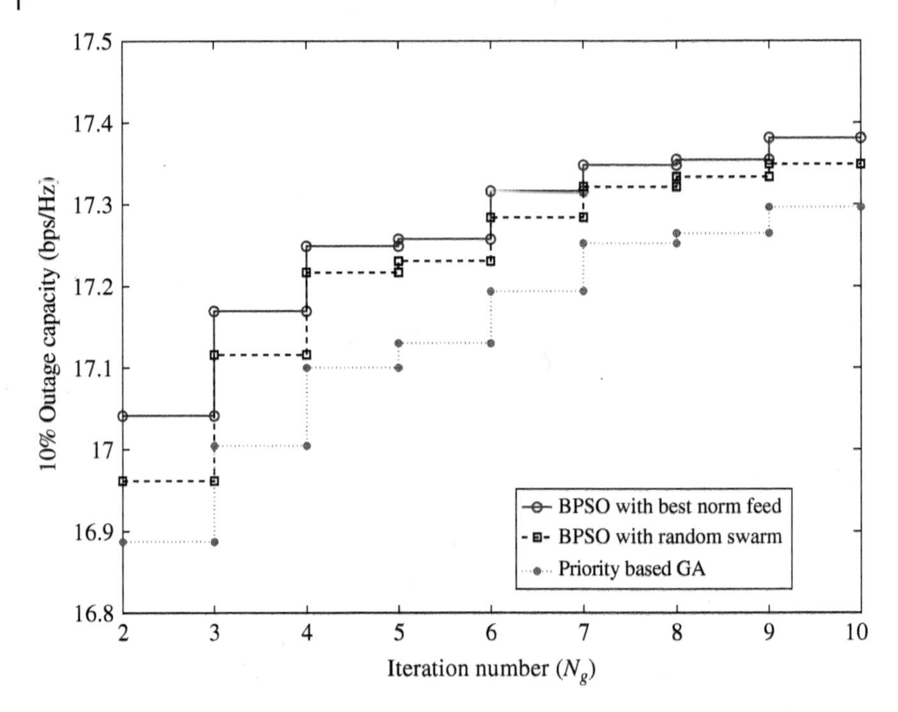

Figure 5.12 Comparison of 10% outage capacity achieved by GA [50] and various BPSO schemes for different number of generations.

It can be observed that BPSO performs better than GA for this JTRAS.

The performance of this BPSO technique has been shown in Figure 5.12 with respect to number of iterations. The convergence of this BPSO technique with a cyclic shift of initial feed performs better than GA and BPSO with random initialized population. The MIMO system parameters considered for Figure 5.12 are $M_T = 8, N_R = 12, M_t = 2, N_r = 5, N_g = 10, P = 25, c_1 = 2, c_2 = 2, \Xi = 25\,\text{dB}$.

5.5.2 JUSRAS in MU-MIMO Systems Using BPSO

Here a multi-user MIMO system is considered. The parameters used earlier will remain valid for this section as well. There are K number of UEs, where each UE has N_R number of antennas. The BS has M_T number of antennas. In this section, the usage of BPSO for JTRAS in single-user MIMO systems will be shown. The authors in [42] have considered the uplink channel for this JUSRAS problem. The total power at the BS has been divided between all the UEs equally, i.e. each UE has $\frac{T}{K}$ amount of power, where T is the total power present at the BS. Then, each of the antennas at UE also has the same power, i.e. $\frac{T}{K \times N_R}$. Further, it has also been assumed that the UEs do not have the uplink CSI whereas the BS has the complete uplink CSI. The instantaneous channel sum-rate can be expressed as 5.3. This is also the utility function for the JUSRAS.

For the JUSRAS, the total number of potential joint user and receive antenna selections are given as:

$$N_{ESS} = \binom{N_T}{N_t} \times \sum_{k=1}^{K_s} \binom{K}{k}, \tag{5.21}$$

where K_s is the total number of users selected. The number of probable candidates (N_{ESS}) increases with number of users and number of antennas. These many candidates cannot be examined within the period of a symbol duration of the current high data rate communication systems. Therefore, the authors of [42] have proposed to use BPSO to have a better solution. The BPSO technique with cyclically shifted population has been used for the JUSRAS as that was used for the JTRAS in Section 5.5.1.

5.5.2.1 Simulation Results and Discussion

In this section, the performance of this BPSO method with a cyclic shifted population has been analyzed by three ways of finding the PDFO, system sum-rate per generation, and the standard deviation achieved. The GA technique, which has been used here for comparison purposes, is the basic GA method where probability of crossover and mutation have been assumed to be 0.9 and 0.1 respectively. The PDFO has been defined as:

$$PDFO = \frac{\left\{C_{sum-rate}^{ESS} - C_{sum-rate}^{technique}\right\}}{C_{sum-rate}^{ESS}} \times 100\%, \tag{5.22}$$

where $C_{sum-rate}^{ESS}$ is the sum-rate achieved by the exhaustive search technique, which yields the maximum system sum-rate and *technique* can be either BPSO or GA.

The MU-MIMO system parameters considered for Figure 5.13 are $N_R = 2$, $M_t = 2$, $M_T = 10$, $K_s = 4$, and $K = 12$. It has been clearly observed that the PDFO for GA is more than BPSO, which underlines that the system sum-rate achieved by BPSO is more than GA for this JUSRAS.

The MU-MIMO system parameters considered for Figure 5.14 are $N_R = 2$, $M_t = 4$, $M_T = 8$, $K_s = 3$, $K = 10$, and $\Xi = 20\,\text{dB}$. The population size used for BPSO and GA for this figure has been assumed to be 10, which is equal to the number of total users in this uplink scenario.

The MU-MIMO system parameters considered for Figure 5.15 are $N_R = 2$, $M_t = 2$, $M_T = 8$, $K_s = 4$, and $K = 14$. Here also, the population size for BPSO and GA has been assumed to be 14, which is the number of users present in the

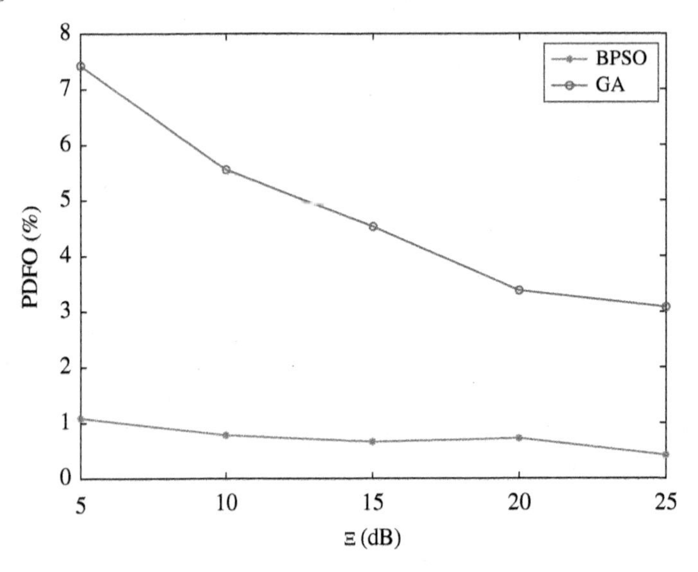

Figure 5.13 PDFO comparison between GA and BPSO. *PDFO* has been plotted for different Ξ values.

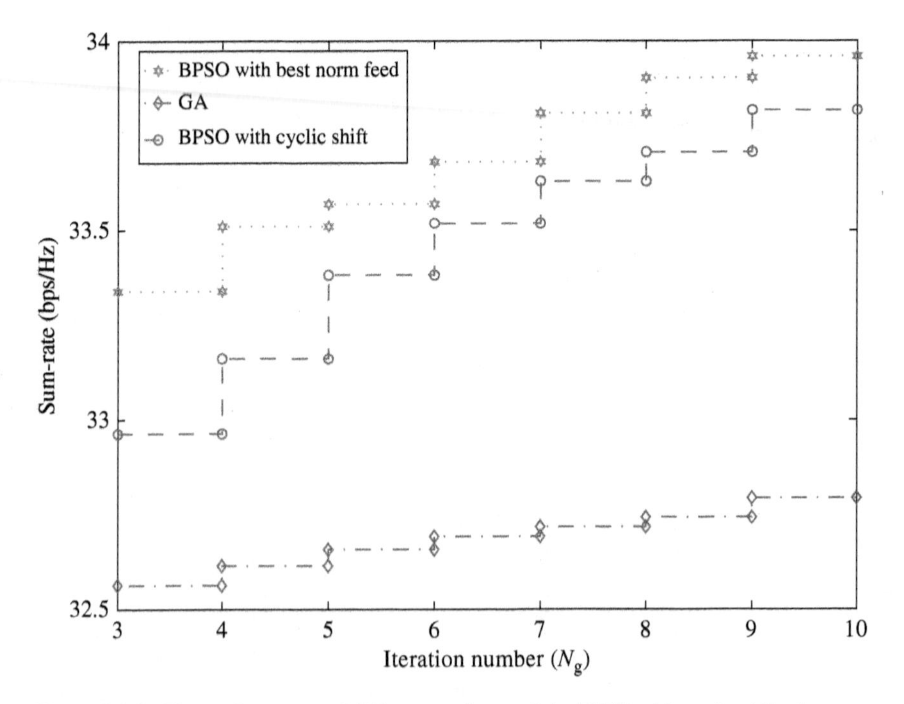

Figure 5.14 The performance of different variants of the BPSO with cyclic shifted population and GA are compared generation wise.

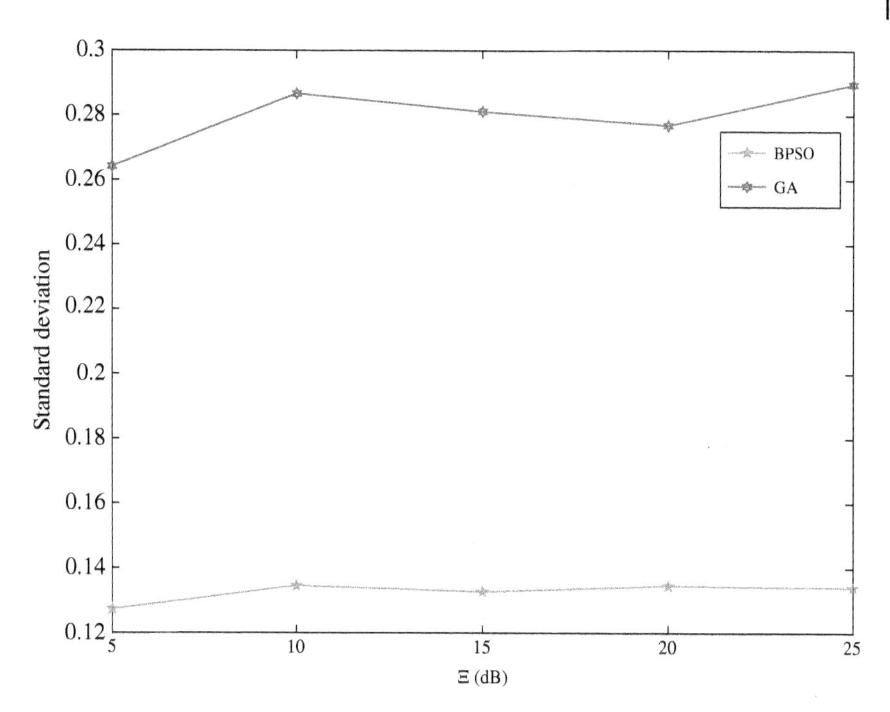

Figure 5.15 The standard deviation for BPSO and GA has been plotted against different system SNR values (Ξ).

MU-MIMO system. The standard deviation achieved by the BPSO is less than that of the GA, which emphasizes that BPSO attains better system sum-rate than the GA technique.

5.6 Conclusion

The use of GA and PSO for user and antenna scheduling of multi-user multiple antenna systems has been discussed in this chapter. The search space for both user and antenna scheduling process is large and this space grows dramatically with the increase in numbers of users and antennas at both transmitter and receiver. The use of GA has resulted in a convincing reduction in the magnitude of such a problem. It has been demonstrated that a reduction in computational complexity has been achieved by using BGA for the MU-MIMO broadcast network. The estimated timing for completion of the scheduling process by BGA with elitism and AM has been shown to be well within the required duration of the current and next generation packet data communications. Moreover, the advantage of using elitism and AM in achieving higher system sum-rate than the limited feedback scheduling scheme has also been shown. The PDFO achieved by BGA elitism and AM is less than that of limited feedback scheduling and BGA without elitism and AM. Quantization of CSI by multiple bits has been proposed and discussed for achievement of feedback overhead reduction of the closed-loop MU-MIMO systems. However, multi-bit quantization is effective only when the optimal quantization thresholds are selected. The search space for this optimal quantization threshold is very large. Therefore, BGA has been used for finding the optimal quantization thresholds for both MU-MIMO and MU-MIMO-OFDM systems. Further, these quantization thresholds have been expressed in terms of system SNR (Ξ) for easier system implementation. Both homogeneous and heterogeneous MIMO systems have been considered. The use of PSO has also been discussed for JTRAS in single-user MIMO systems and JUSRAS for MU-MIMO systems. The performance of PSO with a cyclic shifted initial population after each generation has been shown to be better than GA and basic PSO. It has also been shown that this kind of PSO outperforms GA in terms of the achievable system sum-rate. The standard deviation achieved by PSO for JTRAS and JUSRAS is less than that of GA.

References

1 Knopp, R. and Humblet, P.A. (1995). Information capacity and power control in single cell multiuser communications. *Proceedings of the IEEE Int. Conf. Commun.* (June 1995), 331–335.

2 Letaief, K.B. and Zhang, Y. (2006). Dynamic multiuser resource allocation and adaptation for wireless systems. *IEEE Wirel. Commun. Mag.* 13: 38–47.

3 Costa, M.H.M. (1983). Writing on dirty paper. *IEEE Trans. Inf. Theory* 29: 439–441.

4 Caire, G. and Shamai, S. (2003). On the achievable throughput of a multiantenna gaussian broadcast channel. *IEEE Trans. Inf. Theory* 43: 1691–1706.

5 Mazzarese, D.J. and Krzymien, W.A. (2007). Scheduling algorithms and throughput maximization for the downlink of packet-data cellular systems with multiple antennas at the base station. *IEEE Trans. Veh. Technol.* 43: 215–260.

6 Ajib, W. and Haccoun, D. (2005). An overview of scheduling algorithms in MIMO-based fourth-generation wireless systems. *IEEE Net.* 19 (5): 43–48.

7 Chung, J., Hwang, C.S., Kim, K., and Kim, Y.K. (2003). A random beamforming techniques in MIMO systems exploiting multiuser diversity. *IEEE J. Sel. Areas Commun.* 21: 848–855.

8 Gozali, R., Buehrer, R.M., and Woerner, B.D. (2003). The impact of multiuser diversity on space-time block coding. *IEEE Commun. Lett.* 7: 213–215.

9 Hochwald, B.M., Marzetta, T.L., and Tarokh, V. (2004). Multiple-antenna channel hardening and its implications for rate feedback and scheduling. *IEEE Trans. Inf. Theory* 50: 1893–1909.

10 Sharma, N. and Ozarow, L.H. (2005). A study of opportunism for multipleantenna systems. *IEEE Trans. Inf. Theory* 51: 1804–1814.

11 Elliott, R.C. and Krzymien, W.A. (2009). Downlink scheduling via genetic algorithms for multiuser single-carrier and multicarrier MIMO systems with dirty paper coding. *IEEE Trans. Veh. Technol.* 58: 3247–3262.

12 Lau, V.K.N. (2005). Optimal downlink space-time scheduling design with convex utility functionsmultiple-antenna systems with orthogonal spatial multiplexing. *IEEE Trans. Veh. Technol.* 54: 1322–1333.

13 Pattanayak, P. and Kumar, P. (2015). A computationally efficient genetic algorithm for MIMO broadcast scheduling. *Elsevier Appl. Soft Comput.* 37: 545–553.

14 Bayesteh, A. and Khandani, A. (2005). On the user selection for MIMO broadcast channels. Proceedings of the IEEE Int. Symp. Inf. Theory, Adelaide, Australia (4–9 September), 2325–2329.

15 Jaewoo, S. and Cioffi, J.M. (2009). Multiuser diversity in a MIMO system with opportunistic feedback. *IEEE Trans. Veh. Technol.* 58: 4909–4918.

16 Min, M., Kim, D., Kim, H., and Im, G. (2013). Opportunistic two-stage feedback and scheduling for MIMO downlink systems. *IEEE Trans. Commun.* 61: 312–324.

17 Pattanayak, P. and Kumar, P. (2016). Quantized feedback MIMO scheduling for heterogeneous broadcast networks. *Wirel. Netw.* 23: 1449–1466.

18 Pattanayak, P., Roy, K.M., and Kumar, P. (2015). Analysis of a new MIMO broadcast channel limited feedback scheduling algorithm with user grouping. *Springer Wirel. Pers. Commun.* 80: 1079–1094.

19 Sharif, M. and Hassibi, B. (2005). On the capacity of MIMO broadcast channels with partial side information. *IEEE Trans. Inf. Theory* 51: 506–522.

20 Swannack, C., Uysal-Biyikoglu, E., and Wornell, G.W. (2005). MIMO broadcast scheduling with limited channel state information. Proc. Annual Allerton Conf. on Commun., Control and Computing, September.

21 Wunder, G., Schreck, J., and Jung, P. (2012). Nearly doubling the throughput of multiuser MIMO systems using codebook tailored limited feedback protocol. *IEEE Trans. Wirel. Commun.* 11: 3921–3931.

22 Yoo, T. and Goldsmith, A. (2006). On the optimality of multiantenna broadcast scheduling using zero-forcing beamforming. *IEEE J. Select. Areas Commun.* 24: 528–541.

23 Zhang, W. and Letaief, K.B. (2007). MIMO broadcast scheduling with limited feedback. *IEEE J. Select. Areas Commun.* 25: 1457–1467.

24 Yang, C., Han, J., Li, Y., and Xu, X. (2013). Self-adaptive genetic algorithm based MU-MIMO scheduling scheme. *Proc. IEEE International Conf. on Commun. Technol.*: 180–185.

25 Pattanayak, P. and Kumar, P. (2017). Quantized feedback scheduling for MIMO-OFDM broadcast networks with subcarrier clustering. *Elsevier Ad Hoc Netw.* 65: 26–37.

26 Pattanayak, P. and Kumar, P. (Apr 2018). Combined user and antenna scheduling scheme for MIMO–OFDM networks. *Telecom Systems: Modelling, Analysis, Design and Management* 70: 3–12. https://doi.org/10.1007/s11235-018-0462-0.

27 Pattanayak, P. and Kumar, P. (2019). An efficient scheduling scheme for mimo-ofdm broadcast networks. *AEU Int. J. Electron. Commun.* 101: 15–26. https://doi.org/10.1016/j.aeue.2019.01.017.

28 Al-Saedy, M., Al-Imari, M., Al-Shuraifi, M., and Al-Raweshidy, H. (2017). Joint user selection and multimode scheduling in multicell MIMO cellular networks. *IEEE Trans. Veh. Technol.* 66 (12): 10962–10972. https://doi.org/10.1109/TVT.2017.2717909.

29 Fang, F., Zhang, H., Cheng, J. et al. (2017). Joint user scheduling and power allocation optimization for energy-efficient NOMA systems with imperfect CSI. *IEEE J. Sel. Areas Commun.* 35 (12): 2874–2885. https://doi.org/10.1109/JSAC.2017.2777672.

30 Femenias, G. and Riera-Palou, F. (2016). Scheduling and resource allocation in downlink multiuser MIMO-OFDMA systems. *IEEE Trans. Commun.* 64 (5): 2019–2034. https://doi.org/10.1109/TCOMM.2016.2547424.

31 Hou, J., Yi, N., and Ma, Y. (2015). Joint spacefrequency user scheduling for MIMO random beamforming with limited feedback. *IEEE Trans. Commun.* 63 (6): 2224–2236. https://doi.org/10.1109/TCOMM.2015.2432772.

32 Liu, H., Gao, H., Yang, S., and Lv, T. (2017). Low-complexity downlink user selection for massive MIMO systems. *IEEE Syst. J.* 11 (2): 1072–1083. https://doi.org/10.1109/JSYST.2015.2422475.

33 Ronasi, K., Niu, B., Wong, V.W.S. et al. (2014). Throughput-efficient scheduling and interference alignment for MIMO wireless systems. *IEEE Trans. Wirel. Commun.* 13 (4): 1779–1789. https://doi.org/10.1109/TWC.2014.031314.122040.

34 Wang, H., Wang, W., Lau, V.K.N., and Zhang, Z. (2017). Hybrid limited feedback in 5G cellular systems with massive MIMO. *IEEE Syst. J.* 11 (1): 50–61. https://doi.org/10.1109/JSYST.2015.2455061.

35 Xu, G., Liu, A., Jiang, W. et al. (2014). Joint user scheduling and antenna selection in distributed massive MIMO systems with limited backhaul capacity. *China Commun.* 11 (5): 17–30. https://doi.org/10.1109/CC.2014.6880457.

36 Eslami, M. and Krzymien, W.A. (2011). Net throughput maximization of per-chunk user scheduling for MIMO-OFDM downlink. *IEEE Trans. Veh. Technol.* 60: 4338–4348.

37 Fakhereddin, M.J., Sharif, M., and Hassibi, B. (2009). Reduced feedback and random beamforming for OFDM MIMO broadcast channels. *IEEE Trans. Commun.* 57: 3827–3835.

38 Wong, C.Y., Cheng, R.S., Lataief, K.B., and Murch, R.D. (1999). Multiuser ofdm with adaptive subcarrier, bit and power allocation. *IEEE J. Sel. Areas Commun.* 17: 1747–1758.

39 Stuber, G.L., Barry, J.R., McLaughlin, S.W. et al. (2004). Broadband mimo-ofdm wireless communications. *Proc. IEEE* 92: 271–294.

40 Molisch, A.F., Win, M.Z., Choi, Y.-S., and Winters, J.H. (2005). Capacity of MIMO systems with antenna selection. *IEEE Trans. Wirel. Commun.* 4 (4): 1759–1772. https://doi.org/10.1109/TWC.2005.850307.

41 Naeem, M. and Lee, D.C. (2011). Low-complexity joint transmit and receive antenna selection for MIMO systems. *Eng. Appl. Artif. Intell.* 24: 1046–1051.

42 Naeem, M. and Lee, D.C. (2014). A joint antenna and user selection scheme for multiuser MIMO system. *Appl. Soft Comput.* 23: 366–374.

43 Zhang, H., Xie, J., Hu, Q. et al. (2018). A hybrid dpso with levy flight for scheduling mimo radar tasks. *Appl. Soft Comput.* 71: 242–254. https://doi.org/10.1016/j.asoc.2018.06.028.

44 Makki, B., Svensson, T., and Alouini, M. (2019). On the throughput of large-but-finite mimo networks using schedulers. *IEEE Trans. Wirel. Commun.* 18 (1): 152–166. https://doi.org/10.1109/TWC.2018.2878252.

45 Holland, J.H. (1975). *Adaptation in Natural and Artificial Systems*, 1ee. Ann Arbor, MI: University of Michigan Press.

46 High Speed Downlink Packet Access (HSDPA). Overall Description; Stage 2 (Release 8), Sep. 2008.

47 McKay, M.R., Smith, P.J., Suraweera, H.A., and Collings, I.B. (2008). On the mutual information distribution of OFDM-based spatial multiplexing:exact variance and outage approximation. *IEEE Trans. Inf. Theroy* 54: 3260–3278.

48 Kennedy, J. and Eberhart, R. (1995). Particle swarm optimization. Proceedings of ICNN '95 – International Conference on Neural Networks, 11: 1942–1948. https://doi.org/10.1109/ICNN.1995.488968.

49 Kennedy, J. and Eberhart, R.C. (1997). A discrete binary version of the particle swarm algorithm. *1997 IEEE International Conference on Systems, Man, and Cybernetics. Computational Cybernetics and Simulation*, 5: 4104–4108. https://doi.org/10.1109/ICSMC.1997.637339.

50 Lu, H. and Fang, W. (2007). Joint transmit/receive antenna selection in MIMO systems based on the priority-based genetic algorithm. *IEEE Antennas Wirel. Propag. Lett.* 6: 588–591. https://doi.org/10.1109/LAWP.2007.911387.

6

An Efficient Modified Red Deer Algorithm to Solve a Truck Scheduling Problem Considering Time Windows and Deadline for Trucks' Departure

Amir Mohammad Fathollahi-Fard, Abbas Ahmadi, and Mohsen S. Sajadieh

Department of Industrial Engineering and Management Systems, Amirkabir University of Technology, Tehran, Iran

6.1 Introduction and Literature Review

Recent decades have seen a great deal of interest in today's competitive market and the issue of customers' satisfaction, which has become more crucial for companies than ever [1]. Recent reports confirm that supply chain performance directly relies on customers' satisfaction [2]. Clients generally prefer to receive high-quality products for the lowest cost at the correct place and time [3]. Taking all of this into account, the role of distribution centers is undeniable in creating customers' satisfaction. A high-efficiency distribution center can not only decrease the total cost of supply chain systems, especially in transportation, and control the quality of products receiving from manufacturers, but also its main duty is to deliver products to customers in the correct place at a suitable time [4]. Hence, one of the most beneficial procedures to increase the efficiency of distribution centers is a cross-docking system to achieve the goals described above satisfactorily [5]. One of the main operations of cross-docking is truck scheduling. This complex scheduling often requires extensive effort and time to make a valid plan which respects all the constraints (e.g. departure time, multi-product, ready time and time window, etc.). Based on this motivation, this chapter proposes a new truck scheduling problem considering time windows and the deadline for trucks' departure.

Evolutionary Computation in Scheduling, First Edition. Edited by Amir H. Gandomi, Ali Emrouznejad, Mo M. Jamshidi, Kalyanmoy Deb, and Iman Rahimi.

Generally, the cross-docking system is one way to distribute the products from manufacturers to markets [6]. The finished products are entered into the inbound dock by receiving trucks. Afterwards, these products may be sorted according to the demand of customers [7]. Last but not the least, these products are directly transferred to the outbound dock in order to be loaded into shipping trucks. Notably, the main feature of cross-docking is that long-term storage is not allowed [8]. Accordingly, the cross-docking can remove both long-run storage and the products' retrieval costs. Therefore, an increase in the performance of cross-docking leads to an increase in the efficiency of distribution systems and thus in the satisfaction of customers [9].

From one of basic papers in this research area, Apte and Viswanathan [3] in 2000 explored a group of techniques to make an improvement in the cross-docking center. One of their main recommendations is the use of Information Technology (IT) to automate the handling of material required. Some other advantages of this approach are utilizing the whole capacity of trucks in terms of transferring the products and the efficient design of cross-docking centers. They declared that when the rate of demand is stable and the cost of stock-out is low, using cross-docking is appropriate and highly recommended. Otherwise, using traditional warehousing systems would be more suitable. Generally, an implementation of both systems simultaneously by an interaction between traditional warehousing and cross-docking systems would be more effective [8].

The literature of cross-docking is divided into several concepts, including but not limited to the location of cross-docking, the design and layout of cross-docks, dock door assignment, and supply chain networks based on cross-docking and truck scheduling. Of these, truck scheduling has had more interest shown in investigating it during two last decades [8–10]. From a recent survey written by Ladier and Alpan [10] in 2016, cross-dock scheduling was categorized into five classifications including truck to door assignment, truck to door sequencing, truck to door scheduling, and truck sequencing and truck scheduling. In the majority of cases, the goal is the minimization of the makespan and the distance traveled as the main objective functions in cross-dock scheduling. They also declared that one of main suggestions for future works in the area of truck scheduling is to consider the deadline for shipping trucks as one of real-life constraints. This study is another variant of truck scheduling considering the deadline for trucks' departure and time windows and is among the first studies in this research area.

One of the first studies in the truck scheduling problem is referred to by Yu [11] in 2002. This author developed the truck scheduling problem, aiming to determine the optimal sequence of receiving and shipping trucks. The objective function was to minimize the makespan. Later in 2008, Yu and Egbelu [12] presented a novel mathematical model for truck scheduling in which a receiving door and a shipping door and also a temporary storage in front of the shipping door were considered.

Their main innovation was to develop nine heuristic algorithms. A comparative study was conducted to check the efficiency of algorithms by considering the results of the exact method.

Over the last decade, many studies have made a contribution to the truck scheduling problem. The most important ones are considered in detail below. In 2009, Chen and Lee [13] developed another variant of truck scheduling as a flow-shop machine scheduling problem. To address it, a branch-and-bound algorithm was proposed. Their suggested solution method was able to solve the test problems with up to 60 jobs in an acceptable period of time. In 2010, Boysen [14] introduced a cross-dock scheduling problem for a storage ban mode. The main goal was to minimize the flow time, processing time, and tardiness of outbound trucks, respectively. To tackle their NP-hard problem, a hybrid of dynamic programming and Simulated Annealing (SA) was proposed. Later in 2014, Li et al. [15] offered a two-phase parallel machine problem with earliness and tardiness to formulate the cross-dock scheduling problem. A group of metaheuristic algorithms were applied to solve their problem. In 2015, Mohtashami et al. [16] proposed a bi-objective cross-dock scheduling problem by minimizing both the total cost and the makespan, simultaneously. A Non-dominated Sorting Genetic Algorithm (NSGA-II) was applied to address it. In 2016, Amini and Tavakkoli-Moghaddam [1] discussed a problem in which the trucks might face breakdowns during the service times. They also considered a due date for each shipping truck. Three multi-objective metaheuristic algorithms were utilized to solve their problem. In 2017, Golshahi-Roudbaneh et al. [17] proposed a group of heuristics and metaheuristics such as Differential Evolution (DE) and Genetic Algorithm (GA) to find the optimal receiving and shipping trucks sequence based on Yu [11]. More recently in 2018, Mohammadzadeh et al. [18] applied three recent metaheuristics to solve the truck scheduling problem based on Golshahi-Roudbaneh et al. [17]. Finally, Ye et al. [19] proposed another multi-door truck scheduling problem and solved it by an improved particle swarm optimization (PSO).

To find the pros and cons of each work, Table 6.1 shows the properties of existing works from the aspect of modeling and solution approach. We have classified all papers based on seven properties in order to evaluate them: the main suppositions for the modeling, the multi-door, temporary storage, departure time, multi-product, ready time, and time window limitations. The solution method is characterized as exact, heuristics, and/or metaheuristics.

The main findings and similarities between the works are noted as follows. Considering more than two doors is assumed by a number of recent studies [19, 35, 37]. Temporary storage is a typical constraint for most of the existing studies. Conversely, the time window limitation is rarely considered by the literature [24, 33, 35]. Multi-product and its different groups are considered in a few

Table 6.1 Characteristics related to this work from 2008 to 2018.

Author(s)	Year	Multi-door	Temporary storage	Departure time	Multi-product	Ready time	Time window	Solution method		
								Exact	Heuristic	Metaheuristic
Yu and Egbelu [12]	2008		✓					✓	✓	
Chen and Lee [13]	2009							✓		
Chen and Song [20]	2009		✓					✓	✓	
Boysen [14]	2010		✓					✓		✓
Soltani and Sadjadi [21]	2010		✓					✓		✓
Boysen et al. [22]	2010		✓						✓	
Forouharfard and Zandieh [23]	2010									✓
Larbi et al. [24]	2011				✓		✓	✓	✓	
Arabani et al. [4]	2011		✓					✓		✓
Shakeri et al. [25]	2012	✓			✓				✓	
Berghman et al. [6]	2012		✓					✓		
Davoudpour et al. [26]	2012		✓	✓		✓				✓
Sadykov [27]	2012							✓		
Boysen et al. [28]	2013	✓						✓	✓	✓
Van Belle et al. [11]	2013		✓					✓		✓
Bjelić et al. [29]	2013		✓	✓						✓
Joo and Kim [30]	2013		✓							✓

Konur and Golias [31]	2013	✓			✓				✓	✓
Ladier and Alpan [32]	2013		✓	✓				✓	✓	
Ladier and Alpan [33]	2014		✓	✓		✓	✓	✓	✓	✓
Madani-Isfahani et al. [34]	2014		✓		✓					✓
Amini et al. [7]	2014								✓	✓
Mohtashami et al. [16]	2015			✓		✓				✓
Amini and Tavakkoli-Moghaddam [1]	2016		✓	✓		✓				✓
Golshahi-Roudbaneh et al. [17]	2017		✓						✓	✓
Khalili-Damghani et al. [35]	2017		✓	✓		✓	✓	✓		✓
Wisittipanich and Hengmeechai [36]	2017	✓	✓			✓				✓
Ye et al. [37]	2018	✓	✓							✓
Ye et al. [19]	2018	✓	✓							✓
Mohammadzadeh et al. [18]	2018		✓							✓
This study			✓	✓	✓	✓	✓			✓

studies, and adding departure time, multi-product, and ready time are not presumed simultaneously in any of the studies. Contrary to previous works, this research investigates a new variant of truck scheduling problem considering time windows, departure, and ready time for trucks in a cross-docking system. A mathematical model is introduced on the basis of the rest of the models already existing in this field [11–13, 15]. There are some differences between this work and the papers mentioned in the table. First, a time window is considered for every shipping truck. Second, the products through this system are divided into two groups: perishable and imperishable. Due to the presence of perishable products, a deadline is considered for the shipping trucks. There is no similar study to model all these constraints simultaneously. Last but not the least, the main innovation of this work is to develop a strong modified evolutionary algorithm based on a recently-introduced metaheuristic to solve the proposed model.

Taken together, the truck scheduling problem may be formulated by several new elements to be more realistic. Considering more factors not only alters the feasible area of problem but also increases its complexity. Therefore, a challenging issue similar to the other types of scheduling is selecting an appropriate solution algorithm. As is evident from Table 6.1, the majority of studies mainly innovated by applying metaheuristic algorithms due to the NP-hardness of these problems. In this regard, the theory of No Free Lunch [38] guarantees that there is no optimization algorithm which can solve all complicated optimization problems satisfactorily [2]. It means that there is a chance, albeit of low likelihood, for a new algorithm to show its superiority in comparison with other existing ones [5]. Since the truck scheduling problem as a variant of scheduling model is another complicated optimization problem, this reason has motivated several authors to utilize some recent algorithms to solve this problem better. Nowadays, the high efficiency of recent metaheuristics such as the Imperialistic Competitive Algorithm (ICA) [16, 23], Stochastic Fractal Search (SFS) [17], Keshtel Algorithm (KA) [2, 17], and Social Engineering Optimizer (SEO) [39] have motivated several recent studies to employ them to solve some variants of scheduling problems. This has encouraged our attempts to contribute a new modification of a recently-developed metaheuristic called the Red Deer Algorithm (RDA) [40] to solve the proposed truck scheduling problem properly.

This section is followed by four more. Section 6.2 offers the proposed mathematical model to formulate the developed truck scheduling problem. Section 6.3 introduces the original idea of RDA along with the proposed modified version. Section 6.4 illustrates an extensive comparison for a set of calibrated algorithms based on different criteria. Finally, discussions and conclusions of this research are presented in Section 6.5.

6.2 Proposed Problem

Here, the proposed problem according to the basic and developed mathematical model has been investigated comprehensively. First of all, the original version of truck scheduling adopted from the literature is explained. After that the proposed problem as a new variant of truck scheduling is developed by new suppositions in this research area.

6.2.1 Basic Mathematical Model

The mathematical model shown below is the basic problem of truck scheduling developed by Yu and Egbelu [12]. The following notations are utilized to define it properly:

Indices:

i	Index of receiving trucks (i = 1, 2, ..., I)
j	Index of shipping trucks (j = 1, 2, ..., J)
k	Index of products (k = 1, 2, ...K)

Parameters:

r_{ik}	Number of units of product type k that was initially loaded in receiving truck i
s_{jk}	Number of units of product type k that was initially needed for shipping truck j
D	Truck changeover time
V	Moving time of products from the receiving dock to the shipping dock
M	Big number

Continuous variables:

T	Makespan
c_i	Time at which receiving truck i enters the receiving dock
F_i	Time at which receiving truck i leaves the receiving dock
d_j	Time at which shipping truck j enters the shipping dock
L_j	Time at which shipping truck j leaves the shipping dock

Integer variable:

X_{ijk}	Number of units of product type k that transfer from receiving truck i to shipping truck j

Binary variables:

$$v_{ij} = \begin{cases} 1, \text{If any products transfer from receiving truck i to shipping truck j;} \\ 0, \text{Otherwise;} \end{cases}$$

$$p_{ij} = \begin{cases} 1, \text{If receiving truck i precedes receiving truck j in the receiving truck sequence;} \\ 0, \text{Otherwise;} \end{cases}$$

$$q_{ij} = \begin{cases} 1, \text{If shipping truck i precedes shipping truck j in the shipping truck sequence;} \\ 0, \text{Otherwise.} \end{cases}$$

The mathematical model is formulated as explained below:

$$\text{Min T}$$

s.t.

$$\text{T} \geq \text{L}_\text{j}, \forall j \in J \tag{6.1}$$

$$\sum_{j=1}^{J} x_{ijk} = r_{ik}, \forall i \in I, k \in K \tag{6.2}$$

$$\sum_{i=1}^{I} x_{ijk} = s_{ik}, \forall j \in J, k \in K \tag{6.3}$$

$$x_{ijk} \leq M v_{ij}, \forall i \in I, j \in J, k \in K \tag{6.4}$$

$$F_i \geq c_i + \sum_{k=1}^{K} r_{ik}, \forall i \in I \tag{6.5}$$

$$c_i \geq F_i + D - M(1 - p_{ij}), \forall i \in I, j \in J \tag{6.6}$$

$$c_i \geq F_j + D - M p_{ij}, \forall i \in I, j \in J \tag{6.7}$$

$$p_{ii} = 0, \forall i \in I \tag{6.8}$$

$$L_j \geq d_j + \sum_{k=1}^{K} s_{jk}, \forall j \in J \tag{6.9}$$

$$d_j \geq L_i + D - M(1 - q_{ij}), \forall i \in I, j \in J \tag{6.10}$$

$$d_j \geq L_j + D - M q_{ij}, \forall i \in I, j \in J \tag{6.11}$$

$$q_{ii} = 0, \forall i \in I \tag{6.12}$$

$$L_j \geq c_i + V + \sum_{k=1}^{K} x_{ijk} - M(1 - v_{ij}), \forall i \in I, j \in J \tag{6.13}$$

$$T, c_j, F_j, d_j, L_j, X_{ijk} \geq 0 \forall i \in I, j \in J, k \in K \tag{6.14}$$

$$v_{ij}, q_{ij}, p_{ij} \in \{0,1\} \forall i \in I, j \in J, k \in K \tag{6.15}$$

The objective function (T) aims to minimize the total operational time (makespan) of the cross-docking process. Eq. (6.1) ensures that the departure time of shipping trucks is lower than the total operational time. This time equals the departure time of the last shipping truck. Similarly, Eq. (6.2) guarantees that in the proposed cross-docking system, the total number of arrived products by each receiving truck equals the total number of products loaded by it initially. Eq. (6.3) ensures that the total number of products loaded by each shipping truck equals its demand rate. Eq. (6.4) confirms that the variables x_{ijk} and the variables v_{ij} have the correct relationship. Eq. (6.5) reveals that the arrival and departure times of receiving truck i have a relationship as shown in the equation. Similarly, Eqs. (6.6) and (6.7) confirm that the arrival and departure times of the receiving truck are similar to each other. Eq. (6.8) specifies that there is no received truck which may not be in sequence by preceding itself. The indications behind Eq. (6.9) to (6.12) are the same as Eqs. (6.5) to (6.8). The main difference between these constraints is their relation to the sequence of shipping trucks. Eq. (6.13) illustrates that the departure time of the shipping truck and arrival time of the receiving truck have a specific relationship with each other to determine each factor satisfactorily. Finally, all continuous and binary variables are guaranteed to be bounded as shown in Eq. (6.14) and Eq. (6.15), respectively.

6.2.2 Developed Mathematical Model

The developed model aims to consider a time window as well as a deadline for each shipping truck. There are several types of products which are divided into two groups, namely perishable and imperishable. Accordingly, the main assumptions of the considered problem are as follows:

- If the shipping truck carries perishable products, its departure time can never exceed the determined deadline [41].
- If the shipping truck carries imperishable products, it is possible that its departure time exceeds the determined deadline [42].
- There are a time window and a deadline which both are unique for each shipping truck [43].

Generally speaking, if the departure time of truck j is more than its deadline, a tardiness penalty cost will be allocated to the time difference between tardiness and deadline, and a deadline penalty cost will be assigned to the time difference between departure time and deadline of shipping truck j.

The new parameters and variables added in comparison with the basic model are defined as follows:

Parameters:

DDj	Due date of shipping truck j
l_j	Upper bound of time window for shipping truck j
e_j	Lower bound of time window for shipping truck j
dl_j	Deadline of shipping truck j
α_{1j}	Earliness penalty cost of shipping truck j carrying imperishable products
α_{2j}	Earliness penalty cost of shipping truck j carrying perishable products
β_{1j}	Tardiness penalty cost of shipping truck j carrying imperishable products
β_{2j}	Tardiness penalty cost of shipping truck j carrying perishable products
β_{3j}	Deadline penalty cost of shipping truck j

Continuous variables:

T_j	Tardiness of shipping truck j
E_j	Earliness of shipping truck j

Here, the proposed mathematical formulation of a new variant of truck scheduling is developed as follows. First of all, the objective function is to minimize the total cost of resulting from tardiness and earliness of shipping trucks.

$$\begin{aligned}\min \sum \alpha_{1j} \times \max(0, e_j - L_j) \times (1 - W_j) + \beta_{1j} \times \max(0, L_j - l_j) \times (1 - W_j) \times (1 - Y_j) + \\ \alpha_{2j} \times \max(0, e_j - L_j) \times W_j + \beta_{2j} \times \max(0, L_j - l_j) \times Wj + \\ (\beta_{1j} \times (dl_j - l_j) + \beta_{3j} \times (L_j - dl_j)) \times Y_j\end{aligned} \tag{6.16}$$

To understand the objective function, consider that the first and the second terms respectively calculate the earliness and tardiness penalty of shipping trucks carrying imperishable products. The third and the fourth terms similarly calculate the earliness and tardiness penalty of shipping trucks carrying perishable products. The fifth term computes the penalty amount when the departure time of shipping trucks is greater than the predetermined deadline. It deserves mention that y can accept *1* only for trucks not carrying perishable products. In other words, shipping trucks carrying perishable products are not allowed to bear a departure time greater than the determined deadline. Figure 6.1 demonstrates the

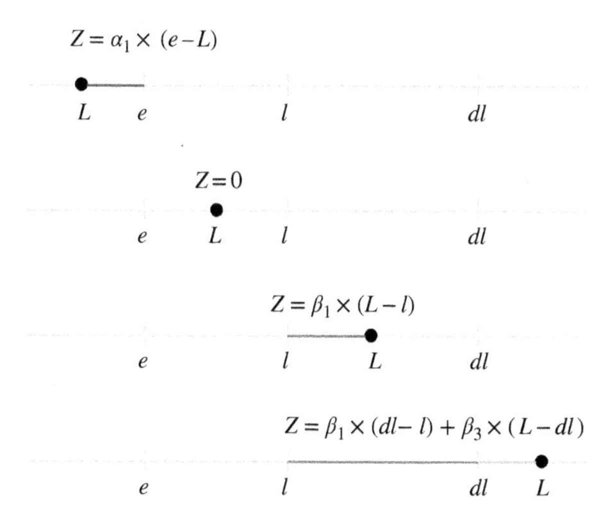

Figure 6.1 An example of computing the objective function.

method of computing the objective function in different moods for a shipping truck lacking perishable products. The horizontal axis shows departure time.

Based on this illustration, the objective function can be revised as follows:

$$\begin{aligned} \min \sum_{j=1}^{J} [&(\alpha_{1j} \times E_j) \times (1 - W_j) + (\beta_{1j} \times T_j) \times (1 - W_j) \times (1 - Y_j) + \\ &(\alpha_{2j} \times E_j) \times W_j + (\beta_{2j} \times T_j) \times Wj + \\ &(\beta_{1j} \times T_j + \beta_{3j} \times (L_j - dl_j)) \times Y_j] \end{aligned} \tag{6.17}$$

In this case, the constraints in Eqs. (6.19), (6.20), and (6.21) are added to the model. Generally, the following constraints define the main limitations of the proposed problem.

As mentioned earlier, the departure time of perishable products should not exceed the determined deadline. The following constraint guarantees this fact:

$$L_j \leq dl_j + M(1 - W_j) \forall j \in J \tag{6.18}$$

In fact, this constraint ensures that if the jth shipping truck is carrying perishable products, its departure time should not be greater than the deadline, whereas if it carries imperishable products, it will be possible that its departure time is greater than the deadline, and it faces a heavy penalty.

Note that by simplifying the objective function in the previous section, we add the following constraints to the model:

$$E_j \geq e_j - L_j \forall j \in J \tag{6.19}$$

$$T_j \geq L_j - l_j \forall j \in J \tag{6.20}$$

$$T_j \geq (L_j - dl_j) \times Y_j \forall j \in J \tag{6.21}$$

Constraints (6.19) and (6.20) respectively compute earliness and tardiness values, if any, for the jth shipping truck. Constraints (6.21) compute the tardiness value if the jth shipping truck does not carry perishable products and its departure time is greater than the deadline. Note that other constraints of the proposed problem are only adopted from the basic model. Finally, the whole model can be written as follows:

$$\begin{aligned} \min z = \sum_{j=1}^{J} [& (\alpha_{1j} \times E_j) \times (1 - W_j) + (\beta_{1j} \times T_j) \times (1 - W_j) \times (1 - Y_j) + \\ & (\alpha_{2j} \times E_j) \times W_j + (\beta_{2j} \times T_j) \times Wj + \\ & (\beta_{1j} \times T_j + \beta_{3j} \times (L_j - dl_j)) \times Y_j] \end{aligned} \tag{6.22}$$

s.t.

$$\sum_{j=1}^{J} x_{ijk} = r_{ik}, \forall i \in I, k \in K \tag{6.23}$$

$$\sum_{i=1}^{I} x_{ijk} = s_{ik}, \forall j \in J, k \in K \tag{6.24}$$

$$x_{ijk} \leq M v_{ij}, \forall i \in I, j \in J, k \in K \tag{6.25}$$

$$F_i \geq c_i + \sum_{k=1}^{K} r_{ik}, \forall i \in I \tag{6.26}$$

$$c_i \geq F_i + D - M(1 - p_{ij}), \forall i \in I, j \in J \tag{6.27}$$

$$c_i \geq F_j + D - M p_{ij}, \forall i \in I, j \in J \tag{6.28}$$

$$p_{ii} = 0, \forall i \in I \tag{6.29}$$

$$L_j \geq d_j + \sum_{k=1}^{K} s_{jk}, \forall j \in J \tag{6.30}$$

$$d_j \geq L_i + D - M(1 - q_{ij}), \forall i \in I, j \in J \tag{6.31}$$

$$d_j \geq L_j + D - M q_{ij}, \forall i \in I, j \in J \tag{6.32}$$

$$q_{ii} = 0, \forall i \in I \tag{6.33}$$

$$L_j \geq c_i + V + \sum_{k=1}^{K} x_{ijk} - M(1 - v_{ij}), \forall i \in I, j \in J \tag{6.34}$$

$$L_j \leq dl_j + M(1 - W_j) \forall j \in J \tag{6.35}$$

$$E_j \geq e_j - L_j \forall j \in J \tag{6.36}$$

$$T_j \geq L_j - l_j \forall j \in J \tag{6.37}$$

$$T_j \geq (L_j - dl_j) \times Y_j \forall j \in J \tag{6.38}$$

$$T, c_j, F_j, d_j, L_j, X_{ijk} \geq 0 \forall i \in I, j \in J, k \in K \tag{6.39}$$

$$v_{ij}, q_{ij}, p_{ij} \in \{0,1\} \forall i \in I, j \in J, k \in K \tag{6.40}$$

6.3 Red Deer Algorithm (RDA)

The RDA first introduced by Fathollahi Fard and Hajiaghaei-Keshteli [40] is one of the recent evolutionary algorithms inspired by the competition of red deer in breeding season. This algorithm has revealed its high-efficiency for solving the combinatorial optimization problems. For example, Golmohamadi et al. [44] addressed a fuzzy fixed charged transportation problem considering batch transferring by RDA and a hybrid version combined by SA. The Variable Neighborhood Search (VNS) and ICA are the other algorithms utilized by their study. In another work in the area of supply chain networks, Samadi et al. [45] proposed a sustainable supply chain network design problem addressed by three heuristics as the initial solutions of their applied evolutionary algorithms including RDA and GA. Their results ensured the performance of heuristics-based RDA through a comparative study. A similar work reported by Hajiaghaei-Keshteli and Fathollahi-Fard [46] developed a sustainable closed-loop supply chain considering discount supposition. They employed different solution methods based on recent metaheuristics including KA [47], ICA, and RDA. In another comparative study, Sahebjamnia et al. [48] illustrated a sustainable tire closed-loop supply chain network design problem for the first time. They solved their model using five hybrid metaheuristics based on recent algorithms, i.e. Water Wave Optimization (WWO) and RDA. Last but not the least, Fathollahi-Fard et al. [49] considered a stochastic multi-objective model to formulate a closed-loop supply chain network design considering social impacts. They tackled this NP-hard problem by three hybrid evolutionary algorithms based on KA and RDA. Overall, the aforementioned papers [44–49] are only some examples among other studies employing RDA and its different variants. All of them confirmed the high efficiency of this recently developed algorithm in solving the combinatorial optimization problems.

Similar to other evolutionary algorithms, the RDA starts with a population of random solutions called Red Deer (RD). This set of solutions has been divided into two types: "male RDs" as the best solutions and "hinds" as the rest of solutions. Generally, roaring, fighting, and mating are the three main behaviors of RDs during the breeding season. First of all, male RDs roar strongly to show their power to other males and attract hinds. After that a number of successful males are selected as the commanders based on their power (fitness of solutions). Another main part of algorithm is the fighting process between the commanders and the rest of males called stags. During this competition, the better solutions as the winners have been chosen again as the commanders to form the harems, which are a group of hinds. Based on the power of each commander, a number of hinds has been adopted to be in the harem. The greater the power of the commander, the more hinds in the harem. After generating the harems, an amazing mating behavior occurs. First of all, the commanders should mate with a number of hinds in the harem and a few in another harem to extend his territory. The stags can mate with the nearest hind without the limitation of harems. Regarding the evolutionary concepts in the RDA, a set of better solutions will be chosen as the next generation of algorithm by roulette wheel selection or tournament selection mechanisms. At the end, the stop condition of this algorithm based on the maximum number of iterations should be satisfied.

Like other metaheuristics, a balance between the exploitation and exploration phases is very important. The RDA does the exploitation properties by roaring and fighting of males as well as the mating of stags with the nearest hind. The main exploration features comprise the generation of some harems for each commander and the mating operator to perform mating with the harem of a commander and a random selected harem. Based on this discussion, Figure 6.2 reveals the flowchart of explained RDA. Accordingly, the light gray boxes (second, third, fourth and eighth after Start) specify the intensification phase, whereas the dark gray boxes (fifth, sixth and seventh after Start) maintain the diversification characteristics of algorithm. As regards the green box (Select the next generation), to escape from the local solutions, the next generation will be selected by using an evolutionary concept. These mechanisms are suited for solving any type of combinatorial optimization problem, such as the proposed truck scheduling in this chapter. To better understand the details of RDA, its pseudocode is also provided in Figure 6.3. More illustration about RDA and related formulas can be found in the main source [40] and other related papers in this area [44–49].

6.3.1 Encoding and Decoding of Algorithm

The first step for solving a mathematical model by using an algorithm such as RDA is to design an appropriate solution representation for the problem under consideration [50–53]. As shown by Golshahi-Roudbaneh et al. [17], Figure 6.4 illustrates the encoding scheme of the problem. For clarification purposes, the figure shows four trucks. The size of this matrix equals the summation of the number of both

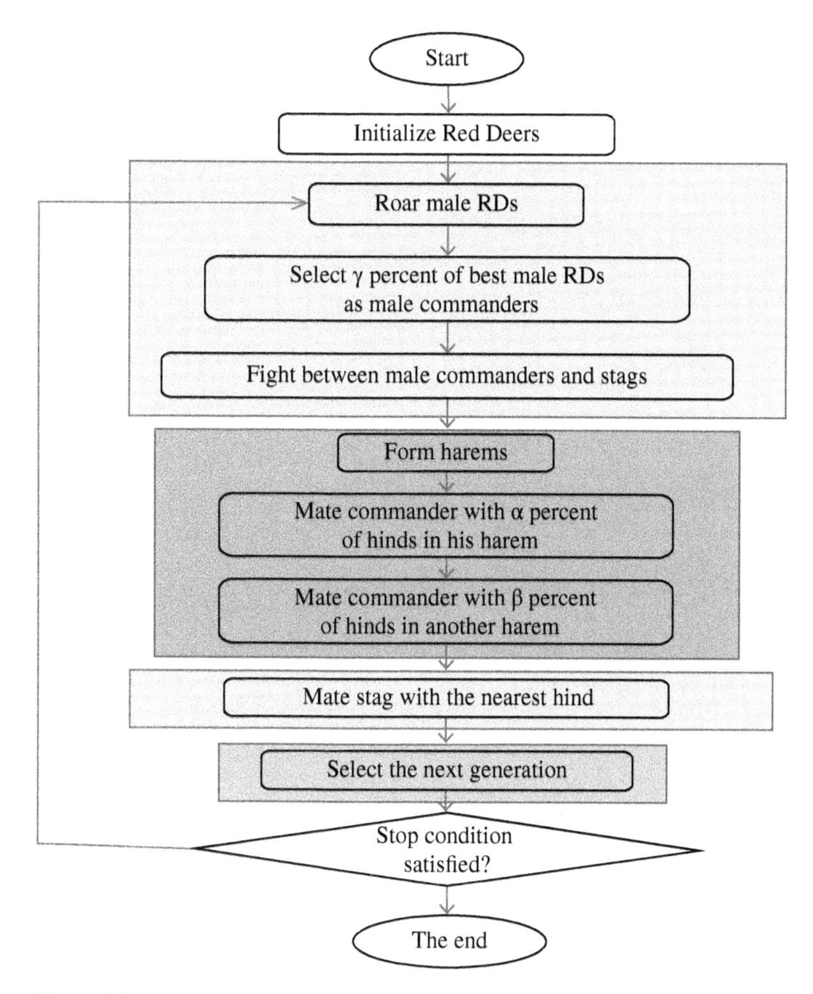

Figure 6.2 Flowchart of employed RDA.

shipping and receiving trucks. Referring to the figure, Part I shows a specific sequence regarding the receiving trucks and Part II represents the sequence for the shipping trucks.

Since the search space of RDA is continuous, a procedure is needed to perform the proposed encoding scheme of the problem [51]. For this purpose, a two stage random technique called Random-Key (RK) has been utilized in this study [52, 53]. Over the last decade, several research studies have applied this technique to run different types of optimizers [52–57]. This technique has two steps. In the first step, random numbers between zero and one from a uniform distribution, shown as $U(0, 1)$, are drawn by the proposed algorithm [54–57]. In the second step, this

```
Initialize the Red Deers population.
Set the parameters (Maxit; nPop; nMale; alpha; beta; gamma)
Calculate the fitness and sort them and form the hinds (Hind) and male RDs (nMale).
X* = the best solution.
it=1;
while (it < Maxit)
  for each male RD
    A local search to update the position if better than the prior ones.
  end for
  Sort the males and also form the stags and the commanders as a gamma percentage of all males.
  for each male commander
     Fight between male commander and stag.
     Update the position of male commander and stag.
  end for
  Form harems.
  for each male commander
     Mate male commander with the selected hinds (alpha percentage) of his harem randomly.
     Select a harem randomly and name it k.
     Mate male commander with some of the selected hinds (beta percentage) of the harem
     randomly.
  end for
  for each stag
     Calculate the distance between the stag and all hinds and select the nearest hind.
     Mate stag with the selected hind.
  end for
  Select the next generation with roulette wheel selection.
  Update the X* if there is better solution.
  it = it+1;
end while
return X*
```

Figure 6.3 Pseudo-code of RDA.

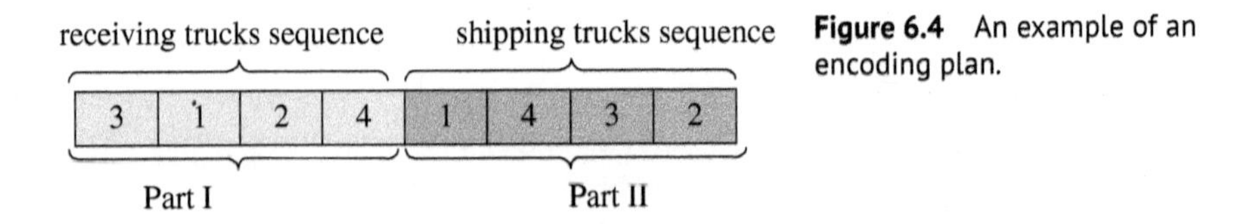

Figure 6.4 An example of an encoding plan.

solution is converted to a feasible representation solution as shown in Figure 6.4. This modification procedure is performed by sorting the vector of this array to consider the sequence of allocation. Figure 6.5 shows an example of the proposed representation method by RK technique. As can be seen, the first row is generated by metaheuristics and the numbers in the second row are determined by procedure.

Figure 6.5 An example of the proposed RK technique.

0.38	0.18	0.35	0.52	0.45	0.91	0.89	0.51
3	1	2	4	1	4	3	2

6.3.2 Proposed Modified RDA

Although there are many modifications of hybridizations of RDA [44–49], the proposed modified version is completely new. In the proposed Modified RDA (MRDA), there are a set of rules to formulate the amount of parameters, including *alpha*, *beta*, and *gamma*. First, the number of commanders should be planned. Each commander generates a harem. From another point of view, the number of commanders is the number of best solutions, which increases the intensification phase. However, this number is directly affected by the diversification properties. Hence, the number of commanders is very important to manipulate both search phases. In the following formula, a procedure to update the percentage of commanders among all males is proposed:

$$gamma = (0.1 + 0.9 \times \frac{it}{Maxit}) \tag{6.41}$$

where *it* is the current iteration of algorithm. By considering the above formula, this number increases during the iterations. This parameter varies between 0.1 and 1. Regarding our experiments, it gives a chance to the user to manipulate both exploitation and exploration properties. To calculate alpha and beta, an elitism strategy is applied. The average of fitness for the hinds in this harem and a randomly selected harem is computed. In the case of minimization, if the average of fitness in the harem of considered commander is lower than that of the other one, it means that the hinds of this harem are better. Therefore, the percentage of mating for the commander is computed as follows:

$$alpha = (0.5 + 0.5 \times \frac{it}{Maxit}) \tag{6.42}$$

Conversely, the beta is considered as follows:

$$beta = 1 - alpha \tag{6.43}$$

Notably, if the average fitness of this harem is higher than that of another harem, the rate of these two parameters should be exchanged. It means that the rate of beta should be calculated by Eq. (6.42). Similarly, Eq. (6.43) should be considered for the rate of alpha. The main reason behind this strategy is to increase the intensification properties of the proposed methodology during an increase in the number of iterations. To ease the details of the proposed MRDA, the associated pseudo-code is reported as can be seen in Figure 6.6. Notably, its flowchart is exactly similar to the original RDA given in Figure 6.2.

```
Initialize the Red Deers population.
Set the parameters (Maxit; nPop; nMale)
Calculate the fitness and sort them and form the hinds (Hind) and male RDs (nMale).
X*=the best solution.
it=1;
while (it< Maxit)
  for each male RD
    A local search to update the position if better than the prior ones.
  end for
  Sort the males and also form the stags and the commanders as a gamma percentage of all males.
  The gamma has been updated by Eq. (39).
  for each male commander
     Fight between male commander and stag.
     Update the position of male commander and stag.
  end for
  Form harems.
  for each male commander
     Select a harem randomly and name it k.
     Calculate the average of fitness for this harem and harem k.
     if the fitness average of this harem is lower than harem k
       Mate male commander with the selected hinds (alpha percentage as updated by Eq. (40)) of his harem randomly.
       Mate male commander with some of the selected hinds (beta percentage as updated by Eq. (41)) of the harem
randomly.
     else
       Mate male commander with the selected hinds (alpha percentage as updated by Eq. (41)) of his harem randomly.
       Mate male commander with some of the selected hinds (beta percentage as updated by Eq. (40)) of the harem
randomly.
     endif
  end for
  for each stag
     Calculate the distance between the stag and all hinds and select the nearest hind.
     Mate stag with the selected hind.
  end for
  Select the next generation with roulette wheel selection.
  Update the X* if there is better solution.
  it=it+1;
end while
return X*
```

Figure 6.6 Pseudo-code of MRDA.

Taken together, the developed MRDA has only three simple parameters, i.e. *Maxit*, *nPop*, and *nMale*. To update the other main parameters of RDA, an adaptive technique is proposed to manage the search mechanism of the algorithm with more efficiency. The main advantage of the proposed MRDA is to have fewer controlling parameters and better designation of search phases to find the global solution. The proposed algorithm is developed in a general way which can be utilized for other complicated optimization problems.

6.4 Computational Results

Here, a comparative study is conducted to assess the developed MRDA. In this regard, the proposed modified algorithm is compared with its original version along with a number of successful evolutionary ones existing in the literature, such as GA, SA, PSO, and ICA. In this section, first of all, the test problems based on the features of the developed mathematical model are generated. Next, to have a fair comparison, all metaheuristics are tuned by the Taguchi method. Finally, an extensive comparison is performed by different measurements to identify the most efficient algorithms.

6.4.1 Instances

In order to check the algorithm's performance, an approach from the literature [12, 17] is adopted to set 10 test problems based on [17] for large sizes. It should be noted that the required time for trucks' changeover equals 75 per time and the needed time to transfer products from receiving dock to shipping dock equals 100 per time. Both loading and unloading time for all products are the same and equal 1 per time. Information related to these 10 problems is shown in Table 6.2.

Based on the novelty of the proposed model, some parameters are generated as follows. Note that the rest of them are taken from the literature [12, 17]. The due date of each shipping truck is obtained through a uniform distribution according to the following equation:

$$DD_j = uniform\left[\sum_{k=1}^{N}(s_{jk})+V \,,\, \sum_{i=1}^{R}\sum_{j=1}^{S}\sum_{k=1}^{N}x_{ijk}+V+(S-1)D\right](1+\lambda) \tag{6.44}$$

From Eq. (6.44), $\sum_{j=1}^{S}\sum_{k=1}^{N}(s_{jk})+V$ is the required operation time for shipping truck j if all its needed products are ready in receiving dock. $\sum_{i=1}^{R}\sum_{j=1}^{S}\sum_{k=1}^{N}x_{ijk}+V+(D-1)S$ is the required operation time for all shipping trucks if their needed products are ready in receiving dock. λ is a random number uniformly distributed between 0 and 0.5.

The lower bound and the upper bound of the time window for each shipping truck are acquired as follows:

$$e_j = uniform\,(0.8,1)\times DD_j \tag{6.45}$$

$$l_j = uniform\,(1,1.2)\times DD_j \tag{6.46}$$

The deadline for each shipping truck is obtained according to the following equation:

$$dl_j = uniform\left[l_j, 2\sum_{i=1}^{R}\sum_{j=1}^{S}\sum_{k=1}^{N}x_{ijk}+V+(D-1)S\right] \tag{6.47}$$

6.4.2 Parameter Tuning

The calibration of parameters is one of the important issues when a metaheuristic is implemented [53–55]. The main reason is the quality of the solutions, which depend directly on the algorithm's parameters to a large extent [57]. The calibration of controlling parameters can help the user to better address the problem under consideration by putting suitable values for these parameters. The tuning procedure, considering all test problems and parameters of algorithms, may need a long time. Accordingly, the design of experiments is a technique that creates the highest payoff with minimal cost and time. The Taguchi [58] method is one of the most well-known and powerful. This methodology reduces the tests and uses the S/N ratio in order to determine the parameters' optimum levels [38, 59, 60]. First, the possible levels of each algorithm are provided as seen in Table 6.3.

Having determined the number of parameters and their levels, a number of required tests should be specified through the proposed Taguchi table for design of experiments known as orthogonal arrays [60, 61]. Accordingly, the standard orthogonal arrays fix most experimental design needs, but sometimes adjustments are unavoidable. Regarding Table 6.3, the Taguchi methodology offers L_{18} for the SA. L_{27} is used for all GA, PSO, ICA, and RDA. For the proposed MRDA, the orthogonal array of L_{16} is employed.

Due to the random nature of algorithms, each experiment is repeated several times. Here, all algorithms are repeated 10 times to be reliable. Analyzing the results using the S/N ratio will be performed using the following equation:

Table 6.2 Information regarding test problems [17].

Test problem	Number of receiving truck	Number of shipping truck	Number of product types	Number of perishable products	Total number of products
1	12	9	9	1	4040
2	12	11	12	1	6340
3	12	13	13	1	5440
4	14	11	13	1	5930
5	13	15	10	1	4627
6	15	16	9	2	3900
7	15	17	15	2	6281
8	14	18	14	2	6190
9	18	19	14	2	6981
10	20	19	16	2	8367

Table 6.3 Factors of metaheuristics and their considered levels.

Algorithm	Factor	Levels			
		1	2	3	4
	A: Maximum iteration (*Maxit*)	1000	1500	—	—
	B: Sub-iteration (*Subit*)	10	20	30	—
SA	C: Used procedure of local search (T_m)	Swap	Reversion	Insertion	—
	D: Initial temperature (*T0*)	1000	1500	2000	—
	E: Rate of reduction (*R*)	0.85	0.9	0.99	—
	A: Maximum iteration (*Maxit*)	300	500	700	—
	B: Population size (*nPop*)	100	200	300	
PSO	C: Inertia weight (W)	0.7	0.8	0.9	
	D: Acceleration coefficient (C1)	1.2	1.5	2	—
	E: Acceleration coefficient (C2)	1.2	1.5	2	—
	A: Maximum iteration (*Maxit*)	300	500	700	—
	B: Population size (*nPop*)	100	150	200	—
	C: Used procedure of mutation (T_m)	Swap	Reversion	Insertion	—
GA	D: Percentage of mutation (*Pm*)	0.05	0.1	0.15	—
	E: Used procedure of cross-over (C_m)	Uniform	Single	Double	—
	F: Percentage of crossover (*Pc*)	0.6	0.7	0.8	—
	A: Maximum iteration (*Maxit*)	200	400	600	1000
MRDA	B: Population size (*nPop*)	50	100	150	200
	C: Number of males (*nMale*)	10	15	20	30
	A: Maximum iteration (*Maxit*)	200	300	400	—
	B: Population size (*nPop*)	100	150	200	—
	C: Number of empires (*Nemp*)	10	15	20	—
ICA	D: Coefficient of total cost (*E*)	0.05	0.07	0.1	—
	E: Rate of assimilation (*Pa*)	0.1	0.2	0.3	—
	F: Rate of revolution (*Pr*)	0.15	0.25	0.4	—
	A: Maximum iteration (*Maxit*)	200	300	500	—
	B: Population size (*nPop*)	100	150	200	—
	C: Number of males (*nMale*)	10	15	20	—
RDA	D: Rate of mating in a harem (*alpha*)	0.6	0.7	0.8	—
	E: Rate of mating in a random harem (*betta*)	0.3	0.4	0.5	—
	F: Rate of male commanders (*gamma*)	0.7	0.8	0.9	—

Table 6.4 Tuned parameters of algorithms.

Algorithm	Parameters
SA	*Maxit* = *1000; Subit* = *20;* T_m = *Swap; T0* = *2000; R* = *0.99;*
GA	*Maxit* = *500; nPop* = *200;* T_m = *Reversion; Pm* = *0.1; Cm* = *Uniform; Pc* = *0.7;*
PSO	*Maxit* = *800; nPop* = *200; W* = *0.9; C1* = *2; C2* = *2;*
ICA	*Maxit* = *300; nPop* = *150; Nemp* = *10; E* = *0.1; Pa* = *0.2; Pr* = *0.4;*
RDA	*Maxit* = *500; nPop* = *150; nMale* = *10; alpha* = *0.7; beta* = *0.4; gamma* = *0.8;*
MRDA	*Maxit* = *600; nPop* = *100; nMale* = *20;*

$$S/N\ ratio = -10\log\sum_i\sum_j f_{ij}^2 \tag{6.48}$$

where f_{ij} is the objective function value acquired in the jth replication of the ith experiment for each problem. Each level of the parameters that has the highest amount of S/N ratio is chosen as the optimal level [49–55]. Due to the page limitation of this book, the results of the S/N ratio are not reported and can be presented upon request of readers. Finally, the calibrated algorithms' parameters are given in Table 6.4.

6.4.3 Comparison of Metaheuristics

An extensive comparison is conducted for applied metaheuristics. To this end, each metaheuristic is run 30 times for each test problem. The best, the worst, and the average as well as the standard deviation of solutions are reported in Table 6.5. Additionally, for each algorithm, the computational time of the algorithms and the hitting time (the first time that the algorithm reaches the best solution) [2] are noted in this table. The behavior of metaheuristics in terms of both computational and hitting time is depicted in Figure 6.7. To identify the best metaheuristic, some statistical analyses are done by the interval plot as seen in Figure 6.8.

From Table 6.5, both RDA and MRDA show the best solution ever found for each test problem. Furthermore, the results based on other criteria confirm that the proposed MRDA is the most efficient algorithm in this comparison.

What may be seen from Figure 6.7a ensures that there are some similarities between the behaviors of algorithms in terms of computational time. It is obvious that the employed GA shows the worst behavior in this item. Conversely, the MRDA in a half of instances shows the best. However, the ICA is also good in this

Table 6.5 Results of applied metaheuristics (B = best, A = average, W = worst, SD = standard deviation, CPU = computational time, HT = hitting time).

		1	2	3	4	5	6	7	8	9	10
SA	B	**4522.5**	5612.4	1333.3	3898	4135.2	1903.2	2259.8	1913.3	1595	2463
	A	5785.4	6818.8	1386.4	4477	4960.9	1990.8	2660.6	3229.2	2325.3	3360.2
	W	6039.9	7048.2	1414.8	4625.5	4994.2	2033.5	2672.5	3291.6	2410.06	3381.4
	SD	2404.3	2241.6	164.41	1088.9	1476.8	274.13	812.6	2339.61	1300.075	1641.2
	CPU	20.64	19.58	20.86	25.37	23.19	24.81	26.39	25.37	30.16	34.52
	HT	14.867	19.15	18.16	21.03	19.99	19.75	25.45	20.55	18.39	26.06
GA	B	**4522.5**	5243	1333.3	3898	3552.3	1903.2	2295	1913.3	1652.6	1062.6
	A	4552.5	5323.4	1336.5	3965.1	3566.7	2046.8	2493.4	1993.2	1924.6	2035.2
	W	4594.7	5537.6	1339.6	4016.7	3648.7	2084.7	2511.9	2055.2	1953.2	2060.9
	SD	125.2	456.99	10.91	178.65	156	287.28	360.73	213.53	497.7845	1707.34
	CPU	21.24	21.72	21.54	26.65	26.28	29.66	29.6	28.13	32.48	35.99
	HT	21.1	18.93	16.06	20.48	26.05	25.06	23.88	26.82	27.29	28.14
PSO	B	4522.5	5366.1	1333.3	3938.4	3811.3	1942.8	2352.9	2217.6	1821.5	2291.3
	A	4728.8	5607.9	1369	4060.5	4022.6	2000.5	2433.7	2352.5	1985.4	2485.05
	W	5347.7	6333.3	1476.4	4427.1	4656.8	2173.8	2676.4	2757.2	2477.3	3066.3
	SD	429.44	503.34	74.48	254.33	440.02	**120.22**	168.36	280.81	341.29	403.32
	CPU	19.735	20.415	20.345	23.03	24.19	27.81	29.41	27.805	29.41	33.56
	HT	19.04	17.24	16.96	19.2	22	23.66	25.54	24.85	24.89	27.86

(*Continued*)

Table 6.5 (Continued)

		1	**2**	**3**	**4**	**5**	**6**	**7**	**8**	**9**	**10**
ICA	B	**4522.5**	**5243**	1333.3	4019.4	3746.6	2022.1	2503.9	2826.2	2217	3348.5
	A	5242.7	6196.6	1609.1	4579.6	5140.8	2332	2829.4	2912.9	3056.9	3620.9
	W	5408.6	6414.1	1674.8	4639.3	5327.6	2403.2	2844.8	2924.8	3068.876	3756.679
	SD	1413.2	1868.7	543.6	1025.9	2591.7	607.9	577.6	161.5	1465.2	623.5
	CPU	18.23	19.11	19.15	19.41	22.1	25.96	29.22	27.48	**26.34**	**31.13**
	HT	17.17	16.08	18.16	18.36	18.66	22.77	27.81	23.16	22.62	28.02
RDA	B	**4522.5**	**5243**	**1333.3**	3898	**3552.3**	**1903**	**2298**	**1650.1**	**1594.9**	**1061.3**
	A	**4522.5**	5369.9	**1333.3**	3964.9	3675.7	1938.1	2376	1882.2	1838.4	1872.3
	W	**4522.5**	**5527.6**	**1333.3**	4004.5	3836.1	**2002.1**	**2419.1**	**1900.1**	**1857.7**	**1890.1**
	SD	**0**	427.8	**0**	**161.6**	427.02	150.89	**184.19**	418.4416	439.45	1420.42
	CPU	21.2	20.83	20.89	21.52	22.76	26.22	29.85	30.77	28.27	**33.91**
	HT	20.6	**14.23**	**14.29**	14.68	17.3	**16.49**	**21.83**	**28.11**	18.52	28.05
MRDA	B	**4522.5**	**5243**	**1333.3**	**3898**	**3552.3**	**1903**	**2259.8**	**1650.1**	**1594.9**	**1061.3**
	A	**4522.5**	**5323.4**	**1333.3**	3964.9	3566.7	1938.1	**2376**	**1882.2**	**1838.4**	**1872.3**
	W	**4522.5**	5589.57	**1333.3**	4163.14	3745.03	2035	2494.8	1976.31	1930.32	1965.91
	SD	**0**	544.18	**0**	413.65	322.07	205.1	352.5	503.6	519.97	1492.38
	CPU	20.23	**18.75**	**19.38**	**18.38**	**20.94**	**23.25**	28.12	28.89	28.17	33.28
	HT	**15.53**	17.29	17.31	**13.01**	**13.95**	22.69	**21.08**	28.53	**17.64**	32.36

The best values for each criteria are shown in bold.

item for two last test problems. SA and PSO give neither the best nor the worst value in the test problems. Overall, the proposed MRDA is better than its general version in terms of computation time.

The convergence rate of applied algorithms is demonstrated by Figure 6.7b. A lower hitting time confirms the efficiency of algorithms to have a better convergence rate as another criterion of comparison. From this figure, there are a set of similarities between both RDA and MRDA. Note that these algorithms are also better than others in the majority of instances. Aside from these algorithms, the SA also outperforms other algorithms in only three test problems. PSO gives neither the best nor the worst value in the test problem. Overall, the proposed MRDA is the most efficient technique in half of instances for this criterion.

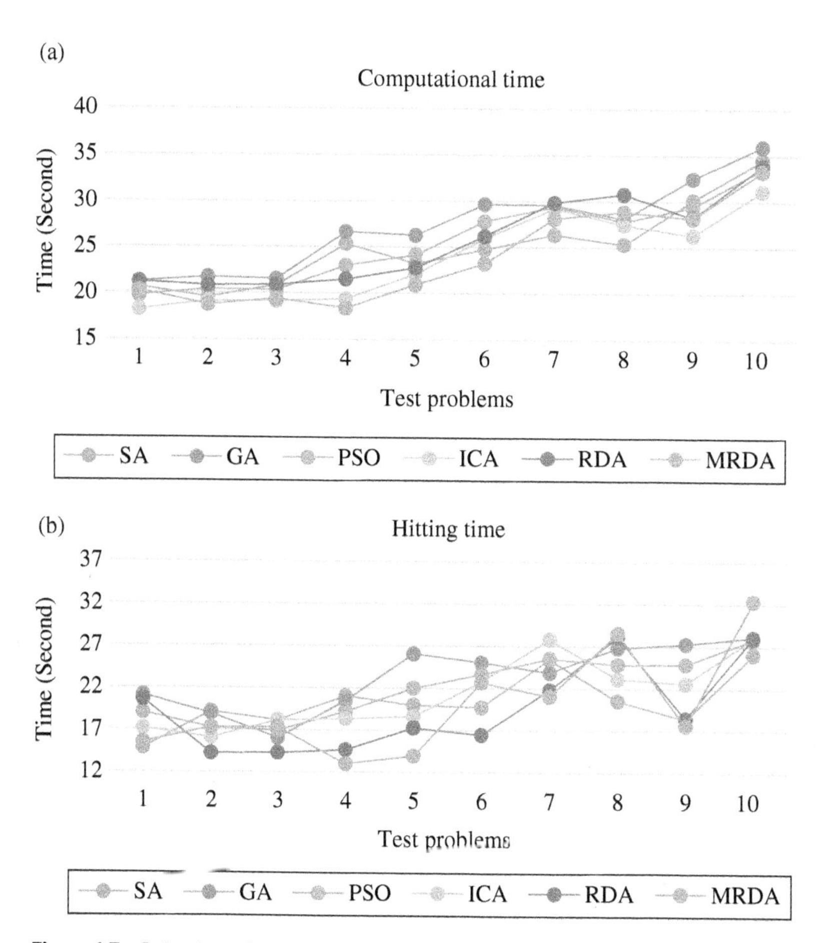

Figure 6.7 Behaviors of metaheuristics in terms of computational time (a) and hitting time (b).

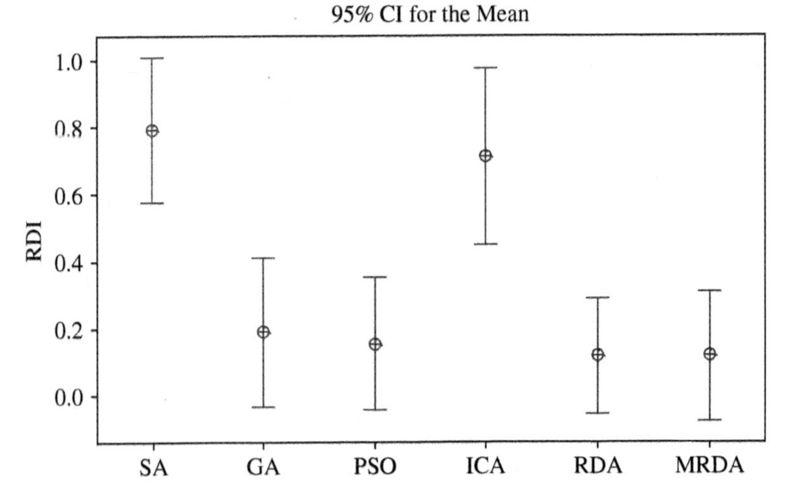

Figure 6.8 Interval plot of employed metaheuristics.

Last but not the least, some statistical comparisons are conducted to find the best metaheuristic among all of them, based on standard deviation of algorithms through 30 run times. The results are transformed to a Relative Deviation Index (RDI) as given below:

$$RDI = \frac{|Alg_{sol} - Best_{sol}|}{max_{sol} - min_{sol}} \tag{6.49}$$

where Alg_{sol} is the objective value given by the algorithm and Max_{sol} and Min_{sol} are respectively the maximum and the minimum values among all solutions of metaheuristics. $Best_{sol}$ is the best solution [56, 57]. Consequently, the means plot and Least Significant Difference (LSD) for all algorithms have been performed. The results are revealed in Figure 6.8. It is evident that the performance of PSO, RDA, and MRDA are clearly better than that of GA, SA, and ICA. Looking more closely, the robustness of the behavior of the proposed MRDA in comparison with that of PSO and RDA is clear.

6.5 Conclusion and Future Works

In this chapter, a new variant of the truck scheduling problem in the cross-docking system was studied. The main contributions of the mathematical model already developed were a time window as well as a deadline associated with each shipping truck in this system. The products were also classified in two groups: perishable and imperishable. Then a new mathematical model was presented, inspired by one of the available models in the literature. Another main innovation of this research was to propose a new modified version of RDA, a recently developed evolutionary algorithm. The proposed algorithm was not only compared with its general idea but also with four successful metaheuristics from the literature, i.e. SA, GA, PSO, and ICA were selected to have an extensive comparison based on different criteria. In this regard, the parameters of each algorithm were set using the Taguchi method. Having set the parameters, each problem was performed 30 times by each algorithm. The consequences demonstrated that the suggested modified RDA called MRDA showed a more desirable performance than the other algorithms.

For future studies, there are many recommendations for ways to continue this line of study. First, other recent metaheuristic and heuristic algorithms could be used to obtain better results. Additionally, multiple receiving and shipping doors could be taken into consideration. Another continuation of this study would involve a time window for arrival trucks as well.

References

1 Amini, A. and Tavakkoli-Moghaddam, R. (2016). A bi-objective truck scheduling problem in a cross-docking center with probability of breakdown for trucks. *Computers and Industrial Engineering* 96: 180–191.

2 Fathollahi-Fard, A.M., Hajiaghaei-Keshteli, M., and Mirjalili, S. (2018). Hybrid optimizers to solve a tri-level programming model for a tire closed-loop supply chain network design problem. *Applied Soft Computing* 70: 701–722.

3 Apte, U.M. and Viswanathan, S. (2000). Effective cross docking for improving distribution efficiencies. *International Journal of Logistics* 3 (3): 291–302.

4 Arabani, A.B., Ghomi, S.F., and Zandieh, M. (2011). Meta-heuristics implementation for scheduling of trucks in a cross-docking system with temporary storage. *Expert Systems with Applications* 38 (3): 1964–1979.

5 Fathollahi-Fard, A.M., Hajiaghaei-Keshteli, M., and Tavakkoli-Moghaddam, R. (2018). A bi-objective green home health care routing problem. *Journal of Cleaner Production* 200: 423–443.

6 Berghman, L., Briand, C., Leus, R., and Lopez, P. (2012). The truck scheduling problem at cross-docking terminals. Paper presented at the International Conference on Project Management and Scheduling (PMS 2012).

7 Amini, A., Tavakkoli-Moghaddam, R., and Omidvar, A. (2014). Cross-docking truck scheduling with the arrival times for inbound trucks and the learning effect for unloading/loading processes. *Production and Manufacturing Research* 2 (1): 784–804.

8 Zuluaga, J.P.S., Thiell, M., and Perales, R.C. (2016). Reverse cross-docking. *Omega* 65: 278–284.
9 Boysen, N. and Fliedner, M. (2010). Cross dock scheduling: classification, literature review and research agenda. *Omega* 38 (6): 413–422.
10 Ladier, A.-L. and Alpan, G. (2016). Cross-docking operations: current research versus industry practice. *Omega* 62: 145–162.
11 Yu, W. (2002). Operational strategies for cross docking systems. *European Journal of Operational Research* 28 (1): 318–326.
12 Yu, W. and Egbelu, P.J. (2008). Scheduling of inbound and outbound trucks in cross docking systems with temporary storage. *European Journal of Operational Research* 184 (1): 377–396.
13 Chen, F. and Lee, C.-Y. (2009). Minimizing the makespan in a two-machine cross-docking flow shop problem. *European Journal of Operational Research* 193 (1): 59–72.
14 Boysen, N. (2010). Truck scheduling at zero-inventory cross docking terminals. *Computers and Operations Research* 37 (1): 32–41.
15 Li, Y., Lim, A., and Rodrigues, B. (2014). Crossdocking – JIT scheduling with time windows. *Journal of the Operational Research Society* 55 (12): 1342–1351.
16 Mohtashami, A., Tavana, M., Santos-Arteaga, F.J., and Fallahian-Najafabadi, A. (2015). A novel multi-objective meta-heuristic model for solving cross-docking scheduling problems. *Applied Soft Computing* 31: 30–47.
17 Golshahi-Roudbaneh, A., Hajiaghaei-Keshteli, M., and Paydar, M.M. (2017). Developing a lower bound and strong heuristics for a truck scheduling problem in a cross-docking center. *Knowledge-Based Systems* 67: 289–308.
18 Mohammadzadeh, H., Sahebjamnia, N., Fathollahi-Fard, A.M., and Hahiaghaei-Keshteli, M. (2018). New approaches in metaheuristics to solve the truck scheduling problem in a cross-docking center. *International Journal of Engineering, Transactions B: Applications* 31 (8): 1258–1266.
19 Ye, Y., Li, J.F., Fung, R.Y. et al. (2018). Optimizing truck scheduling in a cross-docking system with preemption and unloading/loading sequence constraint. 2018 IEEE 15th International Conference on Networking, Sensing and Control (ICNSC), 1–6.
20 Chen, F. and Song, K. (2009). Minimizing makespan in two-stage hybrid cross docking scheduling problem. *Computers and Operations Research* 36 (6): 2066–2073.
21 Soltani, R. and Sadjadi, S.J. (2010). Scheduling trucks in cross-docking systems: a robust meta-heuristics approach. *Transportation Research Part E-Logistics and Transportation Review* 46 (5): 650–666.
22 Boysen, N., Fliedner, M., and Scholl, A. (2010). Scheduling inbound and outbound trucks at cross docking terminals. *OR Spectrum* 32 (1): 135–161.
23 Forouharfard, S. and Zandieh, M. (2010). An imperialist competitive algorithm to schedule of receiving and shipping trucks in cross-docking systems. *The International Journal of Advanced Manufacturing Technology* 51 (9–12): 1179–1193.

24 Larbi, R., Alpan, G., Baptiste, P., and Penz, B. (2011). Scheduling cross docking operations under full, partial and no information on inbound arrivals. *Computers and Operations Research* 38 (6): 889–900.

25 Shakeri, M., Low, M.Y.H., Turner, S.J., and Lee, E.W. (2012). A robust two-phase heuristic algorithm for the truck scheduling problem in a resource-constrained crossdock. *Computers and Operations Research* 39 (11): 2564–2577.

26 Davoudpour, H., Hooshangi-Tabrizi, P., and Hoseinpour, P. (2012). A genetic algorithm for truck scheduling in cross docking systems. *Journal of American Science* 8 (2): 96–99.

27 Sadykov, R. (2012). Scheduling incoming and outgoing trucks at cross docking terminals to minimize the storage cost. *Annals of Operations Research* 201 (1): 423–440.

28 Boysen, N., Briskorn, D., and Tschöke, M. (2013). Truck scheduling in cross-docking terminals with fixed outbound departures. *OR Spectrum* 35 (2): 479–504.

29 Bjelić, N., Popović, D., and Ratković, B. (2013). Genetic algorithm approach for solving truck scheduling problem with time robustness. Paper presented at the Proceedings of the 1st Logistics International Conference LOGIC.

30 Joo, C.M. and Kim, B.S. (2013). Scheduling compound trucks in multi-door cross-docking terminals. *The International Journal of Advanced Manufacturing Technology* 64 (5–8): 977–988.

31 Konur, D. and Golias, M.M. (2013). Cost-stable truck scheduling at a cross-dock facility with unknown truck arrivals: a meta-heuristic approach. *Transportation Research Part E: Logistics and Transportation Review* 49 (1): 71–91.

32 Ladier, A.-L. and Alpan, G. (2013). Scheduling truck arrivals and departures in a crossdock: earliness, tardiness and storage policies, Proceedings of 2013 International Conference on Industrial Engineering and Systems Management (IESM).

33 Ladier, A.-L. and Alpan, G. (2014). Crossdock truck scheduling with time windows: earliness, tardiness and storage policies. *Journal of Intelligent Manufacturing* 29 (3): 569–583.

34 Madani-Isfahani, M., Tavakkoli-Moghaddam, R., and Naderi, B. (2014). Multiple cross-docks scheduling using two meta-heuristic algorithms. *Computers and Industrial Engineering* 74: 129–138.

35 Khalili-Damghani, K., Tavana, M., Santos-Arteaga, F.J., and Ghanbarzad-Dashti, M. (2017). A customized genetic algorithm for solving multi-period cross-dock truck scheduling problems. *Measurement* 79: 548–559.

36 Wisittipanich, W. and Hengmeechai, P. (2017). Truck scheduling in multi-door cross docking terminal by modified particle swarm optimization. *Computers and Industrial Engineering* 68: 339–348.

37 Ye, Y., Li, J., Li, K., and Fu, H. (2018). Cross-docking truck scheduling with product unloading/loading constraints based on an improved particle swarm optimisation algorithm. *International Journal of Production Research* 56 (16): 5365–5385.

38 Wolpert, D.H. and Macready, W.G. (1997). No free lunch theorems for optimization. *IEEE Transactions on Evolutionary Computation* 1 (1): 67–82.

39 Fathollahi-Fard, A.M., Hajiaghaei-Keshteli, M., and Tavakkoli-Moghaddam, R. (2018). The social engineering optimizer (SEO). *Engineering Applications of Artificial Intelligence* 72: 267–293.

40 Fard, A.M.F. and Hajiaghaei-Keshteli, M. (2016). Red deer algorithm (RDA); a new optimization algorithm inspired by red deer's mating. 12th International Conference on Industrial Engineering, 33–34.

41 Van Belle, J., Valckenaers, P., Berghe, G.V., and Cattrysse, D. (2013). A tabu search approach to the truck scheduling problem with multiple docks and time windows. *Computers and Industrial Engineering* 66 (4): 818–826.

42 Fathollahi-Fard, A.M., Hajiaghaei-Keshteli, M., and Tavakkoli-Moghaddam, R. (2018). A Lagrangian relaxation-based algorithm to solve a home health care routing problem. *International Journal of Engineering, Transactions A: Basics* 31 (10): 1734–1740.

43 Ramirez-Nafarrate, A., Lyon, J.D., Fowler, J.W., and Araz, O.M. (2015). Point-of-dispensing location and capacity optimization via a decision support system. *Production and Operations Management* 24 (8): 1311–1328.

44 Golmohamadi, S., Tavakkoli-Moghaddam, R., and Hajiaghaei-Keshteli, M. (2017). Solving a fuzzy fixed charge solid transportation problem using batch transferring by new approaches in meta-heuristic. *Electronic Notes in Discrete Mathematics* 58: 143–150.

45 Samadi, A., Mehranfar, N., Fathollahi Fard, A.M., and Hajiaghaei-Keshteli, M. (2018). Heuristic-based metaheuristic to address a sustainable supply chain network design problem. *Journal of Industrial and Production Engineering* 35 (2): 102–117.

46 Hajiaghaei-Keshteli, M. and Fathollahi Fard, A.M. (2018). Sustainable closed-loop supply chain network design with discount supposition. *Neural Computing and Applications*: 1–35. https://doi.org/10.1007/s00521-018-3369-5.

47 Fathollahi-Fard, A.M. and Hajiaghaei-Keshteli, M. (2018). A stochastic multi-objective model for a closed-loop supply chain with environmental considerations. *Applied Soft Computing* 69: 232–249.

48 Sahebjamnia, N., Fathollahi-Fard, A.M., and Hajiaghaei-Keshteli, M. (2018). Sustainable tire closed-loop supply chain network design: hybrid metaheuristic algorithms for large-scale networks. *Journal of Cleaner Production* 196: 273–296.

49 Fathollahi-Fard, A.M., Hajiaghaei-Keshteli, M., and Mirjalili, S. (2018). Multi-objective stochastic closed-loop supply chain network design with social considerations. *Applied Soft Computing* 71: 505–525.

50 Geismar, H.N., Dawande, M., Murthi, B.P.S., and Sriskandarajah, C. (2015). Maximizing revenue through two-dimensional shelf-space allocation. *Production and Operations Management* 24 (7): 1148–1163.

51 Caglar Gencosman, B., Begen, M.A., Ozmutlu, H.C., and Ozturk Yilmaz, I. (2016). Scheduling methods for efficient stamping operations at an automotive company. *Production and Operations Management* 25 (11): 1902–1918.

52 Gralla, E., Goentzel, J., and Fine, C. (2014). Assessing trade-offs among multiple objectives for humanitarian aid delivery using expert preferences. *Production and Operations Management* 23 (6): 978–989.

53 Fathollahi-Fard, A.M. and Hajiaghaei-Keshteli, M. (2018). Integrated capacitated transportation and production scheduling problem in a fuzzy environment. *International Journal of Industrial Engineering and Production Research* 29 (2): 197–211.

54 Kettunen, J. and Kwak, Y.H. (2018). Scheduling public requests for proposals: models and insights. *Production and Operations Management* 27 (7): 1271–1290.

55 Fard, A.M.F. and Hajiaghaei-Keshteli, M. (2018). A bi-objective partial interdiction problem considering different defensive systems with capacity expansion of facilities under imminent attacks. *Applied Soft Computing* 68: 343–359.

56 Kent, P. and Siemsen, E. (2018). Production process moves: template use and the need to adapt. *Production and Operations Management* 27 (3): 480–495.

57 Hajiaghaei-Keshteli, M. and Fathollahi-Fard, A.M. (2018). A set of efficient heuristics and metaheuristics to solve a two-stage stochastic bi-level decision-making model for the distribution network problem. *Computers and Industrial Engineering* 123: 378–395.

58 Taguchi, G. (1986). *Introduction to Quality Engineering: Designing Quality into Products and Processes*. White Plains, NY; Tokyo: Asian Productivity Organization/UNIPUB.

59 Fu, Y., Tian, G., Fathollahi-Fard, A.M. et al. (2019). Stochastic multi-objective modelling and optimization of an energy-conscious distributed permutation flow shop scheduling problem with the total tardiness constraint. *Journal of Cleaner Production* 226: 515–525.

60 Fathollahi-Fard, A.M., Hajiaghaei-Keshteli, M., and Mirjalili, S. A set of efficient heuristics for a home healthcare problem. *Neural Computing and Applications*: 1–21.

61 Abdi, A., Abdi, A., Fathollahi-Fard, A.M., and Hajiaghaei-Keshteli, M. (2019). A set of calibrated metaheuristics to address a closed-loop supply chain network design problem under uncertainty. *International Journal of Systems Science: Operations and Logistics* https://doi.org/10.1080/23302674.2019.1610197.

7

Application of Sub-Population Scheduling Algorithm in Multi-Population Evolutionary Dynamic Optimization

Javidan Kazemi Kordestani[1] and Mohammad Reza Meybodi[2]

[1] *Department of Computer Engineering, Science and Research Branch, Islamic Azad University, Tehran, Iran*
[2] *Soft Computing Laboratory, Computer Engineering and Information Technology Department, Amirkabir University of Technology (Tehran Polytechnic), Tehran, Iran*

7.1 Introduction

Many problems in real-world applications involve optimizing a set of parameters, in which the objectives of the optimization, some constraints, or other elements of the problems may vary over time. If so, the optimal solution(s) to the problems may change as well. Generally speaking, various forms of dynamic behavior are observed in a substantial part of real-world optimization problems in different domains. Examples of such problems include the dynamic resource allocation in shared hosting platforms [1], dynamic traveling salesman problem that changes traffic over time [2], dynamic shortest path routing in MANETs [3], aerospace design [4], pollution control [5], ship navigation at sea [5], dynamic vehicle routing in transportation logistics [6], autonomous robot path planning [7], optimal power flow problem [8], dynamic load balancing [9], and groundwater contaminant source identification [10].

Evolutionary computation (EC) techniques have attracted a great deal of attention due to their potential for solving complex optimization problems. Even though they are effective for static optimization problems, they should undergo certain adjustments to work well when applied to dynamic optimization problems (DOPs). The reason is that the dynamic behavior of DOPs poses two additional challenges to the EC techniques: (i) *outdated memory*: when a change occurs in the environment, the previously found solutions by the algorithm may no longer be valid. In this case, the EC algorithm will be misled into moving toward false positions. (ii) *Diversity loss*: this issue appears due to the tendency of the

Evolutionary Computation in Scheduling, First Edition. Edited by Amir H. Gandomi, Ali Emrouznejad, Mo M. Jamshidi, Kalyanmoy Deb, and Iman Rahimi.

population to converge to a single optimum. As the result, when the global optimum is shifted away, the number of function evaluations (FEs) required for a partially converged population to relocate the optimum is quite deleterious to the performance.

While both the above challenges can be detrimental to the performance of EC, the second issuc is far more serious.

Over the years, researchers have proposed various techniques to improve the efficiency of traditional EC methods for solving DOPs. According to [11] the existing proposals can be categorized into the following six approaches:

1) increasing the diversity after detecting a change in the environment [12–14];
2) maintaining diversity during the optimization process [15, 16];
3) employing memory schemes to retrieve information about previously found solutions [17, 18];
4) predicting the location of the next optimal solution(s) after a change is detected [19, 20];
5) making use of the self-adaptive mechanisms of ECs [21, 22];
6) using multiple sub-populations to handle separate areas of the search space concurrently [23–35].

Among the above mentioned approaches, the multi-population approach has been shown to be very effective for handling DOPs, especially for multi-modal fitness landscapes. The success of this approach can be attributed to three reasons [36]:

1) As long as different populations search in different sub-areas in the fitness landscape, the overall population diversity can be maintained at the global level.
2) It is possible to locate and track multiple changing optima simultaneously. This feature can facilitate tracking of the global optimum, given that one of the being-tracked local optima may become the new global optimum when changes occur in the environment.
3) It is easy to extend any single-population approach, e.g. diversity increasing/maintain schemes, memory schemes, adaptive schemes, etc. to the multi-population version.

Although effective, the multi-population approach significantly reduces the utilization of FEs and delays the process of finding the global optimum by sharing an equal portion of FEs among sub-populations. In another words, sub-populations located far away from the optimal solution(s) are assigned the same amount of FEs as those located near to optimal solution(s), which in turn exert deleterious effects on the performance of the optimization process.

Since the calculation of FEs is the most expensive component of the EC methods for solving real-world DOPs, dynamic optimization can be considered as scheduling the sub-populations in a way that the major portion of FEs is consumed around the most promising areas of the search space. Therefore, one major challenge is how to suitably assign the FEs to each sub-population to enhance the efficiency of multi-population methods for DOPs.

This chapter is aimed at providing the application of scheduling in enhancing the performance of multi-population methods for tracking optima in DOPs. Eight different sub-population scheduling (SPS) algorithms, which have been applied to a well-known multi-population algorithm called DynDE, will be evaluated and compared using one of the most widely used benchmarks in the literature.

7.2 Literature Review

As mentioned earlier in this chapter, the multi-population is one of the most efficient approaches to tackle the existing challenges in DOPs. The idea behind this is to divide the individuals (candidate solutions) of the main population into multiple sub-populations so they can search in different sub-areas of the fitness landscape in parallel. In this way, the algorithm is able to efficiently handle several issues arising in DOPs: (i) exploration, (ii) optimum tracking, (iii) change detection, and (iv) premature convergence. In the rest of this section, we provide a literature review on multi-population methods for DOPs.

7.2.1 Multi-Population Methods with a Fixed Number of Populations

The main idea of these methods is to divide the task of optimization among a number of fixed-size populations. One way to do this is by establishing a mutual repulsion among a predefined number of sub-populations to place them over different promising areas of the search space. The representative work in this category is that of Blackwell and Branke [37]. They proposed two multi-swarm algorithms based on the particle swarm optimization (PSO), namely mCPSO and mQSO. In mCPSO, each swarm is composed of neutral and charged particles. Neutral particles update their velocity and position according to the principles of pure PSO. On the other hand, charged particles move in the same way as neutral particles, but they are also mutually repelled from other charged particles residing in their own swarm. Therefore, charged particles help to maintain the diversity inside the swarm. In mQSO, instead of having charged particles, each swarm contains quantum particles. Quantum particles change their positions around the center of the best particle of the swarm according to a random uniform distribution with radius r_{cloud}. Consequently, they never converge and provide a suitable level of diversity to swarm in order to follow the shifting optimum. The authors also introduced the *exclusion* operator, which prevents populations from settling on the same peak. In another work [23], the same authors proposed a second operator referred to as *anti-convergence*, which is triggered after all swarms converge and reinitializes the worst swarm in the search space. After the introduction of mQSO, a lot of work has been done to enhance its performance by modifying various aspects of its behavior.

A group of studies analyzed the effect of changing the number and the distribution of quantum particles on the performance of mQSO. For instance, Trojanowski [38] proposed a new class of limited area distribution for quantum particles in which the uniformly distributed candidate solutions within a hyper-sphere with radius r_{cloud} are wrapped using von Neumann's acceptance-rejection method. In another work [39], the same author introduced a two-phased method for generating the cloud of quantum particles in the entire area of the search space based on a direction vector θ and the distance d from the original position using an α-stable random distribution. This approach allows particles to be distributed equally in all directions. The findings of both studies revealed that changing the distribution of quantum particles has a significant effect on the performance of mQSO. del Amo et al. [40] investigated the effect of changing the number of quantum and neutral particles on the performance of mQSO. The three major conclusions of their study can be summarized as follows: (i) an equal number of quantum and neutral particles is not the best configuration for mQSO, (ii) configurations in which the number of neutral particles is higher than the number of quantum particles usually perform better, and (iii) quantum particles are most helpful immediately after a change in the environment.

Some researchers borrowed the general idea of mQSO and developed it using other ECs. The main purpose of these approaches is to benefit from intrinsic positive characteristics of other algorithms to reach a better performance than the mQSO. For instance, Mendes and Mohais [41] introduced a multi-population differential evolution (DE) algorithm, called DynDE. In DynDE several populations are initialized in the search space and explore multiple peaks in the environment incorporating exclusion. DynDE also includes a method for increasing diversity, which enables the partially converged population to track the shifting optimum.

One of the technical drawbacks of the exclusion operator in mQSO is that it ignores the situations when two populations stand on two distinct but extremely close optima (i.e. within the exclusion radius of each other). In these situations, the exclusion operator simply removes the worst population and leaves one of the peaks unpopulated. A group of studies tried to address this issue by adding an extra routine to the exclusion which is executed whenever a collision is detected between two populations. For example, Du Plessis and Engelbrecht [42] proposed re-initialization midpoint check (RMC) to detect whether different peaks are located within the exclusion radius of each other. In RMC, once a collision occurs between two populations, the fitness of the midpoint on the line between the best individuals in each population is evaluated. If the fitness of the midpoint has a lower value than the fitness value of the best individuals of both populations, it implies that the two populations reside on distinct peaks and that neither should be reinitialized. Otherwise, the worst-performing population will be reinitialized in the search space. Although RMC is effective, as pointed by the authors, it is unable to correctly detect all extremely close peaks. In another work, Xiao and Zuo [43] applied hill-valley detection with three checkpoints to determine whether two colliding populations are located on different peaks. In their approach, three points between the best solutions of the collided populations x and y are examined. Thereafter, if there exists a point $z = c \cdot x + (1-c) \cdot y$ for which $f(z) < min\{f(x), f(y)\}$, where $c \epsilon \{0.05, 0.5, 0.95\}$, then two populations are on different peaks and they remain unchanged. Otherwise, they are on the same peak.

Another interesting approach for improving the performance of mQSO is to spend more FEs around the most promising areas of the search space. In this regard, Novoa-Hernández et al. [44] proposed a resource management mechanism, called *swarm control mechanism*, to enhance the performance of the mQSO. The swarm control mechanism is activated on the swarms with *low diversity* and *bad fitness*, and stops them from consuming FEs. In another work, Du Plessis and Engelbrecht [45] proposed an extension to the DynDE referred to as favored populations DE. In their approach, FEs between two successive changes in the environment are divided into three phases: (i) all populations are evolved according to normal DynDE for $\zeta 1$ generations in order to locate peaks, (ii) the weaker populations are frozen and stronger populations are executed for another $\zeta 2$ generations, and (iii) frozen populations are put back to the search process and all populations are evolved for $\zeta 3$ generations in a normal DynDE manner. This strategy adds three parameters $\zeta 1$, $\zeta 2$, and $\zeta 3$ to DynDE, which must be tuned manually.

The SPS, which is the subject of this chapter, falls under this category where the main objective is to distribute FEs among the sub-populations so as to allocate the greatest amount of FEs to the most successful sub-populations. Despite the progress achieved in the past, we believe that there is still room for improving this approach. Therefore, this chapter focuses on different ways to suitably assign FEs to each sub-population using the SPS algorithm.

Different from the above studies, some researchers combined desirable features of various optimization algorithms into a single collaborative method for DOPs. In these methods, each population plays a different role and information can be shared between populations. For example, Lung and Dumitrescu [46] introduced a hybrid collaborative method called collaborative evolutionary-swarm optimization (CESO). CESO has two equal-size populations: a main population to maintain a set of local and global optimum during the search process using crowding-based DE, and a PSO population acting as a local search operator around solutions provided by the first population. During the search process, information is transmitted between both populations via collaboration mechanisms. In another work [47] by the same authors, a third population is incorporated to CESO which acts as a memory to recall some promising information from past generations of the algorithm. Inspired by CESO, Kordestani et al. [48] proposed a bi-population hybrid algorithm. The first population, called QUEST, is evolved by crowding-based DE principles to locate the promising regions of the search space. The second population, called TARGET, uses PSO to exploit useful information in the vicinity of the best position found by the QUEST. When the search process of the TARGET population around the best-found position of the QUEST becomes unproductive, the TARGET population is stopped and all genomes of the QUEST population are allowed to perform an extra operation using hill-climbing local search. They also applied several mechanisms to conquer the existing challenges in the dynamic environments and to spend more FEs around the best-found position in the search space.

7.2.2 Methods with a Variable Number of Populations

Another group of studies has investigated the multi-population schemes with a variable number of sub-populations. These methods can be further categorized into three groups based on how the sub-populations are formed from a main population. They are methods with a parent population and variable number of child populations, methods based on population clustering, and methods based on space partitioning.

7.2.2.1 Methods with a Parent Population and Variable Number of Child Populations

The first strategy to make a variable number of populations is to split off sub-populations from the main population. The major strategy in this approach is to divide the search space into different sub-regions, using a parent population, and carefully exploit each sub-region with a distinct child population. Such a method was first proposed by Branke et al. [49]. Borrowing the concept of forking, they proposed a multi-population genetic algorithm for DOPs called self-organizing scouts (SOS). In SOS, the optimization process begins with a large parent population exploring through the whole fitness landscape with the aim of finding promising areas. When such areas are located by the parent, a child population is split off from the parent population and independently explores the respective sub-space, while the parent population continues to search in the remaining search space for locating new optimum. The search area of each child population is defined as a sphere with radius r and centered at the best individual. In order to determine the number of individuals in each population, including the parent, a quality measure is calculated for each population as defined in [49]. The captured region by the child population is then isolated by re-initializing individuals in the parent population that fall into the search range of the child population. In SOS, overlapping among child populations is usually accepted unless the best individual of a child population falls within the search radius of another population. In this case, the whole child population is removed.

Later, similar ideas were proposed by other authors. For instance, Yang and Li [34] proposed a fast multi-swarm algorithm for DOPs. Their method starts with a large parent swarm exploring the search space with the aim to locate promising regions. If the quality of the best particle in the parent swarm improves, it implies that a promising search area may be found. Therefore, a child swarm is split off from the parent swarm to exploit its own sub-space. The search territory of each child swarm is determined by a radius r which is relative to the range of the landscape and width of the peaks. If the best particle of one child swarm approaches the area captured by another child swarm, the worse child swarm will be removed to prevent child populations residing on the same peak. Besides, if a child swarm fails to improve its quality in a certain number of iterations, the best particle of the child swarm jumps to a new position according to a scaled Gaussian distribution. Another similar approach can be found in [27]. An interesting approach is the *hibernating* multi-swarm optimization algorithm proposed by Kamosi et al. [26], where *unproductive* child swarms are stopped using a hibernation mechanism. In turn, more FEs will be available for *productive* child swarms.

Yazdani et al. [50] employed several mechanisms in a multi-swarm algorithm for finding and tracking the optima over time. In their proposal, a randomly initialized *finder* swarm explores the search space for locating the position of the peaks. In order to enhance the exploitation of promising solutions, a *tracker* swarm is activated by transferring the fittest particles from the finder swarm into the newly created tracker swarm. Besides, the finder swarm is then reinitialized into the search space to capture other uncovered peaks. Afterward, the activated tracker swarm is responsible for finding the top of the peak using a local search along with following the respective peak upon detecting a change in the environment. In order to avoid tracker swarms exploiting the same peak, exclusion is applied between every pair of tracker swarms. Besides, if the finder swarm converges to a populated peak, it is reinitialized into the search space without generating any tracker swarm.

Recently, Sharifi et al. [33] proposed a hybrid approach based on PSO and local search. In this method, a swarm of particles is used to estimate the location of the peaks. Once the swarm has converged, a local search agent is created to exploit the respective region. Moreover, a density control mechanism is used to remove redundant local search agents. The authors also studied three adaptations to the basic approach. Their reported results clearly indicate the importance of managing FEs to reach a better performance.

7.2.2.2 Methods Based on Population Clustering

Differing from the above algorithms where the sub-populations are split off from the main population, another way to create multiple populations is to divide the main population of the algorithm into several clusters, i.e. sub-populations, via different clustering methods. For instance, Parrott and Li [51] proposed a speciation-based PSO for tracking multiple optima in dynamic environments, which dynamically distributes particles of the main swarm over a variable number of so-called *species*. In [52] the performance of speciation-based PSO was improved by estimating the location of the peaks using a least squares regression method.

Similarly, Yang and Li [35] proposed a clustering PSO (CPSO) for locating and tracking multiple optimums. CPSO employed a single linkage hierarchical clustering method to create a variable number of sub-swarms, and assign them to different promising sub-regions of the search space. Each created sub-swarm uses the PSO with a modified *gbest* model to exploit the respective region. In order to avoid different clusters from crowding, they applied a redundancy control mechanism. In this regard, when two sub-swarms are located on a single peak they are merged together to form a single sub-swarm. So, the worst performing individuals of the sub-swarm are removed until its size is equal to a predefined threshold. In another work [53], the fundamental idea of CPSO is extended by introducing a novel framework for covering undetectable dynamic environments.

In [54] a competitive clustering PSO for DOPs is introduced. It employs a multi-stage clustering procedure to split the particles of the main swarm. Then, the particles are assigned to a varying number of sub-swarms based on the particles' positions and their objective function values. In addition to the sub-swarms, there is also a group of free particles that is used to explore the environment to locate new emerging optimums or track the current optimums which are not followed by any sub-swarm. Recently, Halder et al. [55] proposed a cluster-based DE with external archive which uses *k-means* clustering method to create a variable number of populations.

7.2.2.3 Methods Based on Space Partitioning

Another approach for creating multiple sub-populations from the main population is to divide the search space into several partitions and keep the number of individuals in each partition less than a predefined threshold. In this group, we find the proposal of Hashemi and Meybodi [25]. Here, cellular automata are incorporated into a PSO algorithm (cellular PSO). In cellular PSO, the search space is partitioned into some equally sized cells using cellular automata. Then particles of the swarm are allocated to different cells according to their positions in the search space. Particles residing in each cell use their personal best positions and the best solution found in their neighborhood cells for searching an optimum.

Moreover, whenever the number of particles within each cell exceeds a predefined threshold, randomly selected particles from the saturated cells are transferred to random cells within the search space. In addition, each cell has a memory that is used to keep track of the best position found within the boundary of the cell and its neighbors. In another work [24], the same authors improved the performance of cellular PSO by changing the role of particles to quantum particles just at the moment when a change occurs.

Inspired by the cellular PSO, Noroozi et al. [30] proposed an algorithm based on DE, called CellularDE. CellularDE employs the *DE/rand-to-best/1/bin* scheme to provide local exploration capability for genomes residing in each cell. After detecting a change in the environment, the population performs a random local search for several upcoming iterations. In another work [31], the authors improved the performance of CellularDE by using a peak detection mechanism and hill climbing local search.

7.2.3 Methods with an Adaptive Number of Populations

The last group of studies includes methods that control the search progress of the algorithm and adapt the number of populations based on one or more feedback parameters.

The critical drawback of the multi-swarm algorithm proposed in [23] is that the number of swarms should be defined before the optimization process. This is a clear limitation in real-world problems where information about the environment might be not available. The very first attempt to adapt the number of populations in dynamic environments was made by [56]. He developed an adaptive mQSO (AmQSO) which adaptively determines the number of swarms either by spawning new swarms into the search space, or by destroying redundant swarms. In this algorithm, swarms are categorized into two groups: (i) *free* swarms, whose expansion, i.e. the maximum distance between any two particles in the swarm in all dimensions, is larger than a predefined radius r_{conv}, and (ii) *converged* swarms. Once the expansion of a free swarm becomes smaller than a radius r_{conv}, it is converted to the converged swarm. In AmQSO, when the number of free swarms (M_{free}) is dropped to zero, a free swarm is initialized in the search space for capturing undetected peaks. On the other hand, free swarms are removed from the search space if M_{free} is higher than a threshold n_{excess}.

Some researchers adapted the idea of AmQSO and used it in conjunction with other ECs. For instance, Yazdani et al. [57] proposed a dynamic modified multi-population artificial fish swarm algorithm based on the general principles of AmQSO. In this approach, the authors modified the basic parameters, behaviors and general procedure of the standard artificial fish swarm algorithm to fulfill the requirements for optimization in dynamic environments.

Inspired by AmQSO, a modified cuckoo search algorithm was proposed by Fouladgar and Lotfi [58] for DOPs. Their approach uses a modified cuckoo search algorithm to increase the convergence rate of populations for finding the peaks, and exclusion to prevent populations from converging to the same areas of the search space.

Differing from the above algorithms, which use the expansion of the populations as a feedback parameter to adapt the number of populations, Du Plessis and Engelbrecht [59] proposed the dynamic population differential evolution (DynPopDE), in which the populations are adaptively spawned and removed, based on their performance. In this approach, when all of the current populations fail to improve their fitness, i.e. $\forall k \in \kappa \rightarrow \Delta f_k(t) = |f_k(t) - f_k(t-1)| = 0$ where κ is the set of current populations, DynPopDE produces a new population of random individuals in the search space.

DynPopDE starts with a single population and gradually adapts to an appropriate number of populations. On the other hand, when the number of populations surpasses the number of peaks in the landscape, redundant populations can be detected as those which are frequently reinitialized by the exclusion operator. Therefore, a population k will be removed from the search space when it is marked for restart due to exclusion, and it has not improved since its last FEs ($\Delta f_k(t) \neq 0$).

Recently, Li et al. [60] proposed an adaptive multi-population framework to identify the correct number of populations. Their framework has three major procedures: clustering, tracking, and adapting. Moreover, they have employed various components to further enhance the overall performance of the proposed approach, including a hibernation scheme, a peak hiding scheme, and two movement schemes for the best individuals. Their method was empirically shown to be effective in comparison with a set of algorithms.

7.3 Problem Statement

A DOP $\mathcal{F}$ can be described by a quintuple $\{\Omega, \vec{x}, \phi, f, t\}$ where $\Omega \subseteq \mathbb{R}^D$ denotes to the search space, $\vec{x}$ is a feasible solution in Ω, ϕ represents the system control parameters which determine the distribution of the solutions in the fitness landscape, $f: \Omega \rightarrow \mathbb{R}$ is a static objective function, and $t \in \mathbb{N}$ is the time. With these definitions, the problem $\mathcal{F}$ is then formally defined as follows [44]:

$$\begin{aligned} &optimize_x\ f^t(\vec{x}), \vec{x} = (x_1, x_2, \ldots, x_D) \\ &\qquad subject\ to: \vec{x} \in \Omega \end{aligned} \tag{7.1}$$

The goal of optimizing a DOP is to find the set of global optima $X(t)$, in every time t, such as: $X(t) = \{x^* \in \Omega \mid \forall x \in \Omega \Rightarrow f^t(x^*) \succsim f^t(x)\}$. Here, $\succsim$ is a comparison relation which means *is better or equal than*, hence $\succsim \in \{\leq, \geq\}$. As can be seen in the above equation, DOP $\mathcal{F}$ is composed of a series of static instances $f^1(x), f^2(x), \ldots, f^{end}(x)$.

Hence, the goal of the optimization in such problems is no longer just locating the optimal solution(s), but rather tracking the shifting optima over the time.

The dynamism of the problem can then be obtained by tuning the system control parameters as follows:

$$\phi_{t+1} = \phi_t \oplus \Delta\phi \tag{7.2}$$

where $\Delta\phi$ is the deviation of the control parameters from their current values and $\oplus$ represents the way the parameters are changed. The next state of the dynamic environment then can be defined using the current state of the environment as follows:

$$f_{t+1}(\vec{x},\phi) = f_t(\vec{x},\phi_t \oplus \Delta\phi) \tag{7.3}$$

Different change types can be defined, using $\Delta\phi$ and $\oplus$. In [61], the authors proposed a framework of the eight change types including *small step change*, *large step change*, *random change*, *chaotic change*, *recurrent change*, *recurrent change with noise*, and *dimensional change*.

If changes between two successive environments are small enough, one can use the previously found best solutions for accelerating the process of finding the global optimum in the new environment. Otherwise, the best possible strategy would be solving the problem from the scratch.

7.4 Theory of Learning Automata

A learning automaton [62, 63] is an adaptive decision-making unit that improves its performance by learning the way to choose the optimal action from a finite set of allowable actions through iterative interactions with an unknown random environment. The action is chosen at random based on a probability distribution kept over the action-set and at each instant the given action is served as the input to the random environment. The environment responds to the taken action in turn with a reinforcement signal. The action probability vector is updated based on the reinforcement feedback from the environment. The objective of a learning automaton is to find the optimal action from the action set so that the average penalty received from the environment is minimized. Figure 7.1 shows the relationship between the learning automaton and random environment.

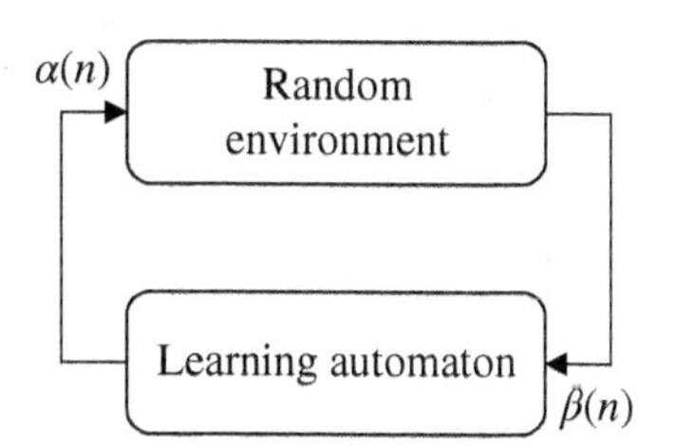

Figure 7.1 The schematic interaction of learning automata with an unknown random environment.

The environment can be described by a triple $E = \langle \alpha, \beta, c \rangle$ where $\alpha = \{\alpha_1, \alpha_2, ..., \alpha_r\}$ represents the finite set of the inputs, $\beta = \{\beta_1, \beta_2, ..., \beta_m\}$ denotes the set of the values that can be taken by the reinforcement signal, and $c = \{c_1, c_2, ..., c_r\}$ denotes the set of the penalty probabilities, where the element c_i is associated with the given action α_i. If the penalty probabilities are constant, the random environment is said to be a stationary random environment, and if they vary with time, the environment is called a non-stationary environment. The environments depend on the nature of the reinforcement signal, β can be classified into P-model, Q-model, and S-model. The environments in which the reinforcement signal can only take two binary values 0 and 1 are referred to as P-model environments. In another class of the environments, a finite number of the values in the interval [0, 1] can be taken by the reinforcement signal. Such an environment is referred to as a Q-model environment. In S-model environments, the reinforcement signal lies in the interval $[a, b]$. Learning automata can be classified into two main families [63]: fixed structure learning automata and variable structure learning automata.

If the probabilities of the transition from one state to another and probabilities of correspondence of action and state are fixed, the automaton is said to be fixed-structure automata and otherwise the automaton is said to be variable-structure automata.

7.4.1 Fixed Structure Learning Automaton

A fixed structure learning automaton (FSLA) is a quintuple $\langle \alpha, \Phi, \beta, F, G \rangle$ where:

- $\alpha = (\alpha_1, \cdots, \alpha_r)$ is the set of actions that it must choose from.
- $\Phi = (\Phi_1, \cdots, \Phi_s)$ is the set of states.
- $\beta = \{0, 1\}$ is the set of inputs where 1 represents a penalty and 0 represents a reward.
- $F : \Phi \times \beta \rightarrow \Phi$ is a map called the transition map. It defines the transition of the state of the automaton on receiving input, F may be stochastic.
- $G : \Phi \rightarrow \alpha$ is the output map and determines the action taken by the automaton if it is in state Φ_j.

A very well-known example of FSLA is the *two-state automaton* ($L_{2,\,2}$). This automaton has two states, Φ_1 and Φ_2 and two actions α_1 and α_2. The automaton accepts input from a set of {0, 1} and switches its states upon encountering an input 1 (unfavorable response) and remains in the same state on receiving an input 0 (favorable response). An automaton that uses this strategy is referred as $L_{2,\,2}$, where the first subscript refers to the number of states and the second subscript is the number of actions.

7.4.2 Variable Structure Learning Automaton

A variable structure learning automaton (VSLA) is represented by a quadruple $\langle\beta, \alpha, p, T\rangle$, where $\beta = \{\beta_1, \beta_2, ..., \beta_m\}$ is the set of inputs, $\alpha = \{\alpha_1, \alpha_2, ..., \alpha_r\}$ is the set of actions, $p = \{p_1, p_2, ..., p_r\}$ is the probability vector which determines the selection probability of each action, and T is learning algorithm which is used to modify the action probability vector, i.e. $p(n+1) = T\,[\alpha(n), \beta(n), p(n)]$. Let $\alpha(n)$ and $p(n)$ denote the action chosen at instant n and the action probability vector on which the chosen action is based, respectively. The recurrence equations shown by Eq. (7.4) and Eq. (7.5) is a linear learning algorithm by which the action probability vector p is updated as follows:

$$p_j(n+1) = \begin{cases} p_j(n) + a.(1 - p_j(n)) & if\ i = j \\ p_j(n).(1 - a) & if\ i \neq j \end{cases} \tag{7.4}$$

when the taken action is rewarded by the environment (i.e. $\beta(n) = 0$), and

$$p_j(n+1) = \begin{cases} p_j(n).(1 - b) & if\ i = j \\ \dfrac{b}{r-1} + (1 - b).p_j(n) & if\ i \neq j \end{cases} \tag{7.5}$$

when the taken action is penalized by the environment (i.e. $\beta(n) = 1$). In the above equations, r is the number of actions. Finally, a and b denote the reward and penalty parameters and determine the amount of increases and decreases of the action probabilities, respectively. If $a = b$, the recurrence Eqs. (7.4) and (7.5) are called a linear reward–penalty (L_{R-P}) algorithm, if $a \gg b$ the given equations are called linear reward–εpenalty ($L_{R-\varepsilon P}$), and finally if $b = 0$ they are called linear reward–inaction (L_{R-I}). In the latter case, the action probability vectors remain unchanged when the taken action is penalized by the environment.

A learning automaton has been shown to perform well in parameter adjustment [64–66], networking [67], social networks [67], etc.

7.5 Scheduling Algorithms

One frequent feature in most of the multi-population methods for DOPs is that the FEs are equally distributed among the sub-populations. It implies that the same quota of the FEs is allocated to all sub-populations, regardless of the quality of their positions in the fitness landscape. However, for several reasons, the equal distribution of FEs among sub-populations is not a good policy [29]:

1) The main goal in dynamic optimization is to locate the global optimum in a minimum amount of time and track its movements in the solution space. Therefore, strategies for quickly locating the global optimum are preferable.

2) A DOP may contain several peaks (local optimum). However, as the heights of the peaks are different, they do not have the same importance from the optimality point of view. Therefore, spending an equal number of FEs on all of them postpones the process of reaching a lower error.
3) Many real-world optimization problems are large-scale in nature. For these problems, the equal distribution of FEs among sub-populations would be detrimental to the performance of the EC algorithms.

To alleviate the above issues, the application of SPS is presented and discussed in this chapter.

The main idea behind the SPS is to change the order by which the sub-populations are executed with the aim to allocate more FEs to profitable sub-populations. This in turn increases the performance of multi-population methods in locating and tracking the optimum in dynamic environments.

The SPS methods can be roughly categorized into three classes. The first class contains strategies in which the sub-populations are executed in a preset or random order based on their index. The second class of SPS are those methods which use one or more feedback parameters to monitor the state of the sub-populations and choose the next sub-population for execution, adaptively. Finally, the third class combines a preset or random SPS method with a feedback. In the rest of this section, we study eight possible SPS algorithms.

7.5.1 Round Robin SPS

The most common and simplest method for scheduling sub-populations, which is currently used in majority of multi-population methods, is using the round robin (RR) strategy. In this strategy, all sub-populations are executed one by one in a circular queue (Figure 7.2). This way, FEs are equally distributed among the sub-populations.

At each time, the sub-population that should be executed is determined as follows:

$$S^t = \begin{cases} 1 & if\, t = 1 \\ (S^{t-1} \% N) + 1 & otherwise \end{cases} \tag{7.6}$$

where % is the modulo operation which returns the remainder after division of one number by another, and N is the number of sub-populations.

7.5.2 Random SPS

The second SPS method is to execute the sub-populations in a random manner. In this method, called random (RND) SPS, at each time, a sub-population is chosen randomly according to the following equation:

$$S^t = randi(N) \tag{7.7}$$

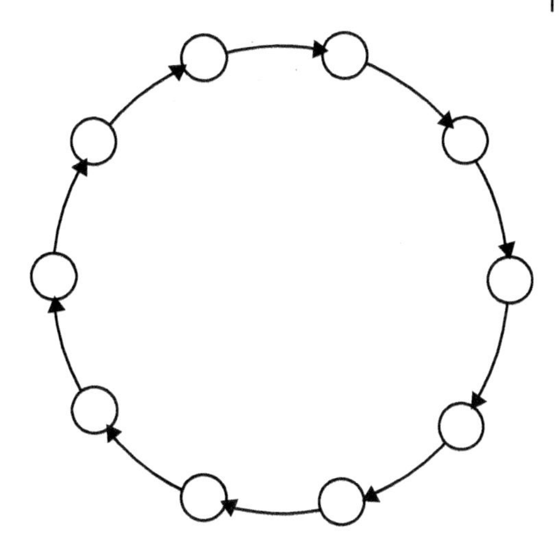

Figure 7.2 The schematic of the round robin policy for sub-population scheduling. Each circle represents a sup-population.

where $\mathcal{S}^t$ is the index of the sub-population that should be executed at time t, and *randi* is a function that returns a random integer in the range 1 and N, and N is equal to the number of sub-populations.

7.5.3 Pure-Chance Two Action SPS

Another option for random SPS is choosing between executing the current sub-population or executing the next sub-population based on pure chance (PC). Let R be a random variable following the discrete uniform distribution over the set $\{current\ sub-population,\ next\ sub-population\}$. At each iteration, the algorithm randomly decides to execute the current sub-population or the next sub-population as follows:

$$\mathcal{S}^t = \begin{cases} \mathcal{S}^{t-1} & if\ rand < 0.5 \\ (\mathcal{S}^{t-1} \% N) + 1 & otherwise \end{cases} \tag{7.8}$$

where *rand* is a random number in [0, 1]. Again, N is the number of sub-populations.

At the beginning of the optimization process, the current sub-population is set to the first sub-population.

7.5.4 SPS Based on Competitive Population Evaluation

As stated before, one way for scheduling the sub-populations is based on receiving one or more feedback parameters from the search progress of the sub-populations. The fourth SPS method, called competitive population evaluation (CPE) [42], uses the combination of the current fitness of the best individual in the sub-populations and the amount that the error of the best individual was reduced during the previous evaluation of the sub-populations as a feedback parameter for SPS. In this method, a performance measure $\mathcal{P}$ is first calculated for all sub-populations as follows:

$$\forall i \in \{1,2,\ldots,N\} \mid \mathcal{P}_i^t = (\Delta f_i^t + 1)\,(R_i^t + 1) \tag{7.9}$$

In the above equation, N is the number of sub-populations and Δf_i^t is the improvement amount of the best individual during the previous evaluation of the sub-population i, which is computed as:

$$\Delta f_i^t = \left| f_i^t - f_i^{t-1} \right| \tag{7.10}$$

R_i^t is also calculated as follows:

$$R_i^t = \begin{cases} f_i^t - \min_{q=1,\ldots,N}\left\{f_q^t\right\} & \textit{for maximization problems} \\ \max_{q=1,\ldots,N}\left\{f_q^t\right\} - f_i^t & \textit{for minimization problems} \end{cases} \tag{7.11}$$

Finally, the sub-population to be executed is selected as follows:

$$\mathcal{S}^t = argmax\left\{\mathcal{P}_i^t\right\},\ \ i = 1,2,\ldots,n \tag{7.12}$$

Regarding Eq. (7.12), at each iteration, the best-performing sub-population takes the FEs and evolves itself until its performance drops below that of another sub-population, where the other one takes the FEs, and this process continues during the run. Figure 7.3 summarizes the working mechanism of CPE.

7.5.5 SPS Based on Performance Index

Another way for adaptive SPS is taking other types of feedbacks into account. In [29], the authors combined the *success rate* and the *quality of the best solution* found by the sub-populations in a single criterion called *performance index* (PI). The success rate (Y) determines the portion of improvement in the individuals of a sub-population which is defined as:

$$\Upsilon = \frac{\textit{number of improved individuals}}{\textit{total number of individuals}} \tag{7.13}$$

For function maximization, the PI ψ_i for each sub-population i is computed as:

$$\psi_i^t = \exp(\Upsilon_i^t)\cdot f_i^t \tag{7.14}$$

where Υ_i^t is the success rate of the sub-population i at time t. *exp* is a mathematical function which returns Euler's number e raised to the power of Υ_i^t, and f_i^t represents the fitness value of the best solution found by sub-population i at time t.

```
1.  Set the parameters N, pop_size; /*Set the multi-population cardinality and sub-population size*/
2.  Randomly initialize N sub-populations of individuals in the search space; /*Initialize sub-
    populations into the search space*/
3.  while termination criterion is not met do
4.      if a change in the environment is detected then
5.          Re-evaluate all sub-populations;
6.          count = 0;
7.      end-if
8.      if count<2 then
9.          Evolve all sub-populations using the inner optimizer; /*all sub-populations are evolved in
            order*/
10.     else
11.         Calculate performance P for all sub-populations according to Eq. (7.9);
12.         Select the best-performing sub-population with respect to Eq. (7.12) and evolve it using the
            inner optimizer;/*e.g. DE, PSO, ACO, GA etc.*/
13.     end-if
14.     count = count + 1;
15. end-while
```

Figure 7.3 Sub-population scheduling based on the competitive population evaluation.

For function minimization, one must convert the function minimization problem into a fitness maximization in which a fitness value $f_i = \overline{f}_{\max} - \overline{f}_i$ is used. Here, $\overline{f}_{\max}$ is the maximum value of the fitness function, while $\overline{f}_i$ is the fitness value of the best-found position by sub-population i.

The FEs at each stage are then allocated to the sub-population with the highest PI, which is selected as follows:

$$S^t = argmax\left\{\psi_i^t\right\},\quad i = 1, 2, \ldots, N \tag{7.15}$$

It is easy to see that the best-performing population will be the population with the highest fitness and success rate.

The pseudo code for the SPS based on the PI is shown in Figure 7.4.

7.5.6 SPS Based on VSLA

Another interesting approach for SPS is by using decision making tools like learning automata. One way to model the SPS using learning automata is by applying an N-action VSLA, where its actions correspond to different sub-populations, i.e. $n_1, n_2, \ldots, n_N$ [29]. In order to select the next sub-population for execution, the learning automaton $LA_{sub-pop}$ selects one of its actions, e.g. α_i, randomly based on its action probability vector. Then, depending on the selected action, the corresponding sub-population is executed.

After execution of the selected sub-population, the automaton $LA_{sub-pop}$ updates its probability vector based on the reinforcement signal received in

1. Set the parameters N, *pop_size*; /*Set the multi-population cardinality and sub-population size*/
2. Randomly initialize N sub-populations of individuals in the search space; /*Initialize populations into the search space*/
3. **while** *termination criterion is not met* **do**
4. | **if** a change in the environment is detected **then**
5. | | Re-evaluate all sub-populations;
6. | **end-if**
7. | Calculate success rate for each sub-population according to Eq. (7.13);
8. | Update the performance index for each sub-population according to Eq. (7.14);
9. | Select the best-performing sub-population with respect to Eq. (7.15) and evolve it according to the inner optimizer; /*e.g. DE, PSO, ACO, GA etc.*/
10. **end-while**

Figure 7.4 Sub-population scheduling based on the performance index.

response to the selected action. Based on the received reinforcement signal, the automaton determines whether its action was a right or wrong, updating its probability vector accordingly. To monitor the overall progress of the SPS process, the reinforcement signal is generated as follows:

$$\beta = \begin{cases} 0 & if\ f(\vec{X}_{g,t}) < f(\vec{X}_{g,t+1}) \\ 1 & otherwise \end{cases} \tag{7.16}$$

where f is the fitness function, and $\vec{X}_{g,t}$ is the global best individual of the algorithm at time step t. According to Eq. (7.16), if the quality of the best-found solution by the algorithm improves, then a favorable signal is received by the automaton $LA_{sub-pop}$. Afterwards, the probability vector of the $LA_{sub-pop}$ is modified based on the learning algorithm of the automaton. The structure of the VLSA-based SPS method is given in Figure 7.5.

7.5.7 SPS Based on FSLA

Apart from the above VSLA approach, one can model the SPS with FSLA. The seventh SPS algorithm is modeled by an FSLA with N states $\Phi = (\Phi_1, \Phi_2, ..., \Phi_N)$, the action set $\alpha = (\alpha_1, \cdots, \alpha_N)$, and the environment response $\beta = \{0, 1\}$ which is related to reward and penalty. Figure 7.6 illustrates the state transition and action selection in the proposed model.

Initially, the automaton is in state $i = 1$. Therefore, the automaton selects action α_1 which is related to evolving the sub-population n_1. Generally speaking, when the automaton is in state $i = 1, ..., N$, it performs the corresponding action α_i which is executing the sub-population n_i. Afterwards, the FSLA receives an input β according to Eq. (7.16). If $\beta = 0$, FSLA stays in the same state, otherwise it goes to the next state.

1. Set the parameters N, *pop_size*; /*Set the multi-population cardinality and sub-population size*/
2. Randomly initialize N sub-populations of individuals in the search space; /*Initialize sub-populations into the search space*/
3. Initialize the $LA_{sub-pop}$ with action set $\alpha = \{1,2,\ldots, N\}$ and action probability vector $p = [(1/N), (1/N),\ldots, (1/N)]$.
4. Set $a = 0.15$ and $b = 0.05$; /* Parameters for updating probability vector in step 12*/
5. **while** *termination criterion not met* **do**
6.   **if** a change in the environment is detected **then**
7.   Re-evaluate all sub-populations;
8.   Reset the action probability vector of $LA_{sub-pop}$;
9.   **end-if**
10.   Select a sub-population i using probability vector p and evolve it according to the inner optimizer; /*e.g. DE, PSO, ACO, GA etc.*/
11.   Compute the reinforcement signal β according to Eq. (7.16);
12.   **for each** action $j \in [1,2,\ldots,r]$ of $LA_{sub-pop}$ **do**
13.   **if** $(\beta = 0)$ **then** //favorable response
14.   Reward selected action according to Eq. (7.4);
15.   **else if** $(\beta = 1)$ //unfavorable response
16.   Penalize selected action according to Eq. (7.5);
17.   **end-if**
18.   **end-for**
19. **end**

Figure 7.5 Sub-population scheduling based on the variable structure learning automaton.

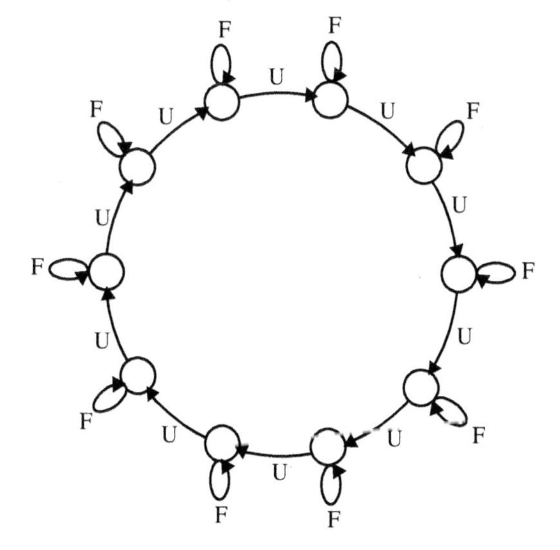

Figure 7.6 The schematic of the proposed FSLA for scheduling the sub-populations. Each circle represents a state of the automaton. The symbols "U" and "F" correspond to unfavorable and favorable responses accordingly.

1. Set the parameters N, *pop_size*; /*Multi-population cardinality and sub-population size*/
2. Randomly initialize N sub-populations of individuals in the search space; /*Initialize sub-populations into the search space*/
3. Initialize the *FSLA* with $\Phi = (\Phi_1,\cdots,\Phi_N)$ and $\alpha = (\alpha_1,\cdots,\alpha_N)$
4. Set the current state of *FSLA* to $i = 1$;
5. **while** *termination criterion not met* **do**
6. | **if** a change in the environment is detected **then**
7. | | Re-evaluate all sub-populations;
8. | **end-if**
9. | Select action i and evolve the corresponding sub-population i according to the Inner optimizer; /*e.g. DE, PSO, ACO, GA etc.*/
10. | Compute the reinforcement signal β according to Eq. (7.16);
11. | **if** ((β=0) **then** //favorable response
12. | | //do nothing and stay in the current state
13. | **else if** (β=1) **then** //unfavorable response
14. | | $i = (i\%N) + 1$; //go to next state
15. | **end-if**
16. **end**

Figure 7.7 Sub-population scheduling based on the fixed structure learning automaton.

Qualitatively, the simple strategy used by FSLA-based scheduling implies that the corresponding multi-population EC algorithm continues to execute whatever sub-population it was evolving earlier as long as the response is favorable but changes to another sub-population as soon as the response is unfavorable. Figure 7.7 shows the different steps of the FSLA model for SPS.

7.5.8 SPS Based on STAR Automaton

Finally, the last SPS method is based on STAR with deterministic reward and deterministic penalty [68]. The automaton can be in any of $N+1$ states $\Phi = (\Phi_0, \Phi_1, \Phi_2, \ldots, \Phi_N)$, the action set is $\alpha = (\alpha_1, \cdots, \alpha_N)$, and the environment response is $\beta = \{0, 1\}$, which is related to reward and penalty. The state transition and action selection are illustrated in Figure 7.8.

When the automaton is in any state $i = 1, \ldots, N$, it performs the corresponding action α_i which is related to evolving the sub-population n_i. On the other hand, the state Φ_0 is called the "neutral" state: when in that state, the automaton chooses any of the N actions with equal probability $1/N$.

Both reward and penalty cause deterministic state transitions, according to the following rules [68]:

1) When in state Φ_0 and chosen action is i ($i = 1, \ldots, N$), if rewarded go to state Φ_i with probability 1. If punished, stay in state Φ_0 with probability 1.
2) When in state Φ_i, $i\neq0$ and chosen action is i ($i = 1, \ldots, N$), if rewarded stay in state Φ_i with probability 1. If punished, go to state Φ_0 with probability 1.

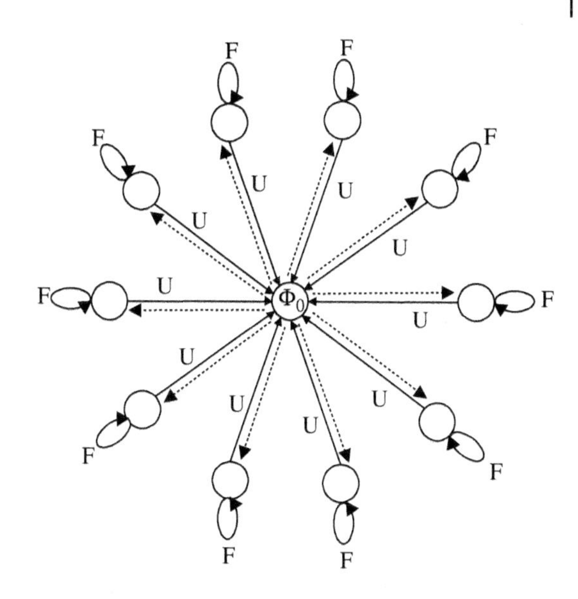

Figure 7.8 The schematic of the STAR for sub-population scheduling. Each circle represents a state of the automaton. The symbols "U" and "F" correspond to unfavorable and favorable responses accordingly.

With these in mind, the STAR-based SPS can be described as follows. At the commencement of the run, the automaton is in state Φ_0. Therefore, the automaton chooses one of its actions, say α_r, randomly. The action chosen by the automaton is applied to the environment, by evolving the corresponding sub-population r, which in turn emits a reinforcement signal β according to Eq. (7.16). The automaton moves to state Φ_r if it has been rewarded and it stays in state Φ_0 in the case of punishment. On the other hand, when in state $\Phi_{i\neq 0}$, the automaton chooses an action based on the current state it resides in. That is, action α_i is chosen by the automaton if it is in the state $\Phi_{i\neq 0}$.

Consequently, sub-population i is evolved. Again, the environment responds to the automaton by generating a reinforcement signal according to Eq. (7.16). On the basis of received signal β, the state of the automaton is updated. If $\beta = 0$, STAR stays in the same state, otherwise it moves to Φ_0. The algorithm repeats this process until the terminating condition is satisfied. Figure 7.9 shows the details of the proposed scheduling method.

A summary of various SPS strategies is given in Table 7.1.

7.6 Application of the SPS Algorithms on a Multi-Population Evolutionary Dynamic Optimization Method

In the following section, we describe how the mentioned SPS schemes can be incorporated into an existing multi-population method for DOPs. Although several options exist, we have considered DynDE, the algorithm proposed in [41].

```
1.  Set the parameters N, pop_size; /*Multi-population cardinality and sub-population size*/
2.  Randomly initialize N sub-populations of individuals in the search space; /*Initialize sub-
    populations into the search space*/
3.  Initialize the STAR with Φ = (Φ0, Φ1,···, ΦN) and α = (α1,···, αN)
4.  Set the current state of STAR(1) to i = 0;
5.  while termination criterion not met do
6.  |  if a change in the environment is detected then
7.  |  |  Re-evaluate all sub-populations;
8.  |  end-if
9.  |  if current state is Φ0 then //select an action randomly
10. |  |  Randomly choose an action from αr ∈ {α1,···, αN} and evolve the corresponding sub-
    |  |  population r according to the inner optimizer; /*e.g. DE, PSO, ACO, GA etc.*/
11. |  |  Compute the reinforcement signal β according to Eq. (7.16);
12. |  |  if ((β=0) then //favorable response
13. |  |  |  i = r; //go to state Φr
14. |  |  else if (β=1) //unfavorable response
15. |  |  |  //do nothing and stay in the state Φ0
16. |  |  end-if
17. |  else if current state is Φi≠0 then//select action i
18. |  |  Select action αi and evolve the corresponding sub-population i according to the inner
    |  |  optimizer; /*e.g. DE, PSO, ACO, GA etc.*/
19. |  |  Compute the reinforcement signal β according to Eq. (7.16);
20. |  |  if (β=0) then //favorable response
21. |  |  |  //do nothing and stay in the current state
22. |  |  else if (β=1) //unfavorable response
23. |  |  |  i = 0; //go to state Φ0
24. |  |  end-if
25. |  end-if
26. end
```

Figure 7.9 Sub-population scheduling based on the STAR.

DynDE has been extensively studied in the past, showing very good results in challenging DOPs.

In order to describe DynDE, we will explain the DE paradigm first, which is the underlying optimization method in DynDE. Later, we give a brief overview of DynDE.

7.6.1 Differential Evolution

DE [69, 70] is a well-known EC paradigm which is one of the most powerful stochastic real-parameter optimization algorithms. Over the past decade, DE has gained much popularity because of its simplicity, effectiveness, and robustness.

The main idea of DE is to use spatial difference among the population of vectors to guide the search process toward the optimum solution. The rest of this section describes the main operational stages of DE in detail.

Table 7.1 Description of different SPS strategies.

Method	Strategy	Scheduling mechanism	Feedback parameter
RR	Preset	All sub-populations are executed one-by-one in a circular queue	—
RND	Random	Sub-populations are executed in a random manner	—
PC	Random	Each sub-population is either executed or skipped randomly	—
CPE	Adaptive	The sub-population with the highest fitness and improvement is executed at each time	Fitness of the best individual and its fitness improvement
PI	Adaptive	The sub-population with the highest fitness and success rate is executed at each time	Success rate and the quality of the best individual
VSLA	Adaptive	Sub-populations are scheduled using VSLA	Success or failure of the global best individual
FSLA	Hybrid	Sub-populations are scheduled using FSLA	Success or failure of the global best individual
STAR	Hybrid	Sub-populations are scheduled using STAR	Success or failure of the global best individual

7.6.1.1 Initialization of Vectors

DE starts with a population of *NP* randomly generated vectors in a *D*-dimensional search space. Each vector *i*, also known as genome or chromosome, is a potential solution to an optimization problem which is represented by $\vec{x}_i = (x_{i1}, x_{i2}, \ldots, x_{iD})$. The initial population of vectors is simply randomized into the boundary of the search space according to a uniform distribution as follows:

$$x_{ij} = lb_j + rand_j\left[0,1\right] \times (ub_j - lb_j) \tag{7.17}$$

where $i \epsilon [1, 2, \ldots, NP]$ is the index of *i*th vector of the population, $j \epsilon [1, 2, \ldots, D]$ represents the *j*th dimension of the search space, and $rand_j[0, 1]$ is a uniformly distributed random number corresponding to the *j*th dimension. Finally, lb_j and ub_j are the lower and upper bounds of the search space corresponding to the *j*th dimension of the search space.

7.6.1.2 Difference-vector-based Mutation

After initialization of the vectors in the search space, a mutation is performed on each genome *i* of the population to generate a donor vector $\vec{v}_i = (v_{i1}, v_{i2}, \ldots, v_{iD})$ corresponding to target vector $\vec{x}_i$. Each component of donor vector $\vec{v}_i$, using a DE/rand/1 mutation strategy, is given by:

$$\vec{v}_i = \vec{x}_{r_1} + \mathcal{F} \cdot (\vec{x}_{r_2} - \vec{x}_{r_3}) \tag{7.18}$$

where $\mathcal{F}$ is the scaling factor used to control the amplification of difference vector, and r_1, r_2 and r_3 are uniformly distributed random integers on the interval $[1, NP]$ such that $r_1 \neq r_2 \neq r_3 \neq i$.

Different mutation strategies have been outlined in [71]. If the generated mutant vector is out of the search boundary, a repair operator is used to bring $\vec{v}_i$ back to the feasible region.

7.6.1.3 Crossover

To introduce diversity to the population of genomes, DE utilizes a crossover operation to combine the components of target vector $\vec{x}_i$ and donor vector $\vec{v}_i$, to form the trial vector $\vec{u}_i$. Two types of crossover are commonly used in the DE community, which are called *binomial crossover* and *exponential crossover*. Binomial crossover is the underlying crossover in DynDE which is defined as follows:

$$u_{ij} = \begin{cases} v_{ij} & \text{if } rand_{ij}[0,1] \leq CR \text{ or } j = jrand \\ x_{ij} & \text{otherwise} \end{cases} \tag{7.19}$$

where $rand_{ij}[0, 1]$ is a random number drawn from a uniform distribution between 0 and 1, $CR \in (0, 1)$ is the crossover probability which is used to control the approximate number of components that are transferred to the trial vector. *jrand* is a random index in the range $[1, D]$, which ensures the transmitting of at least one component from the donor to the trial vector.

7.6.1.4 Selection

Finally, a selection approach is performed on vectors to determine which vector ($\vec{x}_i$ or $\vec{u}_i$) should survive in the next generation. The most fitted vector is chosen to be the member of the next generation.

Different extensions of DE can be specified using the general convention *DE/x/y/z*, where DE stands for "Differential Evolution," *x* represents a string denoting the base vector to be perturbed, *y* is the number of difference vectors considered for perturbation of *x*, and *z* stands for the type of crossover being used, i.e. *exponential* or *binomial*, [71]. Figure 7.10 shows a sample pseudo-code for canonical DE.

DE has several advantages that make it a powerful tool for optimization tasks. Specifically, (i) DE has a simple structure and is easy to implement; (ii) despite its simplicity, DE exhibits a high performance; (iii) the number of control parameters in the canonical DE [69] are very few (i.e. *NP*, *F*, and *CR*); (iv) due to its low space complexity, DE is suitable for handling large-scale problems.

7.6.2 DynDE

DynDE starts with a predefined number of sub-populations, exploring the entire search space to locate the optimum. Each sub-population is composed of *normal* and *Brownian* individuals.

1. Set the parameters *NP*, *F*, and *CR*;
2. Randomly initialize *NP* individuals in the search space;
3. Calculate the fitness of the individuals;
4. **while** *termination criterion not met* **do**
5. **for** each individual $i \in [1,2, \ldots, NP]$ **do**
6. Perform mutation according to Eq. (7.18);
7. Perform binomial crossover according to Eq. (7.19);
8. Calculate fitness of the trial vector;
9. **end-for**
10. Update target vectors according to selection schemea
11. **end**

Figure 7.10 Canonical differential evolution.

7.6.2.1 Normal Individuals

Normal individuals follow the principles of the DE. Although various schemes are typically used for DE, according to the authors [41], DE/best/2/bin produces the best results. In this scheme, the following mutation strategy is used:

$$\vec{v} = \vec{x}_{best} + F \cdot (\vec{x}_1 + \vec{x}_2 - \vec{x}_3 - \vec{x}_4) \tag{7.20}$$

here, $\vec{x}_1 \neq \vec{x}_2 \neq \vec{x}_3 \neq \vec{x}_4$ are four randomly selected vectors from the sub-population. In addition, $\vec{x}_{best}$ is the best vector from the sub-population.

7.6.2.2 Repair Operator

If the generated mutant vector is out of the search boundary, a repair operator is used to bring $\vec{v}$ back to the feasible region. Different strategies have been proposed to repair the out of bound individuals. In this chapter, if the jth element of the ith mutant vector, i.e. v_{ij}, is out of the search region $[lb_j, ub_j]$, then it is repaired as follows:

$$v_{ij} = \begin{cases} lb_j & if \ v_{ij} < lb_j \\ ub_j & if \ v_{ij} > ub_j \end{cases} \tag{7.21}$$

Afterwards, the binomial crossover is applied to form the trial vector $\vec{u}$ according to Eq. (7.19).

7.6.2.3 Brownian Individuals

On the other hand, Brownian individuals are not generated according to the DE/best/2/bin scheme. They are created around the center of the best individual of the sub-population according to the following equation:

$$\vec{x}_{Brownian} = \vec{x}_{best} + \vec{N}(0, \sigma) \tag{7.22}$$

where $\vec{N}(0,\sigma)$ is a Gaussian distribution with zero mean and standard deviation σ. As shown in [41], $\sigma = 0.2$ is a suitable value for this parameter.

7.6.2.4 Exclusion

In order to prevent different sub-populations from settling on the same peak, an exclusion operator is applied. This operator triggers when the distance between the best individuals of any two sub-populations becomes less than the threshold r_{excl}, and reinitializes the worst performing sub-population in the search space. The parameter r_{excl} is calculated as follows:

$$r_{excl} = \frac{X}{2p^{1/d}} \tag{7.23}$$

where X is the range of the search space for each dimension, and p is the number of existing peaks in the landscape.

7.6.2.5 Detection and Response to the Changes

Several methods have been suggested for detecting changes in dynamic environments. See for example [11] for a good survey on the topic. The most frequently used method consists in reevaluating memories of the algorithm for detecting inconsistencies in their corresponding fitness values. In DynDE the best solution of each sub-population was considered. At the beginning of each iteration, the algorithm reevaluates these solutions and compares their current fitness values against those from the previous iteration. If at least one inconsistency is found from these comparisons, then one can assume that the environment has changed. Once a change is detected, all sub-populations are reevaluated.

The procedure for DynDE algorithm is summarized in Figure 7.11.

Bearing in mind the inner working mechanism of DynDE, it is easy to include any SPS scheme in it.

In the rest of this chapter, DynDE with different SPS schemes is specified using the general convention DynDE-XXX, where XXX stands for the type of SPS being used, i.e. RR, RND, etc. For example, DynDE-RND denotes the DynDE with random SPS.

7.7 Computational Experiments

In this section, different SPS algorithms are empirically analyzed by means of computational experiments. The structure of this section is as follows: first, the technical aspects of the artificial DOP used for testing the algorithms is introduced. Then, the performance measures employed for assessing the algorithms are described. Finally, the simulation results are presented and discussed.

1. Set the parameters N, $rexcl$; /*Multi-population cardinality and exclusion radius*/
2. Randomly initialize N sub-populations of individuals in the search space; /*Initialize sub-populations into the search space*/
3. **while** termination condition is not met **do**
4. | Apply exclusion operator;
5. | Detect change of the environment and react to it;
6. | **for** each sub-population n **do**
7. | | Evolve the individuals of the sub-population based on their type (normal or Brownian);
8. | | Evaluate the fitness of the individuals;
9. | | Update the sub-population;
10. | **end-for**
11. **end**

Figure 7.11 DynDE algorithm.

7.7.1 Dynamic Test Function

One of the most widely used dynamic test functions in the literature is the *Moving Peaks Benchmark* (MPB) [17]. MPB is a real-valued dynamic environment with a D-dimensional landscape consisting of m peaks, where the height, width, and position of each peak are changed slightly every time a change occurs in the environment [72]. Different landscapes can be defined by specifying the shape of the peaks and some other parameters. A typical peak shape is conical which is defined as follows:

$$f(\vec{x},t) = \max_{i=1,\ldots,m} H_t(i) - W_t(i)\sqrt{\sum_{j=1}^{D}(x_t(j) - X_t(i,j))^2} \tag{7.24}$$

where $H_t(i)$ and $W_t(i)$ are the height and width of peak i at time t, respectively.

In MPB, both position and objective function of the problem are subject to change. The control parameters of the MPB are $\phi = (H, W, X)$, that deviate from their current values, every time a change occurs, according to the following equation:

$$\Delta\phi = \sigma.\phi_{severity} \tag{7.25}$$

where $\Delta\phi$ is the deviation of the control parameters, σ denotes a normally distributed random number with mean zero and standard deviation one, and $\phi_{severity}$ is a constant number that indicates the change severity of ϕ. The coordinates of each dimension $j \in [1, D]$ of peak i at time t are expressed by $X_t(i, j)$, while D is the problem dimensionality. A typical change of a single peak can be modeled as follows:

$$H_{t+1}(i) = H_t(i) + height_{severity}.\sigma_h \tag{7.26}$$

$$W_{t+1}(i) = W_t(i) + width_{severity}.\sigma_w \tag{7.27}$$

$$\vec{X}_{t+1}(i) = \vec{X}_t(i) + \vec{v}_{t+1}(i) \tag{7.28}$$

$$\vec{v}_{t+1}(i) = \frac{s}{\left|\vec{r} + \vec{v}_t(i)\right|}((1-\lambda)\vec{r} + \lambda\vec{v}_t(i)) \tag{7.29}$$

where σ_h and σ_w are two random Gaussian numbers with zero mean and standard deviation one. Moreover, the shift vector $\vec{v}_{t+1}(i)$ is a combination of a random vector $\vec{r}$, which is created by drawing random numbers in $[-0.5, 0.5]$ for each dimension, and the current shift vector $\vec{v}_t(i)$, and normalized to the length s. Parameter $\lambda \in [0.0, 1.0]$ specifies the correlation of each peak's changes to the previous one. This parameter determines the trajectory of changes, where $\lambda = 0$ means that the peaks are shifted in completely random directions and $\lambda = 1$ means that the peaks always follow the same direction, until they hit the boundaries where they bounce off.

In this chapter we consider the so-called *Scenario 2*, which is the most widely used configuration of MPB. Unless stated otherwise, the MPB's parameters are set according to the values listed in Table 7.2 (*default values*). In addition, to investigate the effect of the environmental parameters (i.e. the change severity, change period, number of peaks, number of dimensions) on the performance of the proposed SPS methods, various experiments were carried out with different combinations of other tested values listed in Table 7.2.

7.7.2 Performance Measure

Several criteria have been already proposed in the literature for measuring the performance of the optimization algorithms in DOPs [73]. In order to validate the efficiency of the tested SPS algorithms, we considered the *offline error*, which is the most well-known metric for dynamic environments [74]. This measure [75] is defined as the average of the smallest error found by the algorithm in every time step:

$$E_{off} = \frac{1}{T}\sum\nolimits_{t=1}^{T} e_t^* \tag{7.30}$$

where T is the maximum number of FEs so far and e_t^* is the minimum error obtained by the algorithm at the time step t. We assume that a FE used by the algorithm corresponds to a single time step.

7.7.3 Experimental Settings

For each experiment of a SPS algorithm on a specific DOP, 50 independent runs with different random seeds were performed. Each run ends when the algorithm has consumed 500 000 FEs. The experimental results are reported in terms of average offline error and standard error, which is computed as standard deviation divided by the squared root of the number of runs.

Table 7.2 Parameter settings for the moving peaks benchmark.

Parameter	Default values (Scenario 2)	Other tested values
Number of peaks (m)	10	1, 2, 5, 7, 20, 30, 40, 50, 100, 200
Height severity	7.0	
Width severity	1.0	
Peak function	Cone	
Number of dimensions (D)	5	10, 15, 20, 25, 30
Height range (H)	$\in$[30.0, 70.0]	
Width range (W)	$\in$ [1.0, 12.0]	
Standard height (I)	50.0	
Search space range (A)	$[0.0, 100.0]^D$	
Frequency of change (f)	5000	1000, 2000, 3000, 4000
Shift severity (s)	1	3, 5
Correlation coefficient (λ)	0.0	
Basic function	No	

To determine the differences among tested algorithms, we applied nonparametric tests. First, we applied the Friedman's test ($P<0.05$) to verify whether significant differences exist at group level, that is, among all algorithms. In the case that such differences exist, then a Wilcoxon test ($\alpha = 0.05$) is applied to compare the best-performing algorithm among DynDE-RND, DynDE-PC, DynDE-CPE, DynDE-PI, DynDE-VSLA, DynDE-FSLA, and DynDE-STAR with DynDE-RR, which is the commonly used SPS algorithm for multi-population methods. The outcomes of the Wilcoxon test revealing that statically significant differences exist between the algorithms are highlighted with an asterisk symbol.

In addition, when reporting the results, the last column(row) of each table includes the improvement rate (%Imp) of the best-performing DynDE variant vs. the DynDE-RR. This measure is computed for every problem instance i as follows:

$$\%Imp_i = 100 \times \left(1 - \frac{e_{i,Best}}{e_{i,DynDE-RR}}\right) \tag{7.31}$$

where $e_{i,\, DynDE-RR}$ is the error value obtained by DynDE-RR for the problem i. Similarly, $e_{i,\, Best}$ is the best (minimum) offline error obtained by the other DynDE variants.

Table 7.3 Default parameter settings for DynDE and VSLA-based sub-population scheduling method.

Parameter	Default value
Number of populations	10
Number of Brownian individuals	2
Number DE individuals	4
σ	0.2
Exclusion radius	31.5
F	0.5
CR	0.5
Reward parameter (a)	0.15
Penalty parameter (b)	0.05

The parameters setting of the DynDE, and the reward and penalty factors for the VSLA-based SPS, are as shown in Table 7.3.

In order to draw valid conclusions regarding the effect of including SPS in DynDE, we have employed the same parameter settings reported by the authors of the algorithm. Moreover, the values for reward and penalty parameters were set according to [29].

7.7.3.1 Experiment 1: Effect of Varying the Shift Severity and Change Frequency

The aim of this experiment is to investigate the performance comparison between different SPS methods on MPB's Scenario 2 with different shift lengths $s \in \{1, 3, 5\}$ and change frequencies $f \in \{1000, 2000, 3000, 4000, 5000\}$. The numerical results of the different algorithms are reported in Table 7.4.

The results of the first set of experiments in Table 7.4 indicate that algorithms including the adaptive and hybrid SPS, i.e. DynDE-PI, DynDE-VSLA, DynDE-FSLA, and DynDE-STAR, reached better results in all DOPs. As can be observed in the last column of Table 7.4, the rate of improvement values of the best-performing algorithm over the DynDE-RR ranges from 12 to 29. Among tested scheduling algorithms, the DynDE-FSLA achieved the best results in 8 out of 15 DOPs. As expected, best results for the fast-changing DOPs ($f = 1000$) were obtained by the DynDE-FSLA. Thanks to FSLA, DynDE-FSLA is able to reduce the computational cost. The reason is that FSLA keeps an account of the rewards and penalties received for each action. It continues performing whatever action it was using earlier as long as it is rewarded, and switches from one action to another only when it receives a penalty. In contrast, VSLA chooses its action randomly based

Table 7.4 Comparison of offline error ± standard error of DynDE with different sub-population scheduling algorithms on DOPs with $D = 5$, $m = 10$ and varying shift severities and change intervals.

MPB		Method								
s	f	DynDE-RR	DynDE-RND	DynDE-PC	DynDE-CPE	DynDE-PI	DynDE-VSLA	DynDE-FSLA	DynDE-STAR	%Imp
1	1000	3.35 ± 0.04	3.61 ± 0.04	3.82 ± 0.05	3.18 ± 0.03	2.96 ± 0.05	2.83 ± 0.04	**2.76 ± 0.04**	2.85 ± 0.04	17.61*
	2000	2.31 ± 0.05	2.47 ± 0.05	2.56 ± 0.04	2.25 ± 0.05	2.05 ± 0.05	1.96 ± 0.04	1.95 ± 0.03	**1.94 ± 0.04**	16.01*
	3000	1.89 ± 0.05	1.97 ± 0.04	2.05 ± 0.04	1.80 ± 0.04	1.69 ± 0.06	1.65 ± 0.05	**1.60 ± 0.04**	1.61 ± 0.05	15.34*
	4000	1.67 ± 0.05	1.78 ± 0.06	1.80 ± 0.05	1.59 ± 0.04	1.56 ± 0.08	1.48 ± 0.06	1.49 ± 0.05	**1.42 ± 0.04**	14.97*
	5000	1.50 ± 0.05	1.64 ± 0.06	1.61 ± 0.05	1.49 ± 0.05	1.47 ± 0.08	1.32 ± 0.06	1.42 ± 0.06	**1.31 ± 0.04**	12.66*
3	1000	8.19 ± 0.06	8.41 ± 0.06	8.68 ± 0.07	8.03 ± 0.06	6.44 ± 0.07	6.48 ± 0.05	**6.06 ± 0.05**	6.37 ± 0.05	26.00*
	2000	5.13 ± 0.06	5.33 ± 0.06	5.44 ± 0.07	5.12 ± 0.06	4.01 ± 0.07	4.13 ± 0.06	**3.90 ± 0.05**	4.06 ± 0.05	23.97*
	3000	3.85 ± 0.07	4.00 ± 0.07	4.02 ± 0.07	3.77 ± 0.06	3.05 ± 0.08	3.18 ± 0.08	**2.98 ± 0.06**	3.08 ± 0.07	22.59*
	4000	3.14 ± 0.07	3.26 ± 0.07	3.30 ± 0.07	3.07 ± 0.06	2.52 ± 0.08	2.63 ± 0.07	**2.47 ± 0.06**	2.54 ± 0.07	21.33*
	5000	2.69 ± 0.07	2.82 ± 0.07	2.81 ± 0.06	2.61 ± 0.07	2.24 ± 0.08	2.23 ± 0.06	2.22 ± 0.06	**2.16 ± 0.06**	19.70*
5	1000	13.67 ± 0.10	13.92 ± 0.09	14.20 ± 0.09	13.37 ± 0.09	10.13 ± 0.09	10.44 ± 0.08	**9.59 ± 0.08**	10.15 ± 0.08	29.84*
	2000	8.66 ± 0.10	8.76 ± 0.09	8.90 ± 0.10	8.56 ± 0.09	**6.13 ± 0.10**	6.57 ± 0.09	6.14 ± 0.08	6.38 ± 0.08	29.21*
	3000	6.29 ± 0.09	6.39 ± 0.09	6.53 ± 0.10	6.19 ± 0.09	**4.45 ± 0.10**	4.88 ± 0.09	4.57 ± 0.08	4.73 ± 0.08	29.25*
	4000	4.98 ± 0.08	5.10 ± 0.09	5.14 ± 0.09	4.88 ± 0.07	**3.59 ± 0.09**	3.87 ± 0.08	3.61 ± 0.07	3.82 ± 0.08	27.91*
	5000	4.26 ± 0.10	4.31 ± 0.10	4.35 ± 0.08	4.18 ± 0.09	3.17 ± 0.10	3.33 ± 0.09	**3.11 ± 0.08**	3.24 ± 0.08	26.99*

Best values are shown in bold.

on the action probability vector kept over the action-set. Therefore, it is probable that the automaton switches to another action even though it has received reward for doing the current action until the selection probability of an action converges to one.

From Table 7.4, it is clear that the SPS approach can significantly improve the performance of DynDE-RR. It should be noted that, except for DynDE-RND and DynDE-PC, which act in random manners, the rest of the SPS algorithms are an improvement over the basic algorithm, i.e. DynDE-RR. As can be inferred from Table 7.4, SPS can improve the performance of DynDE-RR by 22.22% on average.

7.7.3.2 Experiment 2: Effect of Varying Number of Peaks

The experiment aims to investigate the performance comparison between different SPS algorithms on DOPs with different number of peaks $m \in \{1, 2, 5, 7, 10, 20, 30, 40, 50, 100, 200\}$. The other environmental parameters are set according to the MPB's Scenario 2 shown in Table 7.2. The obtained results are reported in Table 7.5.

For a single-peak environment, DynDE-FSLA is remarkably superior to the other methods. The reason lies in the fact that DynDE-FSLA has a faster rate of convergence to the desired action. For those DOPs with $m \leq 10$, best values correspond to the STAR-based approaches, with an improvement rate ranging from 12.66 to 36.51. Finally, for DOPs with a higher number of peaks $m > 10$, DynDE-PI is superior to the others. One possible explanation for such good performance can be due to combining the success rate and the quality of the local best solution into a single feedback parameter, contributing in this way to the algorithm exploration.

It is also worth noticing that when the number of peaks is high ($m = 100, 200$) even the random SPS, i.e. DynDE-RND, could produce better results than DynDE-RR.

7.7.3.3 Experiment 3: Effect of Increasing the Number of Dimensions

In this experiment, the performance of the eight SPS algorithms is investigated on the MPB with different dimensionalities $D \in \{5, 10, 15, 20, 25, 30\}$ and the other dynamic and complexity parameters are the same as the Scenario 2 of the MPB listed in Table 7.2.

As can be clearly observed in Table 7.6, the advantage of using SPS is more marked by increasing the scale of the DOP. As the number of dimensions is increased from 10 to 30, the best rate of improvement values of the SPS methods increases monotonically. This is a further argument to confirm the effectiveness of the SPS.

In order to provide some insights about the performance of the tested algorithms over the all simulated DOPs, a graphical representation using a box plot is presented in Figure 7.12. Each boxplot in the figure corresponds to a DynDE variant with lines

Table 7.5 Comparison of offline error ± standard error of DynDE with different sub-population scheduling algorithms on dynamic environments modeled with MPB.

m	DynDE-RR	DynDE-RND	DynDE-PC	DynDE-CPE	DynDE-PI	DynDE-VSLA	DynDE-FSLA	DynDE-STAR	%Imp
1	3.80 ± 0.17	4.85 ± 0.21	4.88 ± 0.20	2.79 ± 0.16	2.70 ± 0.11	3.07 ± 0.12	**2.37 ± 0.10**	2.49 ± 0.10	37.63*
2	2.41 ± 0.17	2.64 ± 0.16	2.77 ± 0.12	1.93 ± 0.15	1.63 ± 0.13	1.88 ± 0.14	1.54 ± 0.05	**1.53 ± 0.05**	36.51*
5	1.64 ± 0.09	1.70 ± 0.07	1.87 ± 0.08	1.55 ± 0.08	1.32 ± 0.09	1.41 ± 0.08	1.32 ± 0.06	**1.30 ± 0.07**	20.73*
7	1.63 ± 0.08	1.66 ± 0.05	1.82 ± 0.07	1.59 ± 0.06	1.50 ± 0.09	**1.32 ± 0.05**	1.46 ± 0.08	**1.32 ± 0.05**	19.01*
10	1.50 ± 0.05	1.64 ± 0.06	1.61 ± 0.05	1.49 ± 0.05	1.47 ± 0.08	1.32 ± 0.06	1.42 ± 0.06	**1.31 ± 0.04**	12.66*
20	2.74 ± 0.07	2.73 ± 0.06	2.77 ± 0.06	2.66 ± 0.07	**2.46 ± 0.08**	2.60 ± 0.07	2.58 ± 0.07	2.54 ± 0.07	10.21*
30	3.22 ± 0.10	3.24 ± 0.11	3.31 ± 0.09	3.25 ± 0.11	**2.91 ± 0.11**	3.05 ± 0.10	3.15 ± 0.10	3.02 ± 0.08	9.62*
40	3.46 ± 0.08	3.54 ± 0.08	3.56 ± 0.09	3.53 ± 0.09	**3.28 ± 0.09**	3.34 ± 0.07	3.38 ± 0.08	3.41 ± 0.09	5.20
50	3.81 ± 0.10	3.86 ± 0.08	3.90 ± 0.11	3.77 ± 0.09	**3.38 ± 0.10**	3.56 ± 0.09	3.71 ± 0.10	3.62 ± 0.10	11.28*
100	4.21 − 0.12	4.15 ± 0.11	4.11 ± 0.09	4.15 ± 0.12	**3.78 ± 0.11**	3.88 ± 0.11	4.02 ± 0.09	3.97 ± 0.09	10.21*
200	3.99 ± 0.12	3.86 ± 0.09	3.94 ± 0.10	3.98 ± 0.11	**3.62 ± 0.09**	3.71 ± 0.09	3.77 ± 0.09	3.71 ± 0.09	9.27

Best values are shown in bold.

Table 7.6 Average offline error ± standard error of algorithms on DOPs with $m = 10$, $s = 1$, and different dimensionalities.

	Number of dimensions (D)					
Method	**5**	**10**	**15**	**20**	**25**	**30**
DynDE-RR	1.50±0.05	4.41±0.22	6.75±0.23	9.05±0.29	11.37±0.29	14.30±0.46
DynDE-RND	1.64±0.06	4.56±0.24	7.00±0.25	9.27±0.28	11.75±0.34	14.49±0.45
DynDE-PC	1.61±0.05	4.52±0.22	7.18±0.28	9.26±0.27	11.96±0.35	14.87±0.45
DynDE-CPE	1.49±0.05	4.25±0.21	6.21±0.23	8.43±0.28	10.34±0.33	12.93±0.51
DynDE-PI	1.47±0.08	4.03±0.23	5.87±0.26	7.33±0.25	8.90±0.33	10.30±0.36
DynDE-VSLA	1.32±0.06	3.99±0.23	**5.60±0.25**	**6.95±0.23**	**8.11±0.25**	**9.43±0.29**
DynDE-FSLA	1.42±0.06	4.05±0.21	**5.60±0.23**	7.23±0.23	8.53±0.25	9.64±0.28
DynDE-STAR	**1.31±0.04**	**3.94±0.22**	5.73±0.23	7.22±0.25	8.29±0.25	9.65±0.30
%Imp	12.66*	10.65*	17.03*	23.20*	28.67*	34.05*

Best values are shown in bold.

at the lower quartile, median as solid black line, and upper quartile data values, where the whiskers are the lines extending from each end of the box to show the extent of the rest of the data. Values outside the range of the box plot are shown as individual points.

Regarding Figure 7.12, some general observations can be made as follows:

1) The boxplots for PI, VSLA, FSLA, and STAR are comparatively shorter than the other SPS methods. This suggests that PI, VSLA, FSLA, and STAR hold different performances compared to RR, RND, PC, and CPE.
2) The boxplots for PI, VSLA, FSLA, and STAR are comparatively lower than RR, RND, PC, and CPE. This suggests a difference between the two groups of methods.

Moreover, Figure 7.13 shows the mean offline error of each SPS method with 95% confidence intervals for all the reported results from Table 7.4 to Table 7.6. The circles represent the mean offline error obtained by each SPS

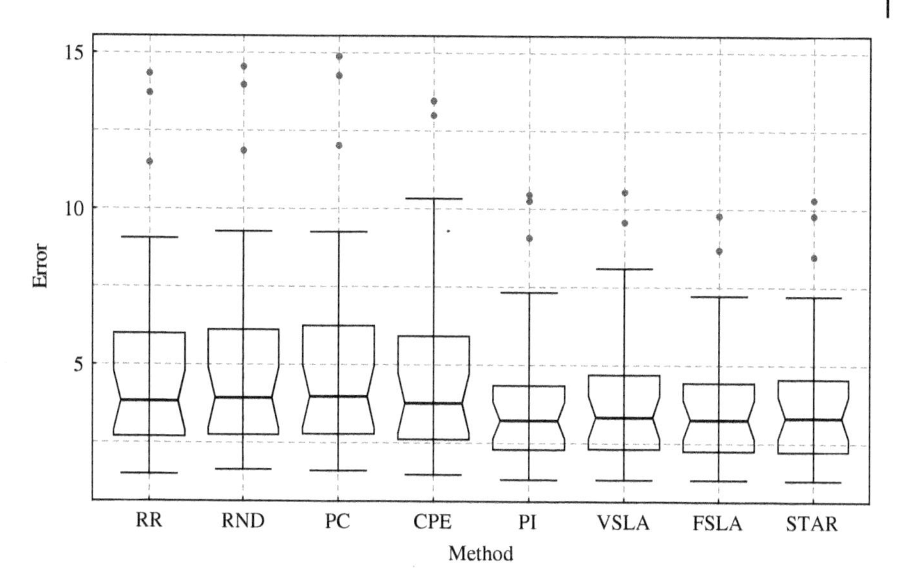

Figure 7.12 Boxplot for the overall performance of DynDE variants with different sub-population scheduling methods.

algorithm and error bars represent 95% confidence intervals for the mean. In addition, the means for every method are connected by a line. From Figure 7.13, it is observed that the results of PI, VSLA, FSLA, and STAR are significantly better than those of RR, RND, PC, CPE. Besides, while PI, VSLA, FSLA, and STAR show a similar performance, the mean error of FSLA is lower than the others.

Finally, in order to fully support the above conclusions, we performed pairwise comparisons with the eight algorithms using the Wilcoxon's rank sum test at 0.05 level ($p < 0.05$) of significance using all the reported means from the previous experiments. Table 7.7 presents whether pairwise performance comparisons between the algorithms are statistically significant or not. Each result given at the Xth row (AX) and the Yth column (AY) is marked with symbols "*S*+," "*S*-," "+" or "−" to indicate that the algorithm on the Xth row is significantly better than, significantly worse than, insignificantly better than, and insignificantly worse than the algorithm on the Yth column, respectively. The rightmost column of the Table 7.7 shows the mean rank of each SPS algorithm using Friedman's multiple comparisons test with $p < 0.05$. It is important to note that the lower the rank the better is the algorithm performance.

From Table 7.7, it is realized that a sophisticated scheduling mechanism can efficiently enhance the performance of multi-population methods for DOPs.

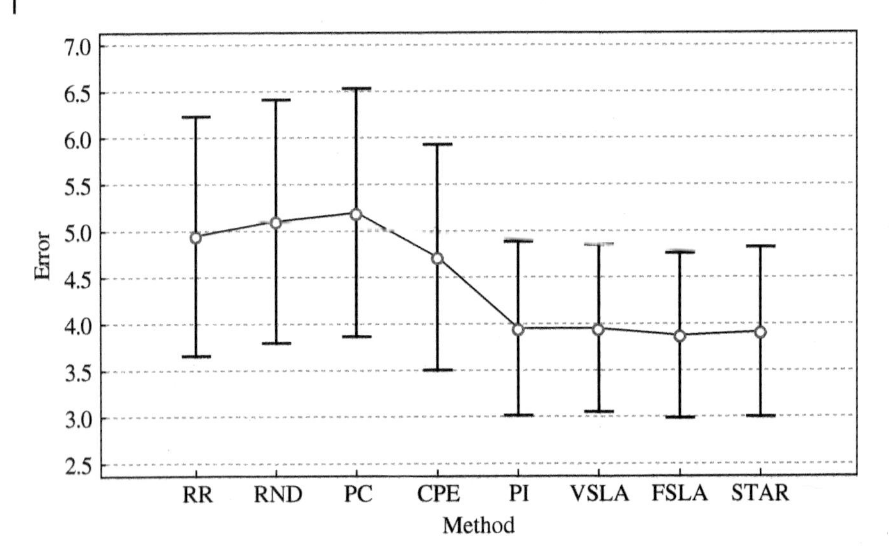

Figure 7.13 Chart for mean error values of different sub-population scheduling algorithms with 95% confidence intervals.

Table 7.7 Results of the pairwise comparisons between different sub-population scheduling algorithms based on offline error values in all MPB's problem instances.

	RR	RND	PC	CPE	PI	VSLA	FSLA	STAR	Mean Rank
RR		S+	S+	S-	S-	S-	S-	S-	6.1000
RND	S-		S+	S-	S-	S-	S-	S-	6.9833
PC	S-	S-		S-	S-	S-	S-	S-	7.7000
CPE	S+	S+	S+		S-	S-	S-	S-	5.2167
PI	S+	S+	S+	S+		+	–	–	2.6000
VSLA	S+	S+	S+	S+	–		–	–	2.8667
FSLA	S+	S+	S+	S+	+	+		+	2.2667
STAR	S+	S+	S+	S+	+	+	–		2.2667

7.7.3.4 Experiment 4: Comparison with Other Methods from the Literature

In this experiment, we study the performance of different DynDE variants in comparison with several well-known algorithms taken from the literature. The algorithms considered include: mQSO [23], SPSO [51], CESO [46], AmQSO [56], CellularPSO [25], ESCA [47], HmSO [26], CellularDE [30], and DynPopDE [59]. Table 7.8 shows the numerical results of different algorithms on MPB scenario 2.

Table 7.8 Comparison of offline error of different algorithms for dynamic environments modeled with MPB (Scenario 2).

Method	Result
mQSO	1.77 ± 0.05
SPSO	2.51 ± 0.09
CESO	1.38 ± 0.02
AmQSO	1.64 ± 0.05
Cellular PSO	1.93 ± 0.08
ESCA	1.53 ± 0.01
HmSO	1.47 ± 0.04
CellularDE	1.64 ± 0.03
DynPopDE	1.39 ± 0.07
DynDE-PI	1.47 ± 0.08
DynDE-VSLA	1.32 ± 0.06
DynDE-FSLA	1.42 ± 0.06
DynDE-STAR	**1.31 ± 0.04**

As can be observed in Table 7.8, DynDE-STAR is the best-performing method. Moreover, despite their simplicity, DynDE variants with SPS display a better or competitive performance compared with other representative multi-population approaches.

From the analysis of the above experiments, we can reach to the conclusion that the SPS algorithm is a fairly simple scheme that can effectively improve the performance of multi-population methods for DOPs.

7.8 Conclusions

This chapter investigates the application of scheduling to improve the performance of evolutionary multi-population approaches in dynamic environments. Eight SPS schemes are integrated into DynDE to address DOPs. Based on dynamic test environments generated by MPB, an experimental study was carried out to test the SPS methods. From the experimental results, the following conclusions can be drawn. First, sophisticated SPS methods are efficient to improve the performance multi-population algorithms in dynamic environments. Second, the performance of DynDE with SPS is directly affected by the selected scheduling strategy and feedback parameters.

The work studied in this chapter can be further extended in several directions. Developing other mechanisms for SPS would be an interesting direction for future work. Studying the effect of different feedback parameters in adaptive or hybrid SPS is also interesting. Evaluating the SPS for noisy dynamic environments is another future work. Finally, it is also very valuable to apply the proposed approach to several real-world dynamic problems.

Acknowledgments

We would like to thank the editors for the invitation to participate in this book. The first author would like to express gratitude to his beloved wife Mahsa Azadeh for being an endless source of happiness, inspiration, support, and peace.

References

1. Shirali, A., Kazemi Kordestani, J., and Meybodi, M.R. (2018). Self-adaptive multi-population genetic algorithms for dynamic resource allocation in shared hosting platforms. *Genet. Program Evolvable Mach.* 19 (4): 505–534.
2. Mavrovouniotis, M. and Yang, S. (2013). Ant colony optimization with immigrants schemes for the dynamic travelling salesman problem with traffic factors. *Appl. Soft Comput.* 13 (10): 4023–4037.
3. Yang, S., Cheng, H., and Wang, F. (2010). Genetic algorithms with immigrants and memory schemes for dynamic shortest path routing problems in mobile ad hoc networks. *IEEE Trans. Syst. Man Cybern. Part C Appl. Rev.* 40 (1): 52–63.
4. Mack, Y., Goel, T., Shyy, W., and Haftka, R. (2007). Surrogate model-based optimization framework: a case study in aerospace design. In: *Evolutionary Computation in Dynamic and Uncertain Environments* (eds. S. Yang, Y.-S. Ong and Y. Jin), 323–342. Berlin, Heidelberg: Springer.
5. Michalewicz, Z., Schmidt, M., Michalewicz, M., and Chiriac, C. (2007). Adaptive business intelligence: three case studies. In: *Evolutionary Computation in Dynamic and Uncertain Environments* (eds. S. Yang, Y.-S. Ong and Y. Jin), 179–196. Berlin, Heidelberg: Springer.
6. Pillac, V., Gendreau, M., Guéret, C., and Medaglia, A.L. (2013). A review of dynamic vehicle routing problems. *Eur. J. Oper. Res.* 225 (1): 1–11.
7. Hossain, M.A. and Ferdous, I. (2015). Autonomous robot path planning in dynamic environment using a new optimization technique inspired by bacterial foraging technique. *Robot. Auton. Syst.* 64: 137–141.
8. Tang, W.J., Li, M.S., Wu, Q.H., and Saunders, J.R. (2008). Bacterial foraging algorithm for optimal power flow in dynamic environments. *IEEE Trans. Circuits Syst. Regul. Pap.* 55 (8): 2433–2442.

9 Sesum, V.-C., and Kuhn, E. (2010). Applying swarm intelligence algorithms for dynamic load balancing to a cloud based call center. 2010 Fourth IEEE Int. Conf. Self-Adapt. Self-Organ. Syst., 255–256.
10 Liu, L. and Ranji Ranjithan, S. (2010). An adaptive optimization technique for dynamic environments. *Eng. Appl. Artif. Intell.* 23 (5): 772–779.
11 Nguyen, T.T., Yang, S., and Branke, J. (2012). Evolutionary dynamic optimization: a survey of the state of the art. *Swarm Evol. Comput.* 6: 1–24.
12 Hu, X. and Eberhart, R.C. (2002). Adaptive particle swarm optimization: detection and response to dynamic systems. *Proc. IEEE Congr. Evol. Comput.* 2: 1666–1670.
13 Vavak, F., Jukes, K.A., and Fogarty, T.C. (1998). Performance of a genetic algorithm with variable local search range relative to frequency of the environmental changes. Genet. Program., 22–25.
14 Vavak, F., Jukes, K., and Fogarty, T.C. (1997). Learning the local search range for genetic optimisation in nonstationary environments. 1997 IEEE Int. Conf. On Evol. Comput., 355–360.
15 Janson, S. and Middendorf, M. (2006). A hierarchical particle swarm optimizer for noisy and dynamic environments. *Genet. Program Evolvable Mach.* 7 (4): 329–354.
16 Mori, N., Kita, H., and Nishikawa, Y. (2001). Adaptation to changing environments by means of the memory based thermodynamical genetic algorithm. *Trans. Inst. Syst. Control Inf. Eng.* 14: 33–41.
17 Branke, J. (1999). Memory enhanced evolutionary algorithms for changing optimization problems. *Proc. Congr. Evol. Comput.*, 3: 1875–1882.
18 Wang, H., Wang, D., and Yang, S. (2007). Triggered memory-based swarm optimization in dynamic environments. In: *Applications of Evolutionary Computing* (ed. M. Giacobini), 637–646. Berlin, Heidelberg: Springer.
19 Hatzakis, I. and Wallace, D. (2006). Dynamic multi-objective optimization with evolutionary algorithms: a forward-looking approach. Proc. Genet. Evol. Comput. Conf., 1201–1208.
20 Rossi, C., Abderrahim, M., and Díaz, J.C. (2008). Tracking moving optima using kalman-based predictions. *Evol. Comput.* 16 (1): 1–30.
21 Grefenstette, J.J. (1999). Evolvability in dynamic fitness landscapes: a genetic algorithm approach. *Proc. IEEE Congr. Evol. Comput.*, 3: 1–2038.
22 Novoa-Hernández, P., Corona, C.C., and Pelta, D.A. (2016). Self-adaptation in dynamic environments – a survey and open issues. *Int. J. Bio-Inspired Comput.* 8 (1): 1–13.
23 Blackwell, T.M. and Branke, J. (2006). Multiswarms, exclusion, and anti-convergence in dynamic environments. *IEEE Trans. Evol. Comput.* 10 (4): 459–472.
24 Hashemi, A.B. and Meybodi, M.R. (2009). A multi-role cellular PSO for dynamic environments. Proc. 14th Int. CSI Comput. Conf., 412–417.

25 Hashemi, A.B. and Meybodi, M.R. (2009). Cellular PSO: A PSO for dynamic environments. 4th International Symposium on Intelligence Computation and Applications, 422–433.

26 Kamosi, M., Hashemi, A.B., and Meybodi, M.R. (2010). A hibernating multi-swarm optimization algorithm for dynamic environments. Proc. Second World Congr. Nat. Biol. Inspired Comput., 363–369.

27 Kamosi, M., Hashemi, A.B., and Meybodi, M.R. (2010). A new particle swarm optimization algorithm for dynamic environments. Proceedings of the First International Conference on Swarm, Evolutionary, and Memetic Computing, 129–138.

28 Kordestani, J.K., Firouzjaee, H.A., and Reza Meybodi, M. (2018). An adaptive bi-flight cuckoo search with variable nests for continuous dynamic optimization problems. *Appl. Intell.* 48 (1): 97–117.

29 Kordestani, J.K., Ranginkaman, A.E., Meybodi, M.R., and Novoa-Hernández, P. (2019). A novel framework for improving multi-population algorithms for dynamic optimization problems: a scheduling approach. *Swarm Evol. Comput.* 44: 788–805.

30 Noroozi, V., Hashemi, A.B., and Meybodi, M.R. (2011). CellularDE: A cellular based differential evolution for dynamic optimization problems. Proceedings of the 10th International Conference on Adaptive and Natural Computing Algorithms, 340–349.

31 Noroozi, V., Hashemi, A.B., and Meybodi, M.R. (2012). Alpinist CellularDE: A cellular based optimization algorithm for dynamic environments. Proc. Genet. Evol. Comput. Conf., 1519–1520.

32 Sharifi, A., Noroozi, V., Bashiri, M. et al. (2012). Two phased cellular PSO: A new collaborative cellular algorithm for optimization in dynamic environments. Proc. IEEE Congr. Evol. Comput., 1–8.

33 Sharifi, A., Kazemi Kordestani, J., Mahdaviani, M., and Meybodi, M.R. (2015). A novel hybrid adaptive collaborative approach based on particle swarm optimization and local search for dynamic optimization problems. *Appl. Soft Comput.* 32: 432–448.

34 Yang, S. and Li, C. (2008). Fast multi-swarm optimization for dynamic optimization problems. *Proceeding Fourth Int. Conf. Nat. Comput.*, 7: 624–628.

35 Yang, S. and Li, C. (2010). A clustering particle swarm optimizer for locating and tracking multiple optima in dynamic environments. *IEEE Trans. Evol. Comput.* 14 (6): 959–974.

36 Li, C., Nguyen, T.T., Yang, M. et al. (2015). Multi-population methods in unconstrained continuous dynamic environments: the challenges. *Inf. Sci.* 296: 95–118.

37 Blackwell, T. and Branke, J. (2004). Multi-swarm optimization in dynamic environments. Appl. Evol. Comput., 489–500.

38 Trojanowski, K. (2008). Tuning quantum multi-swarm optimization for dynamic tasks. Artif. Intell. Soft Comput. – ICAISC 2008, 499–510.

39 Trojanowski, K. (2008). Adaptive non-uniform distribution of quantum particles in mQSO. *Simulated Evolution and Learning: 7th International Conference, SEAL 2008*, Melbourne, Australia, 7–10 December. (eds. Li, X., Kirley, M., Zhang, M. et al.), Springer Berlin Heidelberg, Berlin, Heidelberg: 91–100.

40 del Amo, I.G., Pelta, D.A., González, J.R., and Novoa, P. (2009). An analysis of particle properties on a multi-swarm pso for dynamic optimization problems. Conf. Span. Assoc. Artif. Intell., 32–41.

41 Mendes, R. and Mohais, A.S. (2005). DynDE: A differential evolution for dynamic optimization problems. *Proc. IEEE Congr. Evol. Comput.*, 3: 2808–2815.

42 du Plessis, M.C. and Engelbrecht, A.P. (2012). Using competitive population evaluation in a differential evolution algorithm for dynamic environments. *Eur. J. Oper. Res.* 218 (1): 7–20.

43 Xiao, L. and Zuo, X. (2012). Multi-DEPSO: A DE and PSO based hybrid algorithm in dynamic environments. 1–7.

44 Novoa-Hernández, P., Corona, C.C., and Pelta, D.A. (2011). Efficient multi-swarm PSO algorithms for dynamic environments. *Memetic Comput.* 3 (3): 163.

45 du Plessis, M.C. and Engelbrecht, A.P. (2008). Improved differential evolution for dynamic optimization problems. 2008 IEEE Congr. Evol. Comput. IEEE World Congr. Comput. Intell., 229–234.

46 Lung, R.I. and Dumitrescu, D. (2007). A new collaborative evolutionary-swarm optimization technique. Proc. 9th Annu. Conf. Companion Genet. Evol. Comput., 2817–2820.

47 Lung, R.I. and Dumitrescu, D. (2009). Evolutionary swarm cooperative optimization in dynamic environments. *Nat. Comput.* 9 (1): 83–94.

48 Kordestani, J.K., Rezvanian, A., and Meybodi, M.R. (2014). CDEPSO: a bi-population hybrid approach for dynamic optimization problems. *Appl. Intell.* 40 (4): 682–694.

49 Branke, J., Kaussler, T., Smidt, C., and Schmeck, H. (2000). A multi-population approach to dynamic optimization problems. In: *Evolutionary Design and Manufacture* (ed. I.C. Parmee), 299–307. London: Springer.

50 Yazdani, D., Nasiri, B., Sepas-Moghaddam, A., and Meybodi, M.R. (2013). A novel multi-swarm algorithm for optimization in dynamic environments based on particle swarm optimization. *Appl. Soft Comput.* 13 (4): 2144–2158.

51 Parrott, D. and Li, X. (2006). Locating and tracking multiple dynamic optima by a particle swarm model using speciation. *IEEE Trans. Evol. Comput.* 10 (4): 440–458.

52 Bird, S. and Li, X. (2007). Using regression to improve local convergence. 2007 IEEE Congr. Evol. Comput., 592–599.

53 Li, C. and Yang, S. (2012). A general framework of multipopulation methods with clustering in undetectable dynamic environments. *IEEE Trans. Evol. Comput.* 16 (4): 556–577.

54 Nickabadi, A., Ebadzadeh, M.M., and Safabakhsh, R. (2012). A competitive clustering particle swarm optimizer for dynamic optimization problems. *Swarm Intell.* 6 (3): 177–206.

55 Halder, U., Das, S., and Maity, D. (2013). A cluster-based differential evolution algorithm with external archive for optimization in dynamic environments. *IEEE Trans. Cybern.* 43 (3): 881–897.

56 Blackwell, T. (2007). Particle swarm optimization in dynamic environments. In: *Evolutionary Computation in Dynamic and Uncertain Environments* (eds. S. Yang, Y.-S. Ong and Y. Jin), 29–49. Berlin, Heidelberg: Springer.

57 Yazdani, D., Akbarzadeh-Totonchi, M.R., Nasiri, B., and Meybodi, M.R. (2012). A new artificial fish swarm algorithm for dynamic optimization problems. IEEE Congr. Evol. Comput., 1–8.

58 Fouladgar, N. and Lotfi, S. (2015). A novel approach for optimization in dynamic environments based on modified cuckoo search algorithm. *Soft. Comput.* 7: 2889–2903.

59 du Plessis, M.C. and Engelbrecht, A.P. (2013). Differential evolution for dynamic environments with unknown numbers of optima. *J. Glob. Optim.* 55 (1): 73–99.

60 Li, C., Nguyen, T.T., Yang, M. et al. (2016). An adaptive multipopulation framework for locating and tracking multiple optima. *IEEE Trans. Evol. Comput.* 20 (4): 590–605.

61 Li, C., Yang, S., Nguyen, T. et al. (2008). Benchmark generator for CEC 2009 competition on dynamic optimization. Technical Report, Department of Computer Science, University of Leicester.

62 Narendra, K.S. and Thathachar, M.A. (1974). Learning automata-a survey. *IEEE Trans. Syst. Man Cybern.* 4 (4): 323–334.

63 Narendra, K.S. and Thathachar, M.A. (2012). *Learning Automata: An Introduction*. Courier Corporation.

64 Abedi Firouzjaee, H., Kordestani, J.K., and Meybodi, M.R. (2017). Cuckoo search with composite flight operator for numerical optimization problems and its application in tunnelling. *Eng. Optim.* 49 (4): 597–616.

65 Kordestani, J.K., Ahmadi, A., and Meybodi, M.R. (2014). An improved differential evolution algorithm using learning automata and population topologies. *Appl. Intell.* 41 (4): 1150–1169.

66 Mahdaviani, M., Kordestani, J.K., Rezvanian, A., and Meybodi, M.R. (2015). LADE: learning automata based differential evolution. *Int. J. Artif. Intell. Tools* 24 (6): 1550023.

67 Rezvanian, A., Saghiri, A.M., Vahidipour, S.M. et al. (2018). *Recent Advances in Learning Automata*. Berlin: Springer.

68 Economides, A.A. and Kehagias, A. (2002). The STAR automaton: expediency and optimality properties. *IEEE Trans. Syst. Man Cybern. B Cybern.* 32 (6): 723–737.

69 Storn, R. and Price, K. (1997). Differential evolution - a simple and efficient heuristic for global optimization over continuous spaces. *J. Glob. Optim.* 11 (4): 341–359.

70 Storn, R. and Price, K. (1995). Differential evolution-a simple and efficient adaptive scheme for global optimization over continuous spaces.

71 Das, S. and Suganthan, P.N. (2011). Differential evolution: a survey of the state-of-the-art. *IEEE Trans. Evol. Comput.* 15 (1): 4–31.

72 Branke, J. (2002). *Evolutionary Optimization in Dynamic Environments*. Springer Science and Business Media.

73 Kordestani, J.K., Rezvanian, A., and Meybodi, M.R. (2017). New measures for comparing optimization algorithms on dynamic optimization problems. *Nat. Comput.* 18 (4): 705–720.

74 Cruz, C., González, J.R., and Pelta, D.A. (2011). Optimization in dynamic environments: a survey on problems, methods and measures. *Soft. Comput.* 15 (7): 1427–1448.

75 Branke, J. and Schmeck, H. (2003). Designing evolutionary algorithms for dynamic optimization problems. In: *Advances in Evolutionary Computing: Theory and Applications* (eds. A. Ghosh and S. Shigeyoshi), 239–262. Berlin, Heidelberg: Springer.

8

Task Scheduling in Cloud Environments

A Survey of Population-Based Evolutionary Algorithms

Fahimeh Ramezani[1], *Mohsen Naderpour*[1], *Javid Taheri*[2], *Jack Romanous*[1], *and Albert Y. Zomaya*[3]

[1] *Centre for Artificial Intelligence, Faculty of Engineering and Information Technology, University of Technology Sydney (UTS), Sydney, NSW, Australia*
[2] *Department of Computer Science, Karlstad University, Karlstad, Sweden*
[3] *Centre for Distributed and High Performance Computing, School of Computer Science, University of Sydney, Sydney, NSW, Australia*

8.1 Introduction

Cloud and grid are two distributed, i.e. heterogeneous, computing environments that facilitate large-scale computing. Grid computing shares dispersed, heterogeneous pools of hosts, servers, storage systems, data, networks, and sensors within one integrated system. One of the main strategies of grid computing is to distribute pieces of a program over several computers [1]. Cloud computing is also a large-scale distributed computing paradigm, but these systems are driven by economies of scale. The idea is to serve collections of abstract, virtual, and dynamically scalable power and storage resources as managed platforms and services on demand to external customers over the Internet under a service level agreement (SLA). The basic principle is to relocate computing from a local computer to an online network of devices [1].

There are two main differences between the architectures and principles of cloud and grid computing that need to be considered when designing a task scheduling system. These are pricing schemes and the large size of the resource pools. For example, in a grid, the cost of computation is based on the volume of resources used, such as the number of complete central processing unit (CPU) cycles. However, the running time of the underlying hosting instances may affect any cost calculations based on pay-per-use pricing strategies. In addition, the size

Evolutionary Computation in Scheduling, First Edition. Edited by Amir H. Gandomi, Ali Emrouznejad, Mo M. Jamshidi, Kalyanmoy Deb, and Iman Rahimi.

of the resource pool in a grid is usually limited, whereas a cloud is usually assumed to have an infinite amount of resources [2].

While typically more expansive than grid systems, clouds do not have an unlimited amount of resources despite what customers might believe. These resources can become strained with fluctuating customer needs, which means cloud providers must balance their resource load and utilization. Usually, this balancing involves automatically assigning the resources needed for the current service requests in the most optimal way to meet SLA criteria while minimizing costs. This study focuses on reviewing the existing task scheduling methods for cloud environments.

A cloud environment dynamically receives many jobs from the users of its applications in every millisecond. Individual jobs include several tasks. These tasks accrue in numerous queues and are then transferred to the task schedulers. The task schedulers are responsible for distributing these tasks to virtual machines (VMs) to execute. In turn, the VMs are allocated to physical machines (PM) (see Figure 8.1), each having a different number of virtual CPUs and different amounts of memory. To achieve optimal resource utilization in a cloud environment, optimization measures are applied by the task schedulers to distribute tasks among the VMs. The scheduling process is repeated dynamically to schedule every set of tasks among the VMs as they

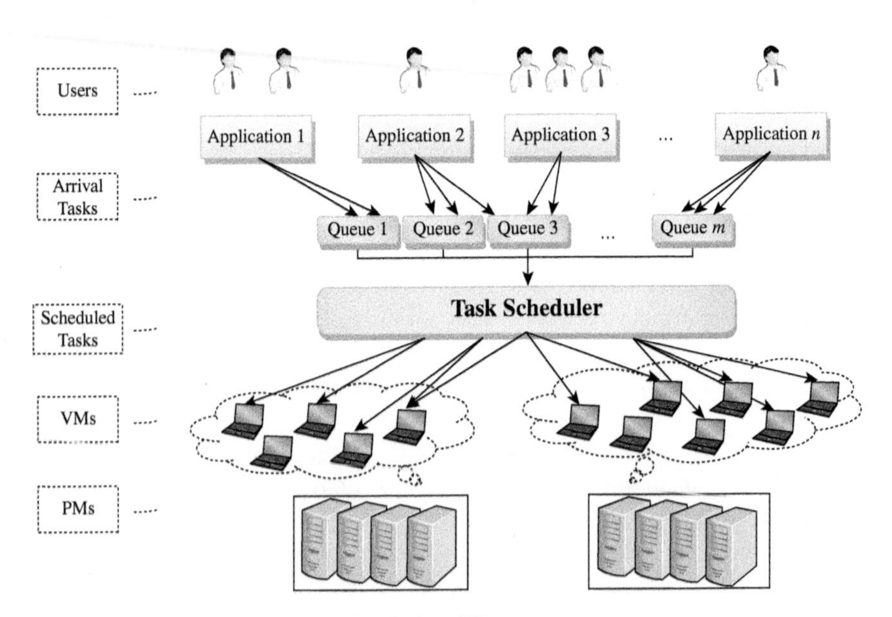

Figure 8.1 Cloud objects and their relations [3].

arrive. Given that each user arbitrarily sends tasks to the cloud environment, the amount and type of tasks in the queues may be significantly different from one schedule batch to the next.

There are two different types of jobs within a cloud environment: computationally intensive jobs and data-intensive jobs. The goal when scheduling computationally intensive jobs is to allocate all jobs among the computational nodes (CNs) or VMs to minimize the overall makespan (i.e. the time to transfer the data files plus execute the job) across the entire system. With computationally intensive jobs, the total transfer time for all the data files (the input and output files for each task of the job) is usually relatively negligible compared to the execution of the job. The speed and number of available resources in different CNs/VMs and the network capacity between the CNs/VMs and SNs are common considerations within these systems. With data-intensive jobs, the time to transfer all data files is generally assumed to take significantly longer than executing the job. Therefore, jobs will demand less time to download the related data files for execution and thus the execution time of the system is lessened. A typical consideration in such allocations is the capacity of the interconnected network links [4].

In complex systems such as this, there are three main phases of scheduling: resource discovery, matchmaking, and job execution. In the resource discovery phase, the schedulers conduct a global search to generate a list of all available resources in the system along with their limitations and history profiles. In the matchmaking phase, the schedulers try to determine the best choices for executing jobs and replicating the data files. The capacity of each CN/VM/SN and the quality of the network connecting them are some of the basic characteristics that schedulers need to account for to perform this phase. In the final phase of a job's execution, the schedulers issue commands to the CNs to execute jobs and to the SNs to replicate the data files [4].

The problem of task scheduling in a distributed computing system is an NP-hard optimization problem. It affects quality of service (QoS) because task scheduling is the process that optimizes QoS criteria like service costs and response times. However, using a heuristic algorithm ensures that the scheduling algorithm keeps to an acceptable runtime as it immensely reduces the complexity of a search space.

The objective of this chapter is to provide a systematic literature review on the existing task scheduling methods that rely on evolutionary computation algorithms. The structure is as follows: a brief introduction to cloud environments and the problem is provided in Section 8.2. The research methodology is described in Section 8.3. Section 8.4 contains the results. Section 8.5 discusses the research gaps and future directions. Section 8.6 concludes the chapter.

8.2 Physical Topology of Cloud

Cloud architectures focus on important difficulties surrounding large-scale data processing. In traditional data processing, it can be difficult to mobilize as many machines as an application requires or to provide a machine when one is needed. Further, it can be hard to distribute and coordinate a large-scale processing job on different machines, provide backup machines for recovery in case of machine failure, or auto-up or downscale a running machine based on dynamic workloads. Moreover, the problems do not end once the job is complete, as the machines either need to be disposed of or redeployed. Cloud architectures were designed to resolve such difficulties [5]. Figure 8.2 illustrates the physical topology of a cloud cluster that consists of physical and logical resources such as: physical hosts (i.e. PMs), data storage (DS), and VMs. Cloud environments also comprise middleware and other components, such as a hypervisor, schedulers, jobs, and data files.

Cloud computing provides a novel infrastructure that focuses on commercial resource provision by incorporating virtualization [6]. Virtualization techniques can consolidate the hardware resources of a PM, such as CPUs or memory, and other physical computing resources, such as storage and networking, into pools of resources within a cloud cluster that can be made available dynamically and flexibly and with several operating systems [7]. In essence, virtualization provides a virtual environment for every operating system allocated to a PM. So, an operating system within a virtual environment can be seen as an independent VM [8]. The virtualization layer schedules and allocates the physical resources, and makes

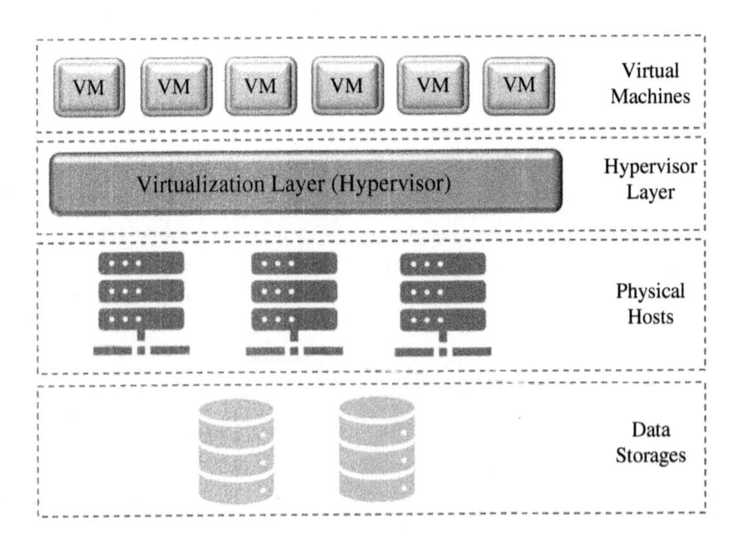

Figure 8.2 Physical topology of a cloud cluster.

each VM think that it totally owns all the physical resources of the underlying hardware [9]. This gives rise to an integral characteristic of a cloud environment: elasticity. Elasticity is the ability to rapidly provide and release, i.e. increase and decrease, the number of physical resources allocated to any single VM [10].

The next section explains physical and logical cloud resources in more detail, as well as their middleware and components.

8.2.1 Cloud Resources

Resources are any physical or virtual components with limited availability in a computer system. There are two types of resources: logical and physical [11]. This section presents the definitions of the main types of resources in a cloud system.

8.2.1.1 Physical Resources

The lowest levels of the cloud stack contain physical infrastructure in the form of clusters, data centers, and desktop computers. The IT infrastructure and virtualized environments are deployed and managed in layers above these foundations. Data centers handle deployment, which might include hundreds or possibly thousands of machines. This level provides the "horsepower" [12].

The physical resources within this infrastructure typically include processors, memory, and peripheral devices. Physical resources can vary dramatically from one computer to the next. Typically, a mainframe system has numerous parallel processors, hundreds of disks, millions of bytes of memory, large numbers of terminals, tapes, and other special-purpose peripheral devices, which is further connected to a global network with thousands of other similar mainframes [11].

- PMs, also known as physical hosts, are a collection of CNs, which are usually configured in clusters. Each CN comprises numerous homogeneous processors that have identical processing speeds and numbers of processors. The processing speed of each CN is a relative number that reflects its processing speed compared to other CNs in the system. The number of processors in a CN determine its ability to execute a job with a certain degree of parallelism in a non-preemptive fashion, i.e. whether it can execute a job without interrupting another job [4].
- CPUs perform most of the processing inside a computer, and one of the prime concerns in cloud systems is high CPU utilization. CPU utilization is the amount of work handled by the CPU, which varies depending on the number and type of computing tasks being performed. Some tasks are CPU-intensive; others are not. Hence, better utilization means more data can be processed with the same number of CPUs [11].
- Memory refers to the devices that store data during processing. Today, memory management is no longer performed using static methods. Rather, as cloud

environments are dynamic and volatile, memory allocation also needs to be dynamic. Hence, most cloud computing architectures are trending toward clustered memory structures in the form of virtual entities [11].

- DS means saving data in a remote storage system maintained by a third party, where the computers and databases are connected via the internet. Cloud storage systems usually comprise hundreds of data servers and, because computers occasionally require maintenance or repair, it is critical to store the same information on multiple machines, i.e. a backup. This is called redundancy. Without a level of redundancy, cloud providers could not guarantee their clients would be able to access their information at any time. Most providers store backup data on servers with a different source of power so clients can access their data even if a power supply fails [11].
- Network elements include a vast array of different components, such as hubs, bridges, switches, etc. Managing many network elements can lead to unsustainable administration costs unless typical system management tasks can be managed with automated methods [11].

8.2.1.2 Logical Resources

Logical resources are system abstractions that have temporary control over physical resources. They can support the development of applications and well-organized communication protocols. Logical resources are important to cloud environments for several reasons [11]:

- Operating systems provide users with mechanisms and policies to manage physical resources (hardware), files, devices, performance, security, faults, etc. which, in turn, help users make the best use of the resources available [11].
- VMs are software computers that, just like physical computers, run operating systems and applications. Each VM comprises its own virtual (i.e. software-based) hardware, including a virtual CPU, memory, hard disk, and network interface card that is decoupled from any underlying physical hardware.
- Energy is also a logical resource. Using less energy is obviously better, and the main technique used to decrease energy consumption is to concentrate the workload on the smallest number of physical nodes while switching off idle nodes. However, this approach does require a trade-off between power and performance as performance can degrade when workloads are consolidated [11].
- Network throughput, also known as network bandwidth, is a measure of the highest amount of data that can be passed in bits per second. A common technique for measuring throughput is to transfer a "large" file between two systems while measuring the time required for the file transfer to complete. Generally, higher throughput means a network is more efficient. Bandwidth management protocols prevent congestion largely by accepting or refusing a

new arrival cell. In a cloud environment, problems with bandwidth allocation are usually resolved by integrating higher link capacities between different types of services to prevent a network overload [11].
- Delay of a second or even a millisecond lag in delay-sensitive traffic can make a notable difference to the QoS for an end user. Therefore, when scaling data centers up or down, cloud service providers must make highly accurate decisions that take a range of utility criteria into account, e.g. the delay in setting up virtual resources, in migrating existing processes, in allocating or reallocating resources, etc. [11].

8.2.2 Resource Utilization

Considering the continued growth in cloud computing and that resource utilization directly impacts costs, resource management techniques enable cloud providers to consolidate workloads to achieve optimal resource utilization while adhering to SLAs at minimum cost [13]. If X is the total hardware capacity of the server, and Y is the current hardware capacity used, then:

- the workload is under-utilized if $\frac{X}{Y} < 1$
- the workload is over-utilized if $\frac{X}{Y} > 1$
- resource utilization is optimal when $X \approx Y$.

8.2.3 Cloud Middleware and Components

This section presents the main middleware and components of a cloud cluster.

8.2.3.1 Hypervisors

In computing, a hypervisor – also known as virtual machine monitor (VMM) – is one of many hardware virtualization techniques that allows multiple operating systems, termed guests, to run simultaneously on a host computer. It is termed a hypervisor because, conceptually, it operates at a level higher than a supervisory program [14]. Hypervisors are installed on physical hardware in a virtualized data center as a platform for running VMs. They are responsible for executing guest operating systems, VMs, and consolidating computing resources. Additionally, the hypervisor dynamically allocates physical hardware resources to the VMs as needed to support their operations. In this way, multiple instances of a variety of VMs can share virtualized hardware resources. Hypervisors allow VMs to operate with some amount of independence from the underlying physical hardware. For example, a hypervisor can move a VM from one physical host to the next, or its virtual disks can be moved from one type of storage to another without impacting the function of the VM [7].

8.2.3.2 Schedulers

Schedulers are independent entities in a distributed cloud system that receive jobs and data files from users and then schedule, assign, or replicate them to destination nodes for processing or storage (i.e. CNs/VMs/SNs). Schedulers are the means by which systems can execute multiple parallel jobs simultaneously, given available nodes, while other jobs are queued until nodes become available. Job schedulers manage queues of waiting jobs and coordinate node allocations [15]. They are the decision makers of the system. They decide where each job and data file should be executed, stored, or replicated. The main goals of a scheduler are to optimize the throughput of a system (number of jobs completed per time unit), provide response time guarantees (finish a job by a deadline), and keep resource utilization high [15]. Every individual scheduler is able to connect to either all CNs/VMs/SNs or only a subset of them. They can be either sub-entities of a CN/SN or an individual broker that accepts jobs and data files from users [4].

8.2.4 Jobs and Tasks

A job is generated by users and given to schedulers to be executed by processing nodes. Each job is made of many dependent tasks known as a directed acyclic graph (DAG). DAGs contain specific characteristics about the tasks, such as execution time, the number of required processors, and so on. Execution time governs the seconds needed for the task to be completed using the slowest CPU in the system. Note that the actual execution time for a task can be significantly decreased if it is allocated to a faster CPU. The number of processors determines the degree of parallelism of a task [4].

In a DAG, jobs are represented as different shapes, which correlate to the various classes of operations. The specifications for these representations include width, height, number of processors, time to execute, shape, and a list of the data files needed. The width is the highest number of tasks able to run concurrently within a job. The height is the number of stages or levels in a job. The number of processors is the largest number of processors needed by any one task in the job. The time to execute specifies the lowest amount of time a job can be run on the slowest CPU in the system. Lastly, the list of required data files contains the files a CPU must download prior to executing the job. The actual data required to execute individual tasks is provided either via: (i) previously existing data files listed as required data files; and/or (ii) the output of the task's immediate predecessors in the DAG. These predecessors could either be local files, temporary files, or inter-processing messages. Job shapes can be series-parallel, homogeneous-parallel, heterogeneous-parallel, and single-task. Figure 8.3 illustrates the different job shapes [4].

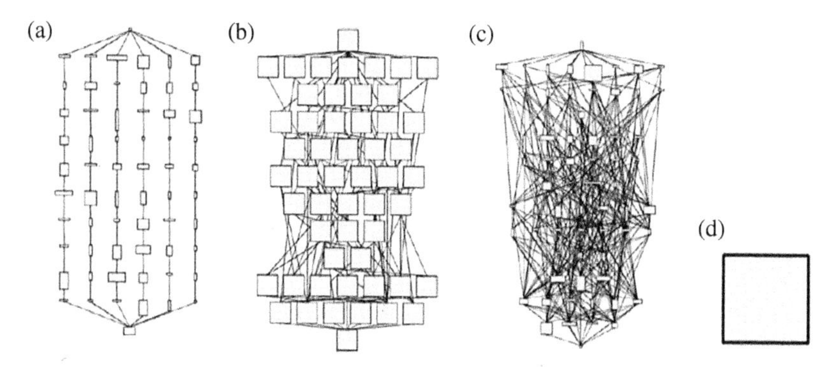

Figure 8.3 Jobs' shapes: (a) series-parallel (b) homogeneous-parallel (c) heterogeneous-parallel (d) single-task [4].

8.2.5 Virtual Cluster

A virtual cluster is a collection of VMs configured in a way that makes them act as though they were a traditional high-performance computing (HPC) cluster. This usually involves installing and configuring job management software, which includes a batch scheduler and a shared storage system (e.g. a network/distributed file system) [16].

8.2.6 Multi-Objective Task Scheduling

A multi-objective optimization problem (MOOP) is a problem that has several conflicting objectives that need to be optimized simultaneously:

$$Minimize\, F(x) = (f_1(x), f_2(x), \ldots, f_m(x))^T$$

where x ∈ X and X is the decision space.

The objectives in a MOOP often conflict with one another. Hence, Pareto dominance is used to perform a comparison between the suggested solutions as follows:

For u, v ∈ X, u dominates v if and only if,

$$\forall i : f_i(u) \leq f_i(v) \wedge \exists j : f_j(u) < f_j(v)$$

If solution x* is not dominated by any other solution, then it is considered Pareto optimal. The term "Pareto front" describes the set containing every Pareto optimal solution within the objective space [2].

In this study, most of the reviewed papers considered task scheduling as a MOOP that optimizes some of the conflicting objectives, e.g. service cost, power consumption, makespan, response time, flow time, execution time, data/task transfer time, task queue length, and resource utilization.

8.3 Research Methodology

The research methodology is based on PRISMA, i.e. the preferred reporting items for systematic reviews and meta-analyses. PRISMA consists of a 27-item checklist and a four-phase flow diagram that focuses on randomized trials. However, it also can be used as a basis for conducting systematic reviews of other types of research [17]. Figure 8.4 illustrates our research methodology followed by a detailed description.

- Rationale and objectives: As described in the introduction, the rationale is task scheduling within cloud computing. The objective is reviewing state-of-the-art evolutionary computation algorithms developed during the last decade.
- Eligibility criteria: To be included in this chapter, the studies must have:
  - been published in peer-reviewed journals
  - been written in English
  - been published between 2008 and 2018
  - developed an evolutionary computation-based scheduling solution.
- Studies were excluded if the full text of the study was unavailable or if it related to:
  - QoS parameter-based scheduling
  - multi-processor-based scheduling

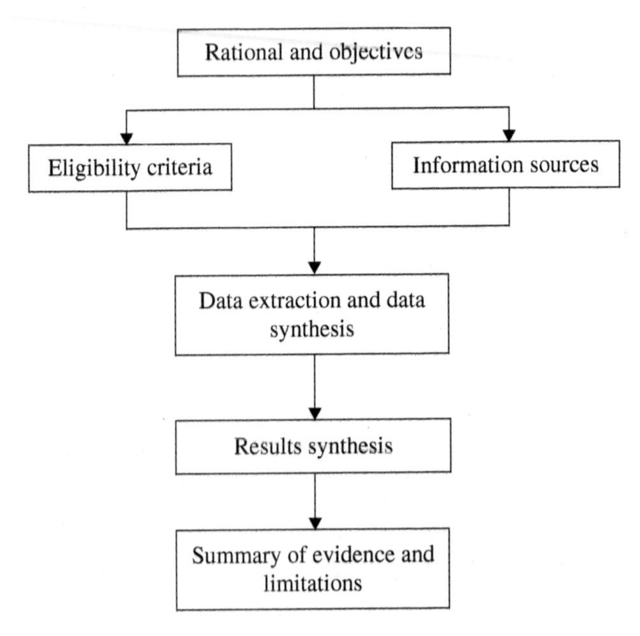

Figure 8.4 Research methodology.

  - fuzzy-based scheduling
  - clustering-based scheduling
  - another type of scheduling approach.
- Information sources: Scopus – a large dataset of publications commonly used in this field.
- Data extraction and data synthesis: Several combinations of keywords were used to search the Scopus dataset: (i) "cloud computing" and "task scheduling"; (ii) "cloud computing" and "job scheduling"; and (iii) "cloud computing" and "workflow scheduling." In addition to these keyword combinations, the study needed to have at least one evolutionary computation technique including: genetic algorithms (GA), particle swarm optimization (PSO), bee colony optimization (BCO), ant colony optimization (ACO), etc. The data extraction process is presented in Figure 8.5. As shown, 187 records were initially identified. After removing duplicated papers, 151 records remained. The titles, keywords, and abstracts of each paper were then evaluated for eligibility by the authors, resulting in a final dataset of 99 studies.
- Results synthesis: The papers were then categorized based on the objectives used and the primary type of evolutionary algorithm. Section 8.4 presents the results of the review.
- Summary of evidence and limitations: The key findings, the strength of the evidence for each key outcome, are discussed in Section 8.5 along with the limitations of this study and recommendations for future research.

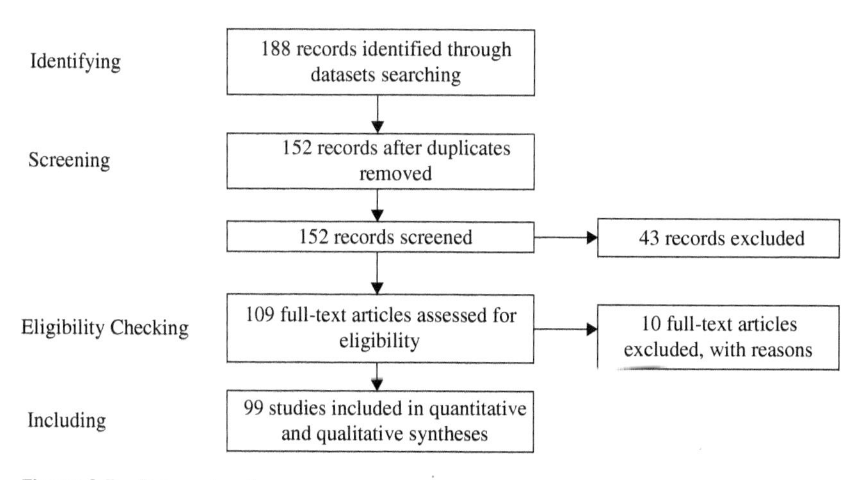

Figure 8.5 Data extraction process.

8.4 Results

This section presents the results of the research methodology categorized into sub-sections. Each sub-section focuses on reviewing the studies that have developed task/job scheduling in cloud environments by applying evolutionary algorithms, such as PSO, GA, ACO, and BCO. The review outlines the basis of how evolutionary algorithms have been improved or used in combination with other algorithms to optimize multi-objective task scheduling problems, as well as the objective functions to be optimized in each scheduling model.

8.4.1 Particle Swarm Optimization

Guo et al. [18] proposed a multi-objective task scheduling model to lessen task execution time and the cost of using the PSO algorithm. Zhan and Huo [19] developed "Improved PSO" (IPSO) by combining PSO and simulated annealing (SA). They leveraged SA's fast convergence time in each iteration of PSO to: increase the convergence rate; improve the efficiency of the scheduling algorithm; and prevent PSO from falling into a local optima. IPSO reduces the average task running time and increases the ratio of resource utilization compared to ACO, GA, and SA.

Taheri et al. [20] considered the data files required for jobs from public or private clouds and proposed a bi-objective job scheduling optimization model using PSO (PSO-ParFnt). Their approach minimizes job execution as well as data file transfer times. They also revised the generic version of PSO in order to create collections of targeted swarms as an alternative to a generic large swarm.

Netjinda et al. [21] developed a workflow scheduling algorithm, called PSO-VNS, by combining PSO with a variable neighborhood search (VNS) technique where the VNS refines the global best from every iteration of PSO to improve fitness convergence. They showed that their method outperforms both greedy methods and infrastructure-as-a-service (IaaS) cloud partial critical paths in reducing costs.

Ouyang et al. [22] proposed a green cloud task scheduling algorithm (GCTA) based on improved binary PSO (BPSO). This work avoids matrix operations (that represent the tasks assigned to VMs) by using a pipeline number for VMs and redefining the position and velocity of the particles. Their simulation results show that the GCTA requires less execution time and consumes fewer resources than sequential scheduling.

Ramezani et al. [23] proposed a PSO-based task scheduling method to minimize task execution and transfer times. The method sits within their task-based load balancing system, which balances the system load by only transferring extra tasks from an overloaded VM rather than migrating the entire overloaded VM.

Yang et al. [24] formulated trust relationships between tasks and resource sets coupled with a trust-based scheduling algorithm (TBHSA) that relies on a set-based PSO (S-PSO) method to select a trustworthy computation service. A set covering problem (SCP) tree search method selects the storage services. These authors show that TBHSA performs better than both PSO and ACO at reducing time and cost and enhancing security and reliability.

Ramezani et al. [3] consider four conflicting objectives, namely minimizing task transfer time, task execution cost, power consumption, and task queue length, to develop a comprehensive multi-objective optimization model using PSO (MOPSO) for task scheduling. The model is essentially an extension to the PSO and GA algorithms to support these four objective functions. Its efficiency is compared against a multi-objective GA (MOGA) using CloudSim. The results show that MOPSO is a faster and more accurate evolutionary algorithm than MOGA for solving such problems.

Alkhashai and Omara [25] proposed a new algorithm called BFPSOTS by combining Best-Fit (BF), PSO, and the Tabu-Search (TS) algorithms. They applied the BF algorithm to generate the initial population for the standard PSO algorithm rather than generating a random population. They also used the TS algorithm to improve PSO's local search. In an evaluation using CloudSim, the results show that BFPSOTS outperforms the standard PSO algorithm in reducing execution times for tasks, cost, and resource utilization.

Other studies focus on hybrid metaheuristic methods for task scheduling [26]. He et al. [27] combined PSO, ACO, and GA to create a more efficient task scheduling method. They improved the PSO algorithm by applying a random inertia weight to generate the initial scheduling results, then used the generated results of the IPSO as the initial pheromones for the ACO algorithm to find the optimal scheduling scheme. They also incorporated an elitist strategy and a crossover operator into the GA to improve the efficiency of the ACO algorithm. The final result is a combined algorithm that can reduce the total task makespan.

Jiang et al. (2016) combined the PSO and ACO algorithms to present a new scheduling algorithm that reduces the processing time and cost of the task. Based on their evaluation results, the combined algorithm is faster than PSO and ACO at generating the scheduling results.

Kamalinia and Ghaffari [26] combined PSO and GA to present a new heterogeneous earliest finish time (HEFT) algorithm, called PSO-GA-HEFT, to reduce makespans and enhance resource efficiency in comparison to HEFT_UpRank, HEFT_DownRank, HEFT_LevelRank, and the multiple priority queues and a memetic (MPQMA) algorithm.

Li et al. [28] applied PSO to propose a security and cost-aware scheduling (SCAS) algorithm for heterogeneous tasks in scientific workflows in the cloud. The algorithm minimizes the execution cost of the total workflow while meeting a prescribed deadline and risk rate constraints.

Nirmala and Bhanu [29] proposed a metaheuristics algorithm called Catfish PSO (C-PSO) that selects the best schedule with the least makespan and execution cost.

Shao [30] proposed a scheduling algorithm for tasks based on a PSO fission mechanism that splits the particles in an appropriate place to generate more kinds of particles so as to create diversity and, hence, avoid premature convergence of the swarm. They showed that this algorithm is faster than first in first out (FIFO) and PSO, and also solves the problem of premature convergence in PSO.

Xie et al. [31] proposed a scheduling algorithm based on PSO and the shuffled frog leaping algorithm (SFLA) that considers efficiency and trustworthiness in the cloud. The proposed algorithm avoids falling into a local optimum and outperforms GA and the TDMin-Min algorithms.

Zhong et al. [32] combined PSO and a greedy algorithm, introducing a greedy PSO (G&PSO) algorithm that converges faster, has a stronger local and global search, and provides a better solution to balancing the workload between VMs.

Chen and Long [33] combined PSO with the ACO algorithm and incorporated parameter determination into the mix. The integrated algorithm can keep particles at a fitness level of a certain concentration. It guarantees diversity in the population, is faster, and performs better in fitting and cost than PSO.

Cui et al. [34] proposed a genetic algorithm based on a chaotic ant swarm (GA-CAS) algorithm, in which four operators and natural selection are applied to solve a constrained multi-objective task scheduling optimization problem that considers reliability, makespan, and flow time. They proved that GA-CAS speeds up convergence and outperforms GA, ACO, and genetic simulated annealing (GSA), which is a hybrid heuristic algorithm comprising GA and a SA algorithm.

Gabi et al. [35] proposed an algorithm called orthogonal Taguchi-based cat swarm optimization (OTB-CSO) to minimize the makespan of tasks. OTB-CSO outperforms minimum and maximum job first (Min-Max), PSO with linear descending inertia weight (PSO-LDIW) and a hybrid PSO with simulated annealing (HPSO-SA) algorithms.

Gan et al. [36] introduced a niche PSO algorithm that overcomes the shortcomings of PSO in local optimization. It also improves convergence precision. The main objective of their model is to reduce execution time and cost.

Ramezani et al. [37] developed a further four-objective PSO algorithm to embed in their bespoke multi-objective load balancing (MO-LB) system. This system avoids VM migration and balances the system load by transferring excessive workloads from a set of VMs allocated to an overloaded PM to other compatible VMs in the cluster with greater capacity. MO-LB achieves better results in reducing response time, memory use, job makespan, power consumption and the time taken for the load balancing process than the VM migration technique in vMotion.

Sadhasivam and Thangaraj [38] developed a different variant of IPSO by updating the position and velocity of each iteration. This algorithm schedules applications to cloud resources while minimizing the total costs including the communication cost between resources, the task dependency cost values, and the execution cost of computer resources.

Yuan et al. [39] proposed simulated annealing PSO (SAPSO) and a temporal task scheduling model, embedded within their proposed profit maximization algorithm (PMA). This task scheduling model dynamically schedules all arriving tasks to be executed in private and public clouds based on the temporal variation of prices in hybrid clouds. They proved that the SAPSO algorithm outperforms PSO and greatly increases the throughput and the profit of a private cloud while guaranteeing the service delay bound. Yuan et al. [40] also proposed a temporal task scheduling algorithm (TTSA) by applying a hybrid SAPSO while modeling the cost as a mixed integer linear program in each iteration.

Ben Alla et al. [41] proposed two hybrid metaheuristic algorithms by combining dynamic dispatch queues (TSDQ) with PSO with fuzzy logic (TSDQ-FLPSO) and combining TSDQ and PSO with simulated annealing (TSDQ-SAPSO). They show that the TSDQ-FLPSO algorithm performs better than PSO, SAPSO, and TSDQ-SAPSO in reducing waiting times, queue length, makespan, and cost as well as optimizing resource utilization and load balancing.

Domanal and Reddy [42] considered VM loads to determine the local best and the global best particles in each iteration of PSO and proposed a modified PSO (MPSO) algorithm to schedule a bag of tasks to a set of VMs. The main goal of the model is efficient use of the cloud's resources (VMs, CPU, and memory) while optimizing the execution cost of the tasks. MOPSO outperforms round robin, first-come-first-served (FCFS), ACO, and GA in executing the bag of tasks in heterogeneous cloud environments in terms of QoS parameters including reliability, time, and cost.

Guo et al. [43] developed an adaptive discrete PSO strategy with genetic algorithm operators (ADPSOGA) for scientific workflow scheduling. ADPSOGA is a PSO-based method that also adopts the randomly two-point crossover operator and the randomly single point mutation operator in GA. They considered minimizing the cost of execution and data transfer while meeting workflow deadlines. The performance of ADPSOGA was evaluated against multi-cloud partial critical paths with pretreatment (MCPCPP), workflow scheduling with PSO (WPSO), and PSO with GA operators (PSOGA).

Midya et al. [44] proposed a hybrid adaptive PSO (HAPSO) algorithm, which is a combination of GA and PSO to optimize resource allocation and task scheduling for task requests arriving from on road users. They showed that HAPSO converges faster than PSO and self-adaptive PSO in reducing the mean square error, task response time, and energy consumption.

A comparison study on the performance evaluation of PSO and GA on workflow scheduling was conducted by Shishido et al. [45] using a security and cost-aware workflow scheduling algorithm. Their findings indicated that GA-based algorithms significantly outperform PSO in terms of both cost-effectiveness and response time.

Tong et al. [46] proposed a heuristic algorithm called hybrid biogeography-based optimization (BBO) algorithm, which integrates a migration strategy into PSO. In their method, the flight strategy under the BBO migration structure is hybridized to accelerate search speeds, and HEFT_D is used to evaluate the task sequence. The objective function was makespan, and the experiments show that, compared to several classical heuristic algorithms, the new hybrid migrating BBO has the advantage in terms of global search ability, fast convergence rates, and high-quality solutions.

Yuan et al. [47] developed a time-aware task scheduling algorithm that considers temporal variations and schedules all admitted tasks that meet the delay bounds. In each iteration, their algorithm solves the formulated profit maximization problem with a hybrid chaotic PSO based on SA. Compared to several other scheduling algorithms, the model increases profit and throughput without violating the delay constraints of all admitted tasks. Subsequently, they used a hybrid optimization algorithm called GSA-based PSO [47]. In the hybrid model, the genetic operators in GA increase the diversity of particles and the efficiency of global searches.

Dordaie and Navimipour [48] combined PSO with the hill climbing algorithm to optimize the scheduling workflow makespan. The algorithm initializes the population randomly using PSO. Each particle is then evaluated and ranked with makespan by HEFT processor mapping method and the hill climbing algorithm is finally applied to some selected particles. The proposed algorithm is repeated until the termination condition is satisfied.

Table 8.1 summarizes the PSO-based methods and reports the implemented improvements, objective functions, algorithms outperformed, and evaluation tool.

8.4.2 Genetic Algorithms

Tayal [51] presented a fuzzy GA-based optimization approach for improving the accuracy of GA's results with job scheduling. Scheduling decisions are made via the evaluation of the entire group of tasks in the job queue. Juhnke et al. [52] presented a multi-objective scheduling algorithm intended for cloud-based workflow applications to lessen costs and execution times by implementing a type of GA called the Pareto archived evolution strategy.

Jang et al. [53] used GA to minimize task execution times resulting in improved user satisfaction and an increase in profit for the service providers.

Table 8.1 Research summary of PSO-based methods.

	Improvement						Objective functions								
	Initial population	Local search	Global best and local best	Conversion	Position and Velocity	Combined/ Used	Power consumption	Makespan, Response time, Flow time, Execution time	Transfer time	Queue length	Cost	Resource utilization	Trust, Reliability, Security	Algorithms Outperformed	Evaluation Tool
[18]						PSO		✓			✓			NA	NA
[19]						Improved PSO (IPSO)		✓				✓		ACO, GA, SA	C1oudSim
[20]	✓	✓				PSO-ParFnt		✓						PSO	Their designed/ modified simulator
[21]	✓		✓	✓		PSO, VNS						✓		Greedy, IC-PCP	NA
[22]					✓	BPSO		✓				✓		SS	C1oudSim
[23]						PSO		✓	✓					NA	C1oudSim
[3]						Extended PSO and GA	✓	✓		✓	✓			NA	C1oudSim
[24]						S-PSO, SCP		✓			✓		✓	PSO, ACO	C1oudSim
[25]	✓	✓				PSO, Best-Fit (BF), Tabu-Search (TS)		✓			✓	✓		PSO	C1oudSim
[27]	✓					PSO, ACO, GA		✓						PSO, ACO, GA	NA
[49]						PSO, ACO		✓			✓			PSO, ACO	C1oudSim
[26]						PSO, GA								HEFT_UpRank, HEFT_DownRank, HEFT_LevelRank, MPQMA	NA

(*Continued*)

Table 8.1 (Continued)

	Improvement						Objective functions							Algorithms Outperformed	Evaluation Tool
	Initial population	Local search	Global best and local best	Conversion	Position and Velocity	Combined/ Used	Power consumption	Makespan, Response time, Flow time, Execution time	Transfer time	Queue length	Cost	Resource utilization	Trust, Reliability, Security		
[28]						PSO					✓		✓	NA	C1oudSim
[29]						Catfish PSO								PSO	CloudSim
[30]	✓	✓		✓		PSO on fission mechanism								FIFO, PSO	NA
[31]		✓				PSO and Shuffled Frog Leaping Algorithm (SFLA)								GA, TD MinMin	CloudSim
[32]		✓		✓		Greedy algorithm and PSO (G&PSO)							✓	PSO	C1oudSim
[33]		✓				PSO, ACO					✓			PSO	NA
[34]				✓		GA, chaotic, PSO		✓					✓	GA, GSA, ACO	Matlab
[35]						Taguchi Orthogonal approach and Cat Swarm Optimization		✓						MinMax, PSO, Linear Descending Inertia Weight (PSO-LDIW), Hybrid PSO, Simulated Annealing (HPSO-SA)	C1oudSim
[36]				✓				✓			✓			PSO	NA

[37]						Four objectives PSO	✓	✓		✓		NA	VMware-vSphere, Condor
[38]					✓	Improved IPSO				✓		PSO	NA
[39]						Improve PSO (SAPSO)		✓		✓		PSO	NA
[40]						PSO, integer linear program: TTSA							NA
[41]						TSDQ, PSO, Fuzzy Logic (TSDQ-FLPSO) TSDQ, PSO, Simulated Annealing (TSDQ-SAPSO)		✓	✓	✓	✓	PSO, SAPSO	C1oudSim
[42]		✓	✓			Modified PSO (MPSO)		✓		✓	✓	Round Robin, FCFS, ACO, GA	NA
[43]						PSO, GA				✓		MCPCPP, WPSO, PSOGA	NA
[44]				✓		PSO, GA	✓	✓				PSO, Self-Adaptive PSO	SUMO 0.30.0, NS 3.26 and Matlab R2014a
[45]						PSO, GA		✓				NA	C1oudSim
[46]						PSO, BBO		✓				NA	C1oudSim
[47]						PSO, GA	✓			✓		NA	Matlab
[48]			✓			PSO, GA	✓			✓		NA	Matlab
[50]	✓					PSO, Hill climbing		✓				PSO, HEFT-B	Cloud Azure

Note: NA means information is not available.

Wang et al. [54] performed research into reducing energy consumption in data centers and developed an improved version of GA with designed genetic operators and practical encoding/decoding methods. A local search operator was introduced with the goal of accelerating their algorithm's convergence speed.

Hu et al. [55] considered task waiting and execution times. They improved GA (IGA) with double fitness functions for genetic manipulation to reduce the probability of mutation in the early stages.

Ghorbannia Delavar and Aryan [56] developed a GA hybrid aimed at workflow scheduling that considers task priorities. The BF and round-robin methods were merged to create an optimal initial population to ensure an ideal solution is found swiftly.

In another study on workflow scheduling, Verma and Kaushal [57] developed a deadline-constrained GA in which the priority of each workflow task is assigned at the bottom level. These priorities are then used to create the initial population. The first individual of the initial population is created by allocating tasks to the available machines according to their level in ascending order. The remaining individuals are created by randomly assigning the tasks to available machines.

Wang et al. [58] proposed an integer bi-level programming model to minimize power consumption. They formulated the problem as two upper-level objective functions: (i) minimizing energy inefficiencies in the data center servers by reducing variances in each server's resource utilization; and (ii) maximizing the data locality ratio. A local search operator was used to accelerates convergence.

Hassan et al. [59] combined GA with list scheduling and earliest completion time (ECT) to develop a GA-based task scheduling algorithm that outperforms GA in minimizing makespan.

Hung et al. [60] presented a cost–time aware genetic scheduling algorithm (CTaG) that optimizes bandwidth and costs over the cloud system.

Mahmoud et al. [61] developed a two-phase GA-based job scheduling model that determines the shortest job completion time with Min-Min, Max-Min, and suffrage techniques. The optimal job scheduling among resources is determined by the GA.

Shojafar et al. [62] presented a hybrid approach called FUGE that is based on fuzzy theory and a GA with the objective of optimizing load balancing considering execution time and cost while taking processing speeds, memory, VM bandwidth, and job length into account. They show the efficiency of the FUGE approach in terms of execution time, execution cost, and the average degree of imbalance compared to ACO and the multiple ACO (MACO) algorithms.

He et al. [27] combined PSO, ACO, and the crossover operator from GA to improve ACO. The experimental results show reduced total task completion time.

Kamalinia and Ghaffari [26] aimed to reduce makespans with the HEFT algorithm in combination with the PSO and GA algorithms. The results of the simulation show improved average makespans over traditional GA.

Peng et al. [63] proposed an improved version of niched Pareto GA (NPGA). Results of this algorithm show an improvement in task scheduling over the classical NPGA without compromising time consumption or cost.

Wang et al. [64] developed a new bi-level GA. The results from various consumption tests show significant improvements in the energy efficiency of servers.

Abdulhamid and Latiff [65] presented a checkpointed league championship algorithm (CPLCA) with an emphasis on secure fault tolerance responsiveness. Simulations show an improvement in total average makespan, and total average response times compared to ACO, GA, and the basic LCA.

Anastasi et al. [66] proposed QBROKAGE which is a genetic approach to cloud brokering. This approach considered costs while satisfying the QoS requirements of cloud applications.

Chirkin et al. [67] aimed to provide a solution for estimating makespans. They also integrated the solution into the earlier-developed scheduling algorithm GAHEFT, which is a hybrid algorithm based on a combination of the GA and the HEFT algorithm. Comparison tests with the Min-Min algorithm show that GAHEFT performs better.

Cui et al. [34] proposed a chaotic ant swarm algorithm (GA-CAS), which is GA-based. It has the objective of optimizing reliability, makespan, and flow time. Results from simulations demonstrate GA-CAS typically increases the speed of convergence and outperforms GA, GSA, and ACO in its objectives.

Kamalinia and Ghaffari [68] presented a hybrid metaheuristic method using HEFT algorithm. Testing shows a significant effect on improving the optimization of makespans and improving resource efficiency when compared to a GA and a differential evolution (DE) algorithm.

Shen et al. [69] proposed a genetic algorithm called E-PAGA with the aim of achieving adaptive regulations for different requirements of energy and performance. When compared to artificial fish swarm optimization (AFSA), FCFS, and the modified best-fit descending algorithm (MBFD), E-PAGA had the lowest energy consumption and makespan.

Sreenu and Malempati [70] proposed the modified fractional gray wolf optimizer for multi-objective task scheduling (MFGMTS), which aims to optimize execution time, execution cost, communication time, communication cost, energy consumption, and resource utilization. Simulation results when compared to PSO, GA, grey wolf optimizer (GWO), and the fractional grey wolf optimizer (FGWO) show MFGMTS as performing better across all measures.

Wei [71] proposed bidirectional optimization GA (BOGA) with the aim of optimizing completion time and cost and simulated the algorithm on CloudSim.

Yao et al. [72] proposed an IGA with a three-stage selection method and a total-division-total genetic strategy. The algorithm improves selection and crossover operations. Using CloudSim, the results show that IGA outperforms the simple GA (SGA) in terms of completion time.

Keshanchi et al. [73] optimized task scheduling via an IGA, called N-GA, that uses GA along with a heuristic-based HEFT search to assign subtasks to processors. A one-point crossover operator is used to avoid violating the precedence constraints. They also apply an elitism technique with unusual selections to avoid premature convergence. Their model outperformed three well-known heuristics algorithms including HEFT-B, HEFT-L, and HEFT-T, in reducing the total execution times and makespan.

Basu et al. [74] combined a GA with ACO to form GAACO where only the fittest task combination survives each stage. This approach was specifically designed for task scheduling and load balancing in the Internet of Things (IoT) applications. Task dependencies are considered with the aim of minimizing the total execution time (makespan) of the tasks. GAACO provides superior performance over GA and ACO alone in heterogeneous multiprocessor environments.

Casas et al. [75] developed a variant of a GA called efficient tune-in of resources (GA-ETI) that analyzes applications prior to execution. The characteristics studied are available computing resources, multiple scheduling configurations, and selecting optimal configurations to execute workflows. GA-ETI adapts the conventional crossover and swap mutation from the classical GA and also includes a modified crossover and new increment and decrement mutation operators. GA-ETI distinguishes itself from other schedulers for its adaptability to the size of jobs and the VMs. It contains a non-monetary cost model and considers complex inter-dependencies between tasks. Tested in a VMware-vSphere private cloud without increasing costs, GA-ETI performed well with an 11–85% reduction in makespan when executing workflows.

Zhang et al. [76] proposed a new task scheduling algorithm called RC-GA (a genetic algorithm for task scheduling based on a resource crowd-funding model). The idea of this model is that idle resources are collected to form a virtual resource pool that can provide cloud services. Its goal is to increase task execution stability and reduce power consumption. It improves upon GA's roulette wheel selection operator, which in turn causes a reduction in random errors during the evolution process. Experiments show RC-GA is better than non-dominated sorting genetic algorithm (NSGA-II) at increasing stability and reducing power consumption. Thus, the convergence speed of RC-GA is faster than classical GA.

Singh [77] proposed a hybrid model called HGVP that involves a GA, VNS, and PSO. HGVP is an improvement over PSO in selecting the initial amounts of particles and in stabilizing non-local searches and local utilization within one evolutionary processing period. Simulation results in Matlab show that HGVP outperforms current techniques with a focus on energy consumption compared to the PSO scheduling technique SPSO-SA and to SLPSO-SA.

Sathya Sofia and GaneshKumar [78] claim that reducing energy consumption in cloud service environments increases makespan, which results in customer dissatisfaction. To provide a non-domination solution for these conflicting aims, they proposed the NSGA-II to perform multi-objective optimization. The results of a comparison test between NSGA-II with and without an artificial neural network (ANN) show that incorporating an ANN decreases both makespan and energy consumption.

Aziza and Krichen [79] developed a scheduling decision support system based on a GA with two objective functions – makespan and cost. The algorithm was implemented in CloudSim.

Given there can be a very large amount of tasks in real clouds, Duan et al. [80] developed a single objective makespan-based method. It relies on incremental GA, which has adaptive probabilities for crossover and mutation. The mutation and crossover rates vary conferring to generations and also between individuals. For the sake of evaluation, they randomly generated a large number of tasks based on the instance types in Amazon EC2 and implemented VMs with different computing capacities on CloudSim. The model performs better than the classical GA, Min-Min, Max-Min, SA, and artificial bee colony (ABC) at finding the optimal scheme.

The classical GA was used by Fernández-Cerero et al. [81] for energy-aware and time-aware scheduling policies. The idea behind their model is to combine energy- and performance-aware scheduling policies to hibernate idle VMs. Four strategies and their combinations were tested including: makespan-centric scheduling, energy-centric scheduling, makespan-centric scheduling with a threshold, and energy-centric scheduling with a threshold.

An adaptive GA was proposed by Ibrahim et al. [82] to be used in a dynamic task scheduling model based on an integer linear programming that satisfies an energy consumption objective. Their algorithm searches for an optimal (or near-optimal) solution starting from an initial population of possible solutions (chromosomes) by selecting the fittest operators and the crossover and mutation reproduction operators. The chromosomes are encoded appropriately, and the crossover operator is specifically designed to suit the given problem. Additionally, the model is hybridized with a procedure that determines whether a chromosome (i.e. a possible solution) is feasible and, if not, the chromosome is penalized accordingly with a lower probability of selection (i.e. survival).

A hybrid GA-PSO algorithm was proposed by Manasrah and Ali [83] to reduce makespans and cost, as well as balance the load of dependent tasks across heterogeneous resources. The idea behind this hybrid model is to distribute the number of iterations between the GA and PSO algorithms to reduce complexity.

Shishido et al. [45] compared the performance of GA and PSO for scientific workflow scheduling giving consideration to security and deadline constraints in a simulated environment. GA performed better than PSO.

Tang et al. [84] proposed a scheduling strategy aimed at industrial Internet of Things applications using a GA. The objective was to reduce energy consumption while taking task dependency, data transmission, and some constraint conditions into account, such as response time deadline and cost.

Taj and Basu [85] hybridized both GA and the group search optimization (GSO) algorithm into the genetic and group search optimization (GGSO) algorithm. This algorithm maximizes income for IaaS providers while simultaneously ensuring QoS for users. GGSO has several advantages over the GA and GSO algorithms such as faster convergence, and less computational time to the optimal solution. The results show it outperforms GSO, GA, and SLPSO algorithms at maximizing profit.

Table 8.2 summarizes the GA-based methods and reports the implemented improvements, objective functions, algorithms outperformed, and evaluation tool.

8.4.3 Ant and Bee Colony Optimization

Early attempts to schedule cloud computing tasks using ACO were based on the work of Shi et al. [87], who tried to improve Hadoop architecture and workload balancing with respect to the cost matrix, hormone matrix, and probability matrix. According to their experiments, ACO improves both response time and throughput. Likewise, using conventional ACO, Mateos et al. [88] developed a scheduler for private clouds. Wu et al. [89] proposed a market-oriented hierarchical scheduling strategy for cloud workflow systems, in which service level scheduling handles task-to-service assignments. Here, the tasks in individual workflow instances are mapped to cloud services in global cloud markets based on their functional and non-functional QoS requirements. The task-level scheduler optimizes task-to-VM assignments in local cloud data centers. Their analysis shows that service level scheduling with GA, ACO, and PSO results in improved performance with ACO in terms of optimizing makespan, cost, and CPU time. In addition to shortening the makespan of task scheduling, Xue et al. [90] used an ACO based load balancing algorithm that considers the load of VMs.

As ACO lacks rapid adaptivity, Baxodirjonovich and Choe [91] modified ACO for cloud workflow applications using probability theory to help the ants decide which machine to target. The modified ACO algorithm outperforms classical ACO.

Pacini et al. [92] used ACO for task scheduling based on several interesting metrics: the number of users receiving services and the total number of VMs created in online (non-batch) scheduling scenarios. Their approach showed better results in a cloud simulation environment than random assignment and GA.

Table 8.2 Research summary of GA-based algorithms.

	Improvement					Objective functions								
	Initial population	Crossover	Mutations	Conversion	Combined/Used	Power consumption	Makespan, Response time, Flow time, Execution time	Transfer time	Resource utilization	Cost	Security	Trust, Reliability	Algorithms Outperformed	Evaluation Tool
Tayal [51]					GA, Fuzzy Logic								GA	NA
Juhnke et al. [52]					PAES (type of GA)		✓			✓			GA	NA
[53]							✓						Round robin, load index, Activity-based Costing	NA
[54]		✓	✓			✓							Hadoop MapReduce	Hadoop
[55]			✓				✓						Random, Rotating, and Greedy	CloudSim
[56]	✓						✓						LAGA HSGA	NA
[57]	✓									✓			NA	Simulation environment in Java

(Continued)

Table 8.2 (Continued)

	Improvement					Objective functions								
	Initial population	Crossover	Mutations	Conversion	Combined/Used	Power consumption	Makespan, Response time, Flow time, Execution time	Transfer time	Resource utilization	Cost	Security	Trust, Reliability	Algorithms Outperformed	Evaluation Tool
[58]				✓		✓							Hadoop Scheduler, Fair Scheduler	Hadoop
[59]					GA, List Algorithm, ECT		✓						GA	NA
[60]					GA				✓	✓			NA	NA
[61]					GA, Min-Min, Max-Min, suffrage				✓				NA	CloudSim
[62]					GA, Fuzzy Logic		✓		✓	✓			ACO, MACO	CloudSim
[27]		✓			PSO, ACO, GA		✓						PSO, ACO	NA
[26]					HEFT GA, PSO		✓						GA	CloudSim
Peng et al. [63]			✓		NPGA		✓			✓			NPGA	NA

Wang et al. [64]	✓	✓	✓	GA	✓							GA	NA
Abdulhamid and Latiff [65]				CPLCA		✓				✓	✓	,GA ,ACO LCA	CloudSim
Anastasi et al. [66]				GA				✓	✓			NA	CloudSim
Chirkin et al. [67]				GA HEFT		✓					✓	Min-Min	NA
Cui et al. [34]			✓	GA-CAS		✓	✓				✓	NA	NA
Kamalinia and Ghaffari [68]	✓			HEFT		✓		✓				GA DE	NA
Shen et al. [69]	✓	✓	✓	E-PAGA	✓	✓						,AFSA ,FCFS MBFD	CloudSim
Sreenu and Malempati [70]				MFGMTS	✓	✓			✓			PSO GA FGWO	CloudSim Java
Wei [71]				GA		✓			✓			NA	CloudSim
Yao et al. [72]	✓			IGA		✓						SGA	CloudSim

(*Continued*)

Table 8.2 (Continued)

	Improvement					Objective functions								
	Initial population	Crossover	Mutations	Conversion	Combined/Used	Power consumption	Makespan, Response time, Flow time, Execution time	Transfer time	Resource utilization	Cost	Security	Trust, Reliability	Algorithms Outperformed	Evaluation Tool
[73]		✓		✓	GA		✓						HEFT-B, HEFT-T, HEFT-L	NA
Basu et al. [74]				✓	GAACO		✓						GA, ACO	NA
Zheng et al. [86]					CBT-MD CBT-ED CFMax CRR	✓				✓			HEFT, CSFS-Max	Java (JDK version 1.8)
Casas et al. [75]		✓	✓		GAETI		✓			✓			NA	VMware-vSphere
Zhang et al. [76]					RC-GA	✓						✓	NSGA-II, GA	NA
Singh [77]	✓			✓	HGVP (GA-PSO-VNS)	✓							SPSO-SA, SLPSO-SA	Matlab 13a
[78]				✓	NSGA-II	✓	✓						GA	NA
[79]							✓			✓			NA	CloudSim

[80]	✓	✓			✓			GA, Min-Min, Max-Min, SA, ABC	CloudSim
[81]				✓	✓			NA	SCORE
[82]	✓	✓		✓				FCFS	Simulation environment developed by C++
[83]			GA-PSO		✓	✓		GA, PSO, HSGA, WSGA, MTCT	WorkflowSim
[45]			GA, PSO		✓	✓		NA	CloudSim
[84]				✓				NA	Coding with C++
Taj and Basu [85]	✓		GGSO		✓	✓	✓	GSO, GA, SLPSO	NA

Note: NA means information is not available.

The improved ACO algorithm presented by Zuo et al. [93] was founded on four metrics – makespan, cost, deadline violation rate, and resource utilization – and was designed to solve multi-objective task scheduling problems. Optimal solutions are found in a timely manner based on constraint functions for performance and cost.

Vinothina and Sridaran [94] used ACO in a public cloud scenario to minimize makespan. Massive amounts of heterogeneous resources are allocated via a pay-per-use model.

Moon et al. [95] adapted diversification and reinforcement strategies with slave ants and proposed an ACO algorithm that solves global task scheduling optimization problems by avoiding long paths whose pheromones are wrongly accumulated by the leading ants.

Wu et al. [96] developed a workflow scheduling algorithm based on ACO to optimize execution times and minimize costs for users. They proposed a meta-heuristic algorithm L-ACO as well as a simple heuristic ProLiS. ProLiS distributes the deadline to each task, proportional to a novel definition of a probabilistic upward rank. Then, a two-step list scheduling methodology ranks the tasks and sequentially allocates each task to a service that meets a sub-deadline and minimizes the cost. ACO is used to carry out deadline-constrained cost optimization.

Xiang et al. [97] developed a greedy-ant algorithm with the aim of improving workflow scheduling in heterogeneous computing environments and minimizing the total execution time of an application. Their approach includes a novel perspective on ant colony system scheduling by guiding ants to explore task priorities and simultaneously assigning tasks to machines. They also defined forward/backward dependence to indicate the global significance of each node. Based on this dependence, a novel heuristic factor helps ants search for task sequences. The end result is a greedy machine allocation strategy.

Zuo et al. [98] used ACO to optimize a finite pool of resources in a hybrid cloud computing environment based on deadlines and cost constraints. It minimizes task completion times and cost using time-first and cost-first single objective optimization strategies. QoS for users and profits for resource providers are maximized using an entropy optimization model.

Chen et al. [99] proposed a novel multi-objective ant colony system for workflow scheduling based on multiple-populations. The model includes execution time and execution cost objectives and a new pheromone update rule based on a set of non-dominated solutions from a global archive. The update rule balances the search for both objectives by guiding each colony in a search for one objective only, and a complementary heuristic strategy ensures that each colony only focuses on its own objective. The quality improvement in the of the solution to the global archive helps to further the approach to the global Pareto front.

Jia et al. [100] developed an intelligent scheduling system from the users' perspective to reduce the expenditure of workflow subject to deadlines and other execution constraints. A new estimation model for task execution time was designed according to virtual machine settings in real public clouds and execution data from practical workflows. An adaptive ACO algorithm with a narrower search space and a self-adaptive weight is used to adaptively meet different deadline settings.

Sebastio et al. [101] developed a new scheduling algorithm for volunteer clouds to support collaborative task execution services. The approach combines ACO and an overlay network partitioning method, called ACO-Partitioned, to improve the QoS of computer-intensive workloads. Colored scout ants discover the computational resources in the network and maintain the computational field. These scout ants are spawned continuously to adapt to the variability of the volunteer cloud. In a comparative efficiency evaluation with ACO, random, round robin, greedy oracle, and local diffusion, ACO-Partitioned showed the highest increase in QoS.

Wang et al. [102] proposed cooperative multi-task scheduling based on ACO (CMSACO) to improve completion time, resource consumption, and executive order. Simulation results show that CMSACO performs better than FCFS in a range of factors, including reduced energy consumption, increasing the number of executed tasks, and load ratio.

Applications that rely on BCO are very rare. Thanka et al. [103] improved BCO for task scheduling with a proposal called improved efficient-artificial bee colony (IE-ABC). Unlike the classical BCO algorithm, where the worker bee must update each time it is in a hive, and an onlooker bee selects the food source with the highest nectar quantity, these authors assigned a dedicated worker bee to the data center. This bee updates the current virtual machine's status, its capacity load, memory availability, storage, and security level in the hive table. So, each time the honeybee that is the task of the cloud environment does not want to search for the best fitness. This saves time because the bee can make a decision about whether its VM is over or underloaded when assigning the task to itself. Compared to the ABC algorithm, IE-ABC improves resource utilization and task migration.

In a recent study, Gomathi et al. [104] used epsilon-fuzzy dominance based on a composite discrete artificial bee colony (EDCABC) to generate Pareto-optimal solutions for multi-objective task scheduling problems. EDCABC applies composite mutation strategies and a fast local search method to improve local searches and avoid premature convergence. EDCABC shows higher reductions in makespans and execution costs and better resource utilization than NSGA-II and MOPSO.

Table 8.3 summarizes the ant and bee colony-based algorithms and reports the improvements, objective functions, algorithms outperformed, and evaluation tool.

Table 8.3 Research summary of ant and bee colony-based algorithms.

	Improvement		Objective Functions								
	Local search convergence	Combined/ Used	Power consumption	Makespan, Response time, Flow time, Execution time	Transfer time	Resource utilization	Cost	Security	Trust, Reliability	Algorithms Outperformed	Evaluation Tool
[87]		ACO		✓			✓			FIFO	Hadoop
[88]		ACO		✓						CloudSim built-in, Random	CloudSim
[89]		ACO		✓		✓	✓			GA, PSO	SwinDeW-C
[90]		ACO		✓		✓				FIFO	CloudSim
[91]	✓	ACO								Basic ACO, Min-Min	WorkflowSim
[92]		ACO		✓		✓				Random, GA	NA
[93]				✓			✓			FIFO, Min-Min	CloudSim
[94]		ACO		✓						HEFT	CloudSim
[95]		ACO		✓		✓				NA	NA
[96]		ACO					✓			NA	NA
[97]		Greedy-Ant		✓						PEFT, HEFT	Matlab
[98]	✓	ACO		✓			✓			Min-Min FIFO	CloudSim

[99]	✓		ACO		✓			✓		HEFT, PSO, NSGA-II	Amazon EC2
[100]	✓		Feasibility-based rule, ACO		✓					HEFT, PSO	CloudSim
[101]			ACO, overlay network partitioning (ACO-Partitioned)							Compared to: ACO, Random, Round robin, Greedy oracle and Local diffusion	The Volunteer Cloud Simulator (AVoCloudy)
[102]			CMSACO	✓	✓		✓			FCFS	NA
[103]			Improved ABC (IE-ABC)			✓	✓		✓	ABC	NA
[104]	✓	✓	BC, Fuzzy logic		✓		✓	✓		NSGA-II, MOPSO	NA

Note: NA means information is not available.

8.4.4 Other Evolutionary Algorithms

Song et al. [105] proposed a general job selection and allocation framework that uses an adaptive filter to select jobs, while a modified heuristic algorithm (Min-Min) allocates them. Two objectives and four criteria are considered in the optimization problem. The two objectives are to maximize the remaining CPU capacity and maximize resource utilization. The four criteria are CPU resource requirements, memory, hard disk space, and network bandwidth.

Bilgaiyan et al. [106] applied a cat-swarm-based multi-objective optimization approach to schedule workflows in a cloud computing environment while minimizing costs, makespans, and CPU idle time. This approach performs better than multi-objective PSO (MOPSO).

Abdullahi and Ngadi [107] presented a discrete symbiotic organism search (DSOS) algorithm for the optimal scheduling of tasks to cloud resources with the aim of minimizing makespans, response times, and the degree of imbalance between VMs. Symbiotic organism search (SOS) mimics symbiotic relationships (i.e. mutualism, commensalism, and parasitism) to improve the quality of the given objective functions. They prove that DSOS converges faster than PSO as the search grows larger, which makes it suitable for large-scale scheduling problems.

Torabi and Safi-Esfahani [108] rely on a bio-inspired metaheuristic algorithm, chicken swarm optimization (CSO), and raven roosting optimization (RRO) for task scheduling. CSO brings efficiency in satisfying the balance between the local and the global search, and RRO overcomes the problem of premature convergence. RRO also has better performance in larger search spaces. Simulation tests performed in a cloud computing environment show improvements in terms of reducing execution times, reducing response times, and increasing throughput in dynamic scenarios.

Choudhary et al. [109] hybridized the popular metaheuristic, gravitational search (GSA), and the HEFT algorithm for workflow scheduling. In terms of its two objectives – makespan and cost – the proposed method outperformed GSA, hybrid GA, and HEFT in experiments with scientific workflows.

8.5 Research Gaps and Future Directions

A review of the literature revealed some research gaps, challenges, and promising future research directions. The following sub-sections discuss these topics. In addition, Tables 8.1–8.3 summarize the following aspects of each paper:

- improvements over original algorithms
- improvements by combining algorithms
- considered objective functions
- superseded algorithms
- evaluation tools applied.

8.5.1 Evaluation Tools

Only two studies have included model evaluations based on real cloud environments. The majority have used simulated clouds like CloudSim to assess the efficacy of their work. Measuring the accuracy of an objective function using simulation tools tends not to be very reliable as some values like transfer and execution times cannot be calculated if they have not actually been performed. Hence, a goal should be to verify the accuracy of the presented models in a real cloud environment, such as VMware-vSphere.

8.5.2 Objective Functions

Most studies only consider two objective functions and only four at most. Moreover, none of the existing models provide any flexibility as to the number and type of objectives. Considering the growing popularity of cloud computing, comprehensive models that incorporate all important objective functions need to be developed. This includes objective functions desired by both industry and users. Models should provide users with the opportunity to select and rank the objective functions according to their own preferences and requirements. For example, the importance of objective functions like time and cost can vary in different situations. In some cases, a user may need a task to complete in the shortest possible time no matter how much it costs. On other occasions, cost may be the most important factor.

8.6 Conclusion

Cloud computing is a paradigm that provides dynamic services using very large scalable and virtualized resources over the Internet. Task, job, and workflow scheduling are some of the major activities performed in all computing environments. However, due to the novelty of cloud computing, there is no unique, standard, or universal task scheduling algorithm. To date, many researchers have attempted to develop and apply different scheduling algorithms for different purposes. Each considers different objectives, such as makespan, cost, execution time, execution cost, etc. As these performance indicators are usually in conflict, the usage of evolutionary computation techniques to overcome multi objective optimization problem has received many attention. The aim of this study was to conduct a systematic review of the literature over the last decade on task scheduling methods in cloud computing environments that rely on conventional or improved evolutionary computation techniques. The PRISM research methodology frames the review by providing a set of eligibility criteria. All studies reviewed were drawn from Scopus. The analysis is segmented according to the type of technique, i.e. GA, PSO, ACO, BCO, and other. Each category is qualitatively reviewed and summarized.

References

1 Pourqasem, J., Karimi, S., and Edalatpanah, S. (2014). Comparison of cloud and grid computing. *Amm. J Soft. Eng.* 2: 8–12.
2 Zhu, Z., Zhang, G., Li, M., and Liu, X. (2016). Evolutionary multi-objective workflow scheduling in cloud. *IEEE Trans. Parallel Distrib. Syst.* 27: 1344–1357.
3 Ramezani, F., Lu, J., Taheri, J., and Hussain, F.K. (2015). Evolutionary algorithm-based multi-objective task scheduling optimization model in cloud environments. *World Wide Web* 18: 1737–1757.
4 Taheri, J., Zomaya, A.Y., Bouvry, P., and Khan, S.U. (2013). Hopfield neural network for simultaneous job scheduling and data replication in grids. *Future Gener. Comput. Syst.* 29: 1885–1900.
5 Varia, J. (2008). Cloud architectures. *White Paper of Amazon*, 16.
6 Grzonka, D., Kołodziej, J., Tao, J., and Khan, S.U. (2015). Artificial neural network support to monitoring of the evolutionary driven security aware scheduling in computational distributed environments. *Future Gener. Comput. Syst.* 51: 72–86.
7 VMWARE (2015). VMware vSphere 5.1 Documentation Center [Online]. https://www.vmware.com/support/pubs.
8 Huang, T., Zhu, Y., Wu, Y. et al. (2016). Anomaly detection and identification scheme for VM live migration in cloud infrastructure. *Future Gener. Comput. Syst.* 56: 736–745.
9 Buyya, R., Broberg, J., and Goscinski, A. (eds.) (2011). *Cloud Computing, Principles and Paradigms*. Hoboken, NJ: Wiley.
10 Moltó, G., Caballer, M., and de Alfonso, C. (2016). Automatic memory-based vertical elasticity and oversubscription on cloud platforms. *Future Gener. Comput. Syst.* 56: 1–10.
11 Manvi, S.S. and Krishna Shyam, G. (2014). Resource management for infrastructure as a service (IaaS) in cloud computing: a survey. *J. Network Comput. Appl.* 41: 424–440.
12 Buyya, R., Pandey, S., and Vecchiola, C. (2009). Cloudbus toolkit for market-oriented cloud computing. Proceeding of the 1st International Conference on Cloud Computing, Beijing, China (December 1–4). Springer.
13 Weingärtner, R., Bräscher, G.B., and Westphall, C.B. (2015). Cloud resource management: a survey on forecasting and profiling models. *J. Netw. Comput. Appl.* 47: 99–106.
14 Sahu, Y. and Pateriya, R. (2013). Cloud computing overview with load balancing techniques. *Int. J. Comput. Appl.* 65: 0975–8887.
15 Riesen, R. and Maccabe, A. (2011). Job scheduling. In: *Encyclopedia of Parallel Computing* (ed. D. Padua). Springer.
16 Wang, Y., Lu, P., and Kent, K.B. (2015). WaFS: a workflow-aware file system for effective storage utilization in the cloud. *IEEE Trans. Comput.* 64: 2716–2729.

17 Moher, D., Liberati, A., Tetzlaff, J. et al. (2009). Preferred reporting items for systematic reviews and meta-analyses: the PRISMA statement. *PLoS Med.* 6: e1000097.

18 Guo, L., Zhao, S., Shen, S., and Jiang, C. (2012). Task scheduling optimization in cloud computing based on heuristic algorithm. *J. Netw.* 7: 547–553.

19 Zhan, S. and Huo, H. (2012). Improved PSO-based task scheduling algorithm in cloud computing. *J. Inform. Comput. Sci.* 9: 3821–3829.

20 Taheri, J., Zomaya, A.Y., Siegel, H.J., and Tari, Z. (2014). Pareto frontier for job execution and data transfer time in hybrid clouds. *Future Gener. Comput. Syst.* 37: 321–334.

21 Netjinda, N., Sirinaovakul, B., and Achalakul, T. (2014). Cost optimal scheduling in IaaS for dependent workload with particle swarm optimization. *J. Supercomput.* 68: 1579–1603.

22 Ouyang, Z., Wang, K., and Xu, L. (2014). Green cloud task-scheduling algorithm based on improved BPSO. *J. Inform. Comput. Sci.* 11: 4301–4308.

23 Ramezani, F., Lu, J., and Hussain, F.K. (2014). Task-based system load balancing in cloud computing using particle swarm optimization. *Int. J. Parallel Programm.* 42: 739–754.

24 Yang, Y., Peng, X., and Cao, J. (2015). Trust-based scheduling strategy for cloud workflow applications. *Informatica (Netherlands)* 26: 159–180.

25 Alkhashai, H.M. and Omara, F.A. (2016). An enhanced task scheduling algorithm on cloud computing environment. *Int. J. Grid Distrib. Comput.* 9: 91–100.

26 Kamalinia, A. and Ghaffari, A. (2016). Hybrid task scheduling method for cloud computing by genetic and PSO algorithms. *J. Inf. Syst. Telecommun.* 4: 271–281.

27 He, X.L., Song, Y., and Binsack, R.V. (2016). The intelligent task scheduling algorithm in cloud computing with multistage optimization. *Int. J. Grid Distrib. Comput.* 9: 313–324.

28 Li, Z., Ge, J., Yang, H. et al. (2016). A security and cost aware scheduling algorithm for heterogeneous tasks of scientific workflow in clouds. *Future Gener. Comput. Syst.* 65: 140–152.

29 Nirmala, S.J. and Bhanu, S.M.S. (2016). Catfish-PSO based scheduling of scientific workflows in IaaS cloud. *Computing* 98: 1091–1109.

30 Shao, Y. (2016). Research on cloud task scheduling algorithm of particle swarm optimization algorithm based on fission mechanism. *Int. J. Multimedia Ubiquitous Eng.* 11: 145–152.

31 Xie, X., Liu, R., Cheng, X. et al. (2016). Trust-driven and PSO-SFLA based job scheduling algorithm on cloud. *Intell. Automation Soft Comput.* 22: 561–566.

32 Zhong, Z., Chen, K., Zhai, X., and Zhou, S. (2016). Virtual machine-based task scheduling algorithm in a cloud computing environment. *Tsinghua Sci. Technol.* 21: 660–667.

33 Chen, X. and Long, D. (2017). Task scheduling of cloud computing using integrated particle swarm algorithm and ant colony algorithm. *Cluster Comput.*: 1–9.

34 Cui, H., Li, Y., Liu, X. et al. (2017). Cloud service reliability modelling and optimal task scheduling. *IET Commun.* 11: 161–167.

35 Gabi, D., Ismail, A.S., Zainal, A., and Zakaria, Z. (2017). Solving task scheduling problem in cloud computing environment using orthogonal taguchi-cat algorithm. *Int. J. Electr. Comput. Eng.* 7: 1489–1497.

36 Gan, N., Huang, Y., and Lu, X. (2017). Niching particle swarm optimization algorithm for solving task scheduling in cloud computing. *Agro Food Ind. Hi Tech* 28: 876–879.

37 Ramezani, F., Lu, J., Taheri, J., and Zomaya, A.Y. (2017). A multi-objective load balancing system for cloud environments. *Comput. J.* 60: 1316–1337.

38 Sadhasivam, N. and Thangaraj, P. (2017). Design of an improved PSO algorithm for workflow scheduling in cloud computing environment. *Intell. Automation Soft Comput.* 23: 493–500.

39 Yuan, H., Bi, J., Tan, W., and Li, B.H. (2017a). Temporal task scheduling with constrained service delay for profit maximization in hybrid clouds. *IEEE Trans. Automation Sci. Eng.* 14: 337–348.

40 Yuan, H., Bi, J., Tan, W. et al. (2017b). TTSA: an effective scheduling approach for delay bounded tasks in hybrid clouds. *IEEE Trans. Cybern.* 47: 3658–3668.

41 Ben Alla, H., Ben Alla, S., Touhafi, A., and Ezzati, A. (2018). A novel task scheduling approach based on dynamic queues and hybrid meta-heuristic algorithms for cloud computing environment. *Cluster Comput.* 21: 1797–1820.

42 Domanal, S.G. and Reddy, G.R.M. (2018). An efficient cost optimized scheduling for spot instances in heterogeneous cloud environment. *Future Gener. Comput. Syst.* 84: 11–21.

43 Guo, W., Lin, B., Chen, G. et al. (2018). Cost-driven scheduling for deadline-based workflow across multiple clouds. *IEEE Trans. Netw. Serv. Manag.* 15: 1571–1585.

44 Midya, S., Roy, A., Majumder, K., and Phadikar, S. (2018). Multi-objective optimization technique for resource allocation and task scheduling in vehicular cloud architecture: a hybrid adaptive nature inspired approach. *J. Netw. Comput. Appl.* 103: 58–84.

45 Shishido, H.Y., Estrella, J.C., Toledo, C.F.M., and Arantes, M.S. (2018). Genetic-based algorithms applied to a workflow scheduling algorithm with security and deadline constraints in clouds. *Comput. Electrical Eng.* 69: 378–394.

46 Tong, Z., Chen, H., Deng, X. et al. (2018). A novel task scheduling scheme in a cloud computing environment using hybrid biogeography-based optimization. *Soft Comput.* 23: 11035–11054.

47 Yuan, H., Bi, J., Zhou, M., and Ammari, A.C. (2018). Time-aware multi-application task scheduling with guaranteed delay constraints in green data center. *IEEE Trans. Automation Sci. Eng.* 15: 1138–1151.

48 Yuan, H., Bi, J., and Zhou, M. (2018). Temporal task scheduling of multiple delay-constrained applications in green hybrid cloud. *IEEE Trans. Serv. Comput.* doi: 10.1109/TSC.2018.2878561.

49 Jiang, T. and Li, J. (2016). Research on the task scheduling algorithm for cloud computing on the basis of particle swarm optimization. *Int. J. Simul.: Syst., Sci. Technol.* 17: 11.1–11.5.

50 Dordaie, N. and Navimipour, N.J. (2018). A hybrid particle swarm optimization and hill climbing algorithm for task scheduling in the cloud environments. *ICT Express* 4: 199–202.

51 Tayal, S. (2011). Tasks scheduling optimization for the cloud computing systems. *Int. J. Adv. Eng. Sci. Technol.* 5: 111–115.

52 Juhnke, E., Dörnemann, T., Böck, D., and Freisleben, B. (2011). Multi-objective scheduling of BPEL workflows in geographically distributed clouds. 4th IEEE International Conference on Cloud Computing. 412–419.

53 Jang, S.H., Kim, T.Y., Kim, J.K., and Lee, J.S. (2012). The study of genetic algorithm-based task scheduling for cloud computing. *Int. J. Control Automat.* 5: 157–162.

54 Wang, X., Wang, Y., and Zhu, H. (2012). Energy-efficient task scheduling model based on MapReduce for cloud computing using genetic algorithm. *J. Comput. (Finland)* 7: 2962–2970.

55 Hu, B., Zhang, X., and Zhang, X. (2013). A scheduling mechanism for independent task in cloud computing environment. *J. Inform. Comput. Sci.* 10: 5945–5954.

56 Ghorbannia Delavar, A. and Aryan, Y. (2014). HSGA: a hybrid heuristic algorithm for workflow scheduling in cloud systems. *Cluster Comput.* 17: 129–137.

57 Verma, A. and Kaushal, S. (2014). Deadline constraint heuristic-based genetic algorithm for workflow scheduling in cloud. *Int. J. Grid Utility Comput.* 5: 96–106.

58 Wang, X., Wang, Y., and Cui, Y. (2014). A new multi-objective bi-level programming model for energy and locality aware multi-job scheduling in cloud computing. *Future Gener. Comput. Syst.* 36: 91–101.

59 Hassan, M.A., Kacem, I., Martin, S., and Osman, I.M. (2015). Genetic algorithms for job scheduling in cloud computing. *Stud. Inform. Control* 24.

60 Hung, P.P., Aazam, M., and Huh, E.N. (2015). CTaG: an innovative approach for optimizing recovery time in cloud environment. *KSII Trans. Internet Inf. Syst.* 9: 1282–1301.

61 Mahmoud, A.A., Zarina, M., Wan Nik, W.N.S., and Ahmad, F. (2015). Multi-criteria strategy for job scheduling and resource load balancing in cloud computing environment. *Indian J. Sci. Technol.* 8.

62 Shojafar, M., Javanmardi, S., Abolfazli, S., and Cordeschi, N. (2015). FUGE: a joint meta-heuristic approach to cloud job scheduling algorithm using fuzzy theory and a genetic method. *Cluster Comput.* 18: 829–844.

63 Peng, Y., Xue, S., and Li, M. (2016). An improved multi-objective optimization algorithm based on NPGA for cloud task scheduling. *Int. J. Grid Distrib. Comput.* 9: 161–176.

64 Wang, X., Wang, Y., and Cui, Y. (2016). An energy-aware bi-level optimization model for multi-job scheduling problems under cloud computing. *Soft Comput.* 20: 303–317.

65 Abdulhamid, S.M. and Latiff, M.S.A. (2017). A checkpointed league championship algorithm-based cloud scheduling scheme with secure fault tolerance responsiveness. *Appl. Soft Comput. J.* 61: 670–680.

66 Anastasi, G.F., Carlini, E., Coppola, M., and Dazzi, P. (2017). QoS-aware genetic cloud brokering. *Future Gener. Comput. Syst.* 75: 1–13.

67 Chirkin, A.M., Belloum, A.S.Z., Kovalchuk, S.V. et al. (2017). Execution time estimation for workflow scheduling. *Future Gener. Comput. Syst.* 75: 376–387.

68 Kamalinia, A. and Ghaffari, A. (2017). Hybrid task scheduling method for cloud computing by genetic and DE algorithms. *Wireless Per. Commun.* 97: 6301–6323.

69 Shen, Y., Bao, Z., Qin, X., and Shen, J. (2017). Adaptive task scheduling strategy in cloud: when energy consumption meets performance guarantee. *World Wide Web* 20: 155–173.

70 Sreenu, K. and Malempati, S. (2017). MFGMTS: epsilon constraint-based modified fractional Grey wolf optimizer for multi-objective task scheduling in cloud computing. *IETE J. Res.* 65: 1–15.

71 Wei, G. (2017). Task scheduling algorithm based on bidirectional optimization genetic algorithm in cloud computing environment. *Agro Food Ind Hi Tech* 28: 3062–3067.

72 Yao, H., Fu, X., Li, H. et al. (2017). Cloud task scheduling algorithm based on improved genetic algorithm. *Int. J. Performability Eng.* 13: 1070–1076.

73 Keshanchi, B., Souri, A., and Navimipour, N.J. (2017). An improved genetic algorithm for task scheduling in the cloud environments using the priority queues: formal verification, simulation, and statistical testing. *J. Syst. Software* 124: 1–21.

74 Basu, S., Karuppiah, M., Selvakumar, K. et al. (2018). An intelligent/cognitive model of task scheduling for IoT applications in cloud computing environment. *Future Gener. Comput. Syst.* 88: 254–261.

75 Casas, I., Taheri, J., Ranjan, R. et al. (2018). GA-ETI: an enhanced genetic algorithm for the scheduling of scientific workflows in cloud environments. *J. Comput. Sci.* 26: 318–331.

76 Zhang, N., Yang, X., Zhang, M. et al. (2018). A genetic algorithm-based task scheduling for cloud resource crowd-funding model. *Int. J. Commun. Syst.* 31.

77 Singh, R. (2018). Hybrid genetic, variable neighbourhood search and particle swarm optimisation-based job scheduling for cloud computing. *Int. J. Comput. Sci. Eng.* 17: 184–191.

78 Sathya Sofia, A. and Ganeshkumar, P. (2018). Multi-objective task scheduling to minimize energy consumption and Makespan of cloud computing using NSGA-II. *J. Netw. Syst. Manage.* 26: 463–485.

79 Aziza, H. and Krichen, S. (2018). Bi-objective decision support system for task-scheduling based on genetic algorithm in cloud computing. *Computing* 100: 65–91.

80 Duan, K., Fong, S., Siu, S.W.I. et al. (2018). Adaptive incremental genetic algorithm for task scheduling in cloud environments. *Symmetry* 10: 168–180.

81 Fernández-Cerero, D., Jakóbik, A., Grzonka, D. et al. (2018). Security supportive energy-aware scheduling and energy policies for cloud environments. *J. Parallel Distrib. Comput.* 119: 191–202.

82 Ibrahim, H., Aburukba, R.O., and El-Fakih, K. (2018). An integer linear programming model and adaptive genetic algorithm approach to minimize energy consumption of cloud computing data centers. *Comput. Electr. Eng.* 67: 551–565.

83 Manasrah, A.M. and Ali, H.B. (2018). Workflow scheduling using hybrid GA-PSO algorithm in cloud computing. *Wireless Commun. Mobile Comput.* 2018: 1934784.

84 Tang, C., Wei, X., Xiao, S. et al. (2018). A Mobile cloud based scheduling strategy for industrial internet of things. *IEEE Access* 6: 7262–7275.

85 Taj, N. and Basu, A. (2019). Hybridization of genetic and group search optimization algorithm for deadline-constrained task scheduling approach. *J. Intell. Syst.* 28: 153–171.

86 Zheng, W., Qin, Y., Bugingo, E. et al. (2018). Cost optimization for deadline-aware scheduling of big-data processing jobs on clouds. *Future Gener. Comput. Syst.* 82: 244–255.

87 Shi, H., Bai, G., and Tang, Z. (2011). ACO algorithm-based parallel job scheduling investigation on Hadoop. *Int. J. Digit. Content Technol. Its Appl.* 5: 283–288.

88 Mateos, C., Pacini, E., and Garino, C.G. (2013). An ACO-inspired algorithm for minimizing weighted flowtime in cloud-based parameter sweep experiments. *Adv. Eng. Software* 56: 38–50.

89 Wu, Z., Liu, X., Ni, Z. et al. (2013). A market-oriented hierarchical scheduling strategy in cloud workflow systems. *J. Supercomput.* 63: 256–293.

90 Xue, S., Li, M., Xu, X., and Chen, J. (2014). An ACO-LB algorithm for task scheduling in the cloud environment. *J. Software* 9: 466–473.

91 Baxodirjonovich, K.N. and Choe, T.Y. (2015). Dynamic task scheduling algorithm based on ant colony scheme. *Int. J. Eng. Technol.* 7: 1163–1172.

92 Pacini, E., Mateos, C., and García Garino, C. (2015). Balancing throughput and response time in online scientific clouds via ant Colony optimization (SP2013/2013/00006). *Adv. Eng. Software* 84: 31–47.

93 Zuo, L., Shu, L., Dong, S. et al. (2015). A multi-objective optimization scheduling method based on the ant colony algorithm in cloud computing. *IEEE Access* 3: 2687–2699.

94 Vinothina, V. and Sridaran, R. (2016). Scheduling scientific workflow based application using ACO in public cloud. *Int. J. Eng. Technol.* 7: 1994–2000.

95 Moon, Y.J., Yu, H.C., Gil, J.M., and Lim, J.B. (2017). A slave ants based ant colony optimization algorithm for task scheduling in cloud computing environments. *HCIS* 7: 1–10.

96 Wu, Q., Ishikawa, F., Zhu, Q. et al. (2017). Deadline-constrained cost optimization approaches for workflow scheduling in clouds. *IEEE Trans. Parallel Distrib. Syst.* 28: 3401–3412.

97 Xiang, B., Zhang, B., and Zhang, L. (2017). Greedy-ant: ant Colony system-inspired workflow scheduling for heterogeneous computing. *IEEE Access* 5: 11404–11412.

98 Zuo, L., Shu, L., Dong, S. et al. (2017). A multi-objective hybrid cloud resource scheduling method based on deadline and cost constraints. *IEEE Access* 5: 22067–22080.

99 Chen, Z., Zhan, Z., Lin, Y. et al. (2018). Multiobjective cloud workflow scheduling: a multiple populations ant Colony system approach. *IEEE Trans. Cybern.* 49: 2912–2926.

100 Jia, Y., Chen, W., Yuan, H. et al. (2018). An intelligent cloud workflow scheduling system with time estimation and adaptive ant colony optimization. *IEEE Trans. Syst., Man, Cybern.* doi: 10.1109/TSMC.2018.2881018.

101 Sebastio, S., Amoretti, M., Lafuente, A.L., and Scala, A. (2018). A holistic approach for collaborative workload execution in volunteer clouds. *ACM Trans. Model. Comput. Simul.* 28: 1–27.

102 Wang, T., Wei, X., Tang, C., and Fan, J. (2018). Efficient multi-tasks scheduling algorithm in mobile cloud computing with time constraints. *Peer-to-Peer Netw. Appl.* 11: 793–807.

103 Thanka, M.R., Uma Maheswari, P., and Edwin, E.B. (2017). An improved efficient: artificial bee Colony algorithm for security and QoS aware scheduling in cloud computing environment. *Cluster Comput.* 22: 1–9.

104 Gomathi, B., Krishnasamy, K., and Saravana Balaji, B. (2018). Epsilon-fuzzy dominance sort-based composite discrete artificial bee colony optimisation for multi-objective cloud task scheduling problem. *Int. J. Bus. Intell. Data Mining* 13: 247–266.

105 Song, B., Hassan, M.M., and Huh, E. (2010). A novel heuristic-based task selection and allocation framework in dynamic collaborative cloud service platform. 2nd IEEE International Conference on Cloud Computing Technology and Science (CloudCom) (30 November–3 December). 360–367.

106 Bilgaiyan, S., Sagnika, S., and Das, M. (2015). A multi-objective cat swarm optimization algorithm for workflow scheduling in cloud computing environment. *Int. J. Soft Comput.* 10: 37–45.

107 Abdullahi, M. and Ngadi, M.A. (2016). Symbiotic organism search optimization based task scheduling in cloud computing environment. *Future Gener. Comput. Syst.* 56: 640–650.

108 Torabi, S. and Safi-Esfahani, F. (2018). A dynamic task scheduling framework based on chicken swarm and improved raven roosting optimization methods in cloud computing. *J. Supercomput.* 74: 2581–2626.

109 Choudhary, A., Gupta, I., Singh, V., and Jana, P.K. (2018). A GSA based hybrid algorithm for bi-objective workflow scheduling in cloud computing. *Future Gener. Comput. Syst.* 83: 14–26.

9

Scheduling of Robotic Disassembly in Remanufacturing Using Bees Algorithms

Jiayi Liu[1], *Wenjun Xu*[1], *Zude Zhou*[1], *and Duc Truong Pham*[2]

[1] *School of Information Engineering, Wuhan University of Technology, Wuhan, China*
[2] *Department of Mechanical Engineering, University of Birmingham, Birmingham, UK*

9.1 Introduction

Low resource utilization and high environmental pollution are the major deficiencies of traditional manufacturing. It is said that in China, there are approximately 20 million mobile phones and 5 million cars scrapped every year [1]. Improper handling of end-of-life (EOL) products will cause resource-wasting and environmental pollution [2]. Given this situation, remanufacturing is studied to protect the environment and to save manufacturing resources by making full use of EOL products [3]. It takes both economic development and environmental protection into considerations by reusing the EOL products [4]. In the remanufacturing process, it is necessary to consider the disassembly process to obtain the high-value components. However, the disassembly process is usually executed by manual labor due to its high complexity. Robotic disassembly is recently being paid great attention because of the high automation potential. The cognitive robotics [5], its basic behavior control strategy [6], and learning/revision strategies [7] have been studied to realize the automated process of disassembling the screens.

Scheduling of robotic disassembly can reduce the cost and improve the efficiency of disassembly. Robotic disassembly cells (RDCs) and robotic disassembly lines (RDLs) are two major ways to disassemble EOL products. Scheduling of robotic disassembly mainly contains robotic disassembly sequence planning (RDSP) for an RDC and robotic disassembly line balancing problem (RDLBP) for an RDL. Considering disassembly precedence relationships of EOL products, finding the optimal disassembly sequences for RDCs is the purpose of RDSP [8]. In an RDL, the aim of RDLB is to assign appropriate disassembly tasks to robotic workstations to ensure even distribution of work among them [9]. Finding optimal solutions for RDSP and RDLB problems is important to improve disassembly efficiency.

Evolutionary Computation in Scheduling, First Edition. Edited by Amir H. Gandomi, Ali Emrouznejad, Mo M. Jamshidi, Kalyanmoy Deb, and Iman Rahimi.

RDSP and RDLBP are different from manual disassembly sequence planning (MDSP) and the manual disassembly line balancing problem (MDLBP). The moving time between different disassembly points (DPs) is always ignored in manual disassembly while in robotic disassembly it should not be ignored and can be determined by the end-effector's moving speed and the length of the movement. In addition, the end-effector should also avoid collisions with the product being disassembled.

The Bees algorithm mimics honeybees' foraging behavior and it has been applied to plenty of research fields, such as service aggregation optimization [10], image segmentation [11], and project scheduling [12]. It has been proved that Bees algorithms (BA) are highly competitive compared to the other methods [10]. However, little research has been conducted to solve the scheduling of robotic disassembly using a Bees algorithm.

In this chapter, scheduling of robotic disassembly (both RDSP [8] and RDLBP [9]) was solved by a Bees algorithm. The major contribution of this chapter is to consider the end-effector's movement characteristics, and the moving time caused by the end-effector's movement of different DPs is considered in the scheduling of robotic disassembly (both RDSP and RDLBP). Compared with the traditional methods, the optimal solutions of proposed methods obtained are more suitable for use in robotic disassembly. The related works are described in Section 9.2. Section 9.3 establishes the disassembly model. Consideration of the safe distance, the length of the obstacle-avoiding path built, and the optimization objectives of RDSP and RDLBP are also described in Section 9.4. In Section 9.5, BA are used to search for the optimal solutions of RDSP and RDLBP. Case studies are conducted to verify the proposed methods in Section 9.6. Finally, Section 9.7 concludes this chapter.

9.2 Related Works

Scheduling of robotic disassembly mainly includes RDSP for an RDC and RDLBP for an RDL. Much research has been conducted in disassembly sequence planning (DSP) and disassembly line balancing problem (DLBP) fields.

Under given optimization objectives, finding the optimal disassembly sequence of EOL products is the purpose of DSP. Xing et al. [13] considered the disassembly distance, the disassembly direction changes, and the length of disassembly sequence, then they used the ant colony optimization algorithm to get the optimal solutions. Kheder et al. [14] used the genetic algorithm to get the optimal solutions of a rear axle. Jin et al. [15] used a brute-force method to get the optimal solutions of disassembling the waste electrical and electronic equipment. Meng et al. [16] used an improved co-evolutionary algorithm to simultaneously optimize the recovery options and disassembly level. For parallel and selective disassembly, Kim et al. [17] minimized the sequence-dependent set-ups and the operation cost of disassembly process using a branch and bound algorithm. When RDSP is studied, the end-effector's movement characteristics should be considered. The total disassembly time consists of the basic disassembly time, the direction-change time, the tool-change time, and the moving time of robot's end-effector between different DPs [18]. For robotic disassembly, the optimal solutions of disassembling personal computers were generated using a genetic algorithm [19]. Then DSP was solved using an online genetic algorithm to disassemble the same personal computers [20]. However, the end-effector's moving time was computed using the end-effector's moving speed and Euclidean distance between different DPs. For the practical robotic disassembly, the end-effector's moving path cannot be a straight-line path. The reason is that the end-effector needs to move along the obstacle-avoiding path to avoid physical collisions with the contour of EOL products. Thus, the existing methods used for DSP are not applicable to RDSP.

A disassembly line contains some disassembly workstations, and balancing the workstations' workload in a disassembly line is the optimization objective of DLBP. McGovern et al. [21] have proven that the decision version of DLBP is NP-complete problems and its optimization version is NP-hard problem. The linear physical programming was utilized by Ilgin et al. [22] to balance the disassembly line used for disassembling the smartphones. To disassemble cell phones, Ding et al. [23] minimized the demand rating, workstation number, and the balance measure using the ant colony algorithm. To optimize the design cost of a disassembly line, Ayyuce et al. [24] utilized a genetic algorithm to get the optimal solutions of stochastic DLBP. Bentaha et al. [25] solved DLBP, of which the disassembly task time was described by known probability distributions using the sample average approximation method and stochastic linear mixed-integer programming. To get optimal solutions of DLBP using reasonable computation time, Mete et al. [26] minimized the workstation number of a disassembly line using the beam search algorithm. Nowadays, more attention has been paid to sequence-dependent DLBP in which, if component A interacts with component B, extra time should be added. To solve sequence-dependent DLBP, ant colony optimization [27], particle swarm optimization algorithm [28], improved artificial bee colony algorithm [29], hybrid genetic algorithm [30], and tabu search [31] were used to minimize the workstation number, the hazardous rating, the demand rating, and the balancing measures. The RDL is different from manual disassembly line in which the moving time between different DPs for each workstation is always ignored. In RDLBP, for each robotic workstation, EOL products can be the obstacles along the end-effector's moving path. However, the existing research always ignores this and is not applicable to RDLBP.

Many metaheuristic algorithms are utilized to get the optimal solutions of DSP and DLBP. To disassemble machine vises, Xu et al. [32] solved DSP using the adaptive particle swarm optimization algorithm of which the mutation probability and inertia weight were adaptively determined. Luo et al. [33] obtained the optimal disassembly sequences of single-speed reduction gearbox using multi-layer representation method and ant colony optimization algorithm. For the U-shaped disassembly line, optimal solutions of RDLBP were obtained by Agrawal et al. [34] using collaborative ant colony algorithm. To solve DLBP with a fixed workstation number, Kalaycılar et al. [35] maximized the total net revenue using upper and lower bounding procedures to assign suitable disassembly tasks to the workstations. The profit-oriented partial DLBP was studied by Ren et al. [2] using an improved gravitational search algorithm. BA mimics honeybees' foraging behavior to get the optimal solutions and it is successfully applied to many research fields. Xu et al. [10] solved the service aggregation optimization problems using an improved BA. BA has also been applied to a fast scheme for multi-level image thresholding of image segmentation [11]. Oztemel et al. [12] solved the resource-constrained project scheduling problem in the mold industry using BA. The Bees algorithm has shown its great competitiveness compared with the other optimization algorithms [10], but it has not been applied to the scheduling of robotic disassembly yet.

9.3 Disassembly Model

To build the disassembly model, the disassembly precedence relationships of EOL products should be described. Many researchers either ignore the disassembly directions of components [23] or used the fixed disassembly direction to describe the disassembly directions of components [16]. Disassembly directions of components should be provided for the robots. The disassembly direction of components is not fixed and it should depend on the disassembly status of the other components [8]. Space interference matrix can be used to generate feasible disassembly directions of components according to the disassembly status of the other components [8]. In this chapter, the disassembly model is established.

9.3.1 Space Interference Matrix

The space interference matrix S_{od} along orthogonal directions should be built first [15]. In S_{od}, each element s_{mn} represents when component n exists, whether component m could be disassembled along od direction or not. $s_{mn} = 0$ means component m could be disassembled along od direction if component n is not disassembled. Otherwise, $s_{mn} = 1$. The space interference matrices are built using an example as represented in Figure 9.1.

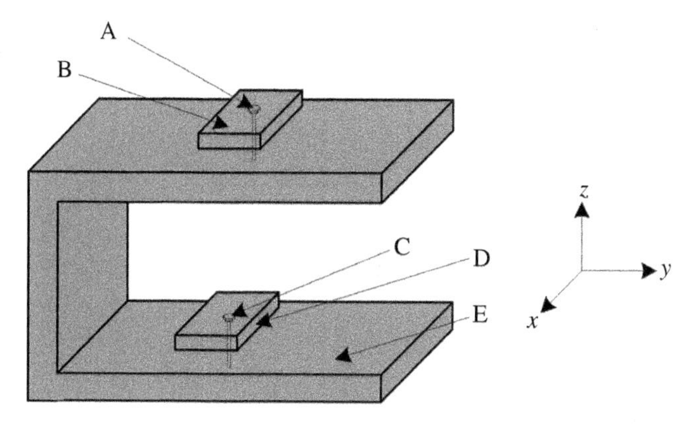

Figure 9.1 An example for space interference matrix.

This assembly's space interference matrices are represented by Eq. (9.1). In matrix S_{z+}, s_{CE} is equal to 0. Although bolt C contacts component E, unscrewing operations along z + direction could be applied to bolt C to realize disassembly. In matrix S_{z-}, s_{EC} is equal to 1. The reason is that before bolt C is disassembled, component E cannot not be disassembled in the z − direction.

$$
S_{x+}=\begin{matrix} & \begin{matrix}A & B & C & D & E\end{matrix} \\ \begin{matrix}A\\B\\C\\D\\E\end{matrix} & \begin{bmatrix}0&1&0&0&1\\1&0&0&0&0\\0&0&0&1&1\\0&0&1&0&0\\1&0&1&0&0\end{bmatrix}\end{matrix};
S_{y+}=\begin{matrix} & \begin{matrix}A & B & C & D & E\end{matrix} \\ \begin{matrix}A\\B\\C\\D\\E\end{matrix} & \begin{bmatrix}0&1&0&0&1\\1&0&0&0&0\\0&0&0&1&1\\0&0&1&0&0\\1&0&1&1&0\end{bmatrix}\end{matrix};
S_{z+}=\begin{matrix} & \begin{matrix}A & B & C & D & E\end{matrix} \\ \begin{matrix}A\\B\\C\\D\\E\end{matrix} & \begin{bmatrix}0&0&0&0&0\\1&0&0&0&0\\0&0&0&0&0\\0&0&1&0&0\\1&1&1&1&0\end{bmatrix}\end{matrix}
$$

$$
S_{x-}=\begin{matrix} & \begin{matrix}A & B & C & D & E\end{matrix} \\ \begin{matrix}A\\B\\C\\D\\E\end{matrix} & \begin{bmatrix}0&1&0&0&1\\1&0&0&0&0\\0&0&0&1&1\\0&0&1&0&0\\1&0&1&0&0\end{bmatrix}\end{matrix};
S_{y-}=\begin{matrix} & \begin{matrix}A & B & C & D & E\end{matrix} \\ \begin{matrix}A\\B\\C\\D\\E\end{matrix} & \begin{bmatrix}0&1&0&0&1\\1&0&0&0&0\\0&0&0&1&1\\0&0&1&0&0\\1&0&1&0&0\end{bmatrix}\end{matrix};
S_{z-}=\begin{matrix} & \begin{matrix}A & B & C & D & E\end{matrix} \\ \begin{matrix}A\\B\\C\\D\\E\end{matrix} & \begin{bmatrix}0&1&0&1&1\\1&0&1&1&1\\0&0&0&1&1\\0&0&1&0&1\\1&0&1&1&0\end{bmatrix}\end{matrix} \quad (9.1)
$$

9.3.2 Interference Matrix Analysis

Feasible disassembly sequences are generated using interference matrix analysis. First, as represented by Eq. (9.2), space interference matrices ($S_{x\pm}$, $S_{y\pm}$, and $S_{z\pm}$) along orthogonal directions are combined into matrix $S_{x,y,z}$. From Eq. (9.2), the first element of column *result* in matrix $S_{x,y,z}$ is 111101. The "OR" operator is applied to the first row of all the six interference matrices to get this element, which contains 6 bits. Only the fifth bit of this element is 0, it represents the fact that the corresponding component could be removed along the z+ direction. It is apparent that both bolts A and C could be removed along the z+ direction as represented by Eq. (9.2). If bolt A is randomly selected and disassembled along the z+ direction, the first row and the first column of matrix $S_{x,y,z}$ should be deleted as shown by Eq. (9.3). It is apparent that component B could be disassembled along any direction except the z− direction, as represented by Eq. (9.3). Component C could be removed along only the z+ direction. Here, it is randomly selected and disassembled along the z+ direction. After that, from Eq. (9.4), it is apparent that both components B and D could be disassembled along any direction except the z− direction, component E could be disassembled along x± and y− directions. If component B is randomly selected and disassembled along y+ direction, according to the same rule as represented by Eq. (9.5), matrix $S_{x,y,z}$ should be updated. As represented by Eq. (9.5), component D could be disassembled along any direction except the z- direction, component E can be disassembled along x± and y− directions. If component E is randomly selected and disassembled along x+ direction, component D could be disassembled along any direction (x+ direction is randomly selected). Thus, A/C/B/E/D and z+/z+/y+/x+/x+ are the feasible solutions.

$$
S_{x,y,z} = \begin{array}{c} \\ A \\ B \\ C \\ D \\ E \end{array}
\begin{array}{c} \begin{array}{ccccc} A & B & C & D & E \end{array} \\ \begin{bmatrix} 000000 & 111101 & 000000 & 000001 & 111101 \\ 111111 & 000000 & 000001 & 000001 & 000001 \\ 000000 & 000000 & 000000 & 111101 & 111101 \\ 000000 & 000000 & 111111 & 000000 & 000001 \\ 111111 & 000010 & 111111 & 001011 & 000000 \end{bmatrix} \end{array}
\begin{array}{c} result \\ 111101 \\ 111111 \\ 111101 \\ 111111 \\ 111111 \end{array} \tag{9.2}
$$

$$
S_{x,y,z} = \begin{array}{c} \\ B \\ C \\ D \\ E \end{array}
\begin{array}{c} \begin{array}{cccc} B & C & D & E \end{array} \\ \begin{bmatrix} 000000 & 000001 & 000001 & 000001 \\ 000000 & 000000 & 111101 & 111101 \\ 000000 & 111111 & 000000 & 000001 \\ 000010 & 111111 & 001011 & 000000 \end{bmatrix} \end{array}
\begin{array}{c} result \\ 000001 \\ 111101 \\ 111111 \\ 111111 \end{array} \tag{9.3}
$$

$$
S_{x,y,z} = \begin{array}{c} \\ B \\ D \\ E \end{array}
\begin{array}{c} \begin{array}{ccc} B & D & E \end{array} \\ \begin{bmatrix} 000000 & 000001 & 000001 \\ 000000 & 000000 & 000001 \\ 000010 & 001011 & 000000 \end{bmatrix} \end{array}
\begin{array}{c} result \\ 000001 \\ 000001 \\ 001011 \end{array} \tag{9.4}
$$

$$S_{x,y,z} = \begin{array}{c} \\ D \\ E \end{array} \begin{array}{cc} D & E \\ \left[\begin{array}{cc} 000000 & 000001 \\ 001011 & 000000 \end{array}\right] \end{array} \begin{array}{c} result \\ 000001 \\ 001011 \end{array} \tag{9.5}$$

9.4 Optimization Objective

For RDC, finding the optimal disassembly sequences of EOL products helps to improve the disassembly efficiency of RDC, which is the core purpose of RDSP. Besides, RDL is the flow-oriented product system, which includes some robotic workstations. It continuously launches the disassembly tasks down the disassembly line and EOL products are sequentially handled from one to another. Disassembly operations are repeatedly executed in each robotic workstation within the cycle time of RDL. RDLBP is the decision problem of balancedly assigning disassembly tasks to robotic workstations. Optimization objectives of both RDSP and RDLBP should be determined in advance.

In this chapter, the following assumptions are made: (i) the EOL products' structure and geometric information are known in advance; (ii) all parts in the EOL products can be disassembled; (iii) the EOL products are used for complete disassembly; (iv) the basic disassembly time, direction-change time, and tool-change time are constant and deterministic; (v) each disassembly task is assigned to only one robotic workstation in the RDL; and (vi) the total working time of robotic workstation is no greater than the cycle time of RDL.

9.4.1 The End-effector's Moving Time

The end-effector's moving time should not be ignored in robotic disassembly. Strictly, the obstacle-avoiding path and the collision-free trajectory should be considered [36]. The end-effector's moving time should be determined by the position and complexity of products and the robot types. In this chapter, the moving time is generated by the moving path length and the end-effector's velocity.

A simple example is illustrated here to represent the calculation process of the moving time between different DPs as shown in Figure 9.2. First, it should be ensured that there are no collisions. The black dotted line is the end-effector's moving path as represented in Figure 9.2. Besides, the length matrix MP which represents the moving path length is denoted by Eq. (9.6). For example, the moving path length between DPs A and B is expressed as a_{12}, which is calculated by Eq. (9.7). After matrix MP is decided, Eq. (9.8) is utilized to compute the moving time, where v_{ee} is the end-effector's linear velocity.

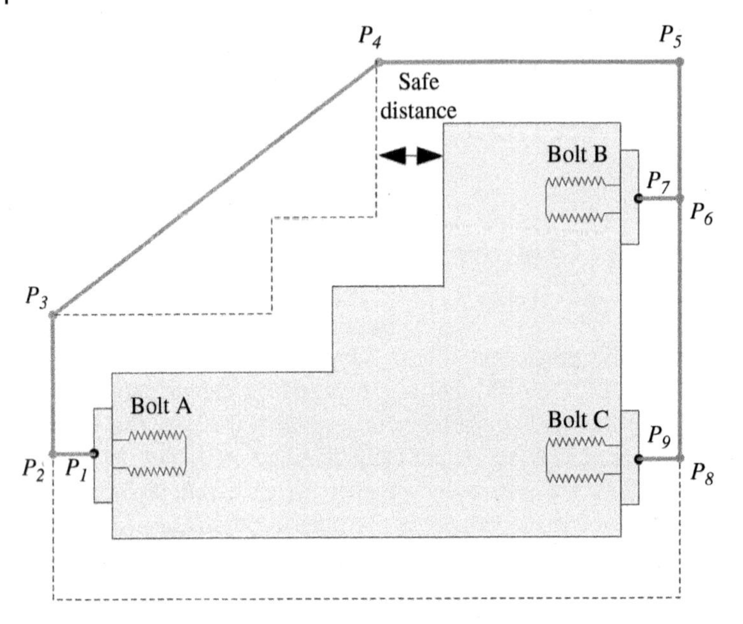

Figure 9.2 An example to calculate the moving path between different DPs.

$$MP = \begin{array}{c} \\ A \\ B \\ C \end{array} \begin{array}{c} \begin{array}{ccc} A & B & C \end{array} \\ \begin{bmatrix} 0 & a_{12} & a_{13} \\ a_{12} & 0 & a_{23} \\ a_{13} & a_{23} & 0 \end{bmatrix} \end{array} \tag{9.6}$$

$$a_{12} = P_1P_2 + P_2P_3 + P_3P_4 + P_4P_5 + P_5P_6 + P_6P_7 \tag{9.7}$$

$$mt(i,j) = MP(i,j) / v_{ee} \tag{9.8}$$

9.4.2 Optimization Objectives of RDSP

For RDC, minimizing the disassembly time of an EOL product is the optimization objective of RDSP. It mainly consists of the basic disassembly time, direction-change time, tool-change time, and the end-effector's moving time.

The time used for disassembling a component using the robots is the basic disassembly time. To cope with the direction changes for the robots, extra time is needed to adjust its posture. The direction-change time is expressed by Eq. (9.9). Similarly, the robot also needs extra disassembly time to handle disassembly tool changes. As shown in Figure 9.3, different disassembly tools correspond to different disassembly operations. The tool-change time is represented by Eq. (9.10)

Disassembly operations	Disassembly tools	Disassembly tools (Subclass)
Unscrewing	Spanners/Screwdriver/...	M1/M2/M3/M4/... Phillips/Slotted/...
Rotating	Spanners...	Size I/Size II/Size III/...
...	...	...
Removing	Grippers/Pliers/...	Size I/Size II/Size III/...
Shredding	Hammer/Cutter/...	Size/Ball, Nail Hammer/...
Cutting	Electric cutting/...	Cutting disc/Thickness/...

Figure 9.3 Disassembly operations and tools.

where *Scd, Spn, Pli, Grp, Elc*, and *Ham* respectively indicate screwdriver, spanner, plier, gripper, electrical cutting and hammer. For the same operations, different disassembly tools (subclass) should also be included to disassemble different component types as represented by Eq. (9.11). For instance, different types of bolts should be disassembled by suitable types of spanners. The end-effector's moving time is represented in Section 9.4.1.

$$dt(x_i, x_{i+1}) = \begin{cases} 0 & \text{direction changes are } 0^{\circ} \\ 1 & \text{direction changes are } 90^{\circ} \\ 2 & \text{direction changes are } 180^{\circ} \end{cases} \tag{9.9}$$

$$TT = \begin{array}{c} \\ Spn \\ Scd \\ Grp \\ Pli \\ Elc \\ Ham \end{array} \begin{array}{c} \begin{array}{cccccc} Spn & Scd & Grp & Pli & Elc & Ham \end{array} \\ \begin{bmatrix} TT_1 & tt_{1,2} & tt_{1,3} & tt_{1,4} & tt_{1,5} & tt_{1,6} \\ tt_{2,1} & TT_2 & tt_{2,3} & tt_{2,4} & tt_{2,5} & tt_{2,6} \\ tt_{3,1} & tt_{3,2} & TT_3 & tt_{3,4} & tt_{3,5} & tt_{3,6} \\ tt_{4,1} & tt_{4,2} & tt_{4,3} & TT_4 & tt_{4,5} & tt_{4,6} \\ tt_{5,1} & tt_{5,2} & tt_{5,3} & tt_{5,4} & TT_5 & tt_{5,6} \\ tt_{6,1} & tt_{6,2} & tt_{6,3} & tt_{6,4} & tt_{6,5} & TT_6 \end{bmatrix} \end{array} \tag{9.10}$$

$$TT_1 = \begin{array}{c} \\ M1 \\ M2 \\ M3 \\ \cdots \\ Mn \end{array} \begin{array}{c} \begin{array}{cccccc} M1 & M2 & M3 & \ldots & Mn \end{array} \\ \begin{bmatrix} 0 & tta_{1,2} & tta_{1,3} & \ldots & tta_{1,n} \\ tta_{2,1} & 0 & tta_{2,3} & \ldots & tta_{2,n} \\ tta_{3,1} & tta_{3,2} & 0 & \ldots & tta_{3,n} \\ \ldots & \ldots & \ldots & \ldots & \ldots \\ tta_{n,1} & tta_{n,2} & tta_{n,3} & \ldots & 0 \end{bmatrix} \end{array} \tag{9.11}$$

Minimizing the disassembly time of an EOL product is the optimization objective which is expressed by Eq. (9.12), where X is the feasible disassembly sequence, n is the part number of EOL products, and $bt(x_i)$ represents the basic disassembly time of part x_i. The tool-change time, direction-change time, and the end-effector's moving time between parts x_i and x_{i+1} are respectively represented by $tt(x_i, x_{i+1})$, $dt(x_i, x_{i+1})$ and $mt(x_i, x_{i+1})$.

$$f(X) = \sum_{i=0}^{n-1} bt(x_i) + \sum_{i=0}^{n-2} mt(x_i, x_{i+1}) + \sum_{i=0}^{n-2} tt(x_i, x_{i+1}) + \sum_{i=0}^{n-2} dt(x_i, x_{i+1}) \tag{9.12}$$

The process of calculating the fitness value of a given feasible disassembly sequence is described using an example here. The feasible disassembly solution is A/C/B/E/D and x+/z+/x+/x+/z+, the corresponding disassembly tool is *Td*/*Ta*/*Ta*/*Te*/*Td* (spanner-I, spanner-II, spanner-III, gripper-I, and gripper-II are respectively represented by *Ta*, *Tb*, *Tc*, *Td*, and *Te*). The tool-change time and direction-change time are respectively represented by Eqs. (9.9) and (9.13). Eq. (9.14) calculates the end-effector's moving time. In this chapter, the basic disassembly time is assumed to be constant. The variable parameters are the latter three factors of Eq. (9.12). The fitness value of this feasible disassembly sequence could be calculated by Eq. (9.15).

$$TT = \begin{array}{c} \\ Ta \\ Tb \\ Tc \\ Td \\ Te \end{array} \begin{array}{c} \begin{array}{ccccc} Ta & Tb & Tc & Td & Te \end{array} \\ \begin{bmatrix} 0 & 1 & 1 & 2 & 2 \\ 1 & 0 & 1 & 2 & 2 \\ 1 & 1 & 0 & 2 & 2 \\ 2 & 2 & 2 & 0 & 1 \\ 2 & 2 & 2 & 1 & 0 \end{bmatrix} \end{array} \tag{9.13}$$

$$mt = \begin{array}{c} \\ A \\ B \\ C \\ D \\ E \end{array} \begin{array}{c} \begin{array}{ccccc} A & B & C & D & E \end{array} \\ \begin{bmatrix} 0 & 2 & 3 & 4 & 5 \\ 2 & 0 & 1 & 2 & 3 \\ 3 & 1 & 0 & 1 & 4 \\ 4 & 2 & 1 & 0 & 2 \\ 5 & 3 & 4 & 2 & 0 \end{bmatrix} \end{array} \tag{9.14}$$

$$
\begin{aligned}
fitness &= \sum_{i=0}^{n-2} dt(x_i, x_{i+1}) + \sum_{i=0}^{n-2} tt(x_i, x_{i+1}) + \sum_{i=0}^{n-2} mt(x_i, x_{i+1}) \\
&= (1+1+0+1) + (2+0+2+1) + (3+1+3+2) \\
&= 17
\end{aligned}
\tag{9.15}
$$

9.4.3 Optimization Objectives of RDLBP

As represented by Eqs. (9.16)–(9.18), minimizing the robotic workstation number (f_1), balancing the robotic workstation's workload (f_2), and disassembling high demand parts as early as possible (f_3) are the optimization objectives used in this chapter. Minimizing optimization objectives f_1 and f_2 respectively can reduce the cost and balance the robotic workstation's workload. Minimizing the optimization objective f_3 helps to protect high-demand parts from damage in the robotic disassembly process.

$$\min f_1 = m \tag{9.16}$$

$$\min f_2 = \sum_{j=1}^{m} (CT - t_{j,total})^2 \tag{9.17}$$

$$\min f_3 = \sum_{i=1}^{n} i * h_{x_i} \tag{9.18}$$

subject to:

$$X = (x_1, x_2, \ldots, x_n) \in FX \tag{9.19}$$

$$\left\lceil \frac{\sum_{i=1}^{n} t_{x_i}}{CT} \right\rceil \leq m \leq n \tag{9.20}$$

$$t_{j,total} \text{ "} CT \tag{9.21}$$

$$\text{component } x_i \text{ is only assigned to } Rob\ Work_j, \forall i \in [0,n], \forall j \in [0,m] \tag{9.22}$$

In Eqs. (9.16) (9.22), m, n, h_{xi}, and $t_{j,total}$ respectively indicate the robotic workstation number, part number of EOL products, the demand value of part xj and the jth robotic workstation's total working time. The cycle time CT is a predefined parameter in this chapter. X and FX respectively indicate the feasible disassembly sequence and feasible disassembly sequence set. Eqs. (9.19) and

(9.20) respectively ensure the disassembly sequence is feasible and the robotic workstation number is limited within permitted values. Eqs. (9.21) and (9.22) respectively indicate that the total of any robotic workstation's working time should be no greater than the cycle time of RDL and a task is only assigned to a workstation.

Feasible solutions of RDL should be obtained by the following method. Based on a given feasible disassembly sequence, an example is illustrated to generate the feasible disassembly solution and calculate its multi-objective values. Table 9.1 lists the properties of the given disassembly sequence "3-5-6-1-4-2-7-8" and the CT of RDL is 25 s. The direction-change time, tool-change time, and the length matrix MP are respectively represented by Eqs. (9.9), (9.23) and (9.24). As represented in Figure 9.4, robotic workstations are assigned different disassembly tasks using the allocation matrix $A = [a_i]$. As shown in step 1, a_4 and a_6 are respectively calculated by Eqs. (9.25) and (9.26). It is obvious that a_3 is less than CT (25 s) and a_4 is greater than CT, which means robotic workstation 1 is assigned parts 3, 5, and 6. Then parts 3, 5, and 6 should be deleted and the disassembly sequence should be "1-4-2-7-8". In step 2, the allocation matrix is obtained in the same manner. The former three elements in the allocation matrix are less than CT, which means robotic workstation 2 is assigned parts 1, 4, and 2. Similarly, in step 3, robotic workstation 3 is assigned parts 7 and 8. Finally, the multi-objective values of this solution are calculated by Eqs. (9.27)–(9.29).

$$TT = \begin{array}{c} \\ Ta \\ Tb \\ Td \\ Te \end{array} \begin{array}{c} \begin{array}{cccc} Ta & Tb & Td & Te \end{array} \\ \begin{bmatrix} 0 & 1 & 2 & 2 \\ 1 & 0 & 2 & 2 \\ 2 & 2 & 0 & 1 \\ 2 & 2 & 1 & 0 \end{bmatrix} \end{array} \tag{9.23}$$

$$MP = \begin{bmatrix} 0 & 14 & 21 & 18 & 12 & 15 & 17 & 22 \\ 14 & 0 & 25 & 20 & 15 & 13 & 19 & 23 \\ 21 & 25 & 0 & 15 & 19 & 24 & 21 & 30 \\ 18 & 20 & 15 & 0 & 20 & 30 & 35 & 28 \\ 12 & 15 & 19 & 20 & 0 & 25 & 10 & 10 \\ 15 & 13 & 24 & 30 & 25 & 0 & 20 & 15 \\ 17 & 19 & 21 & 35 & 10 & 20 & 0 & 17 \\ 22 & 23 & 30 & 28 & 10 & 15 & 17 & 0 \end{bmatrix} \tag{9.24}$$

Table 9.1 Properties of a given disassembly sequence.

Disassembly sequence	3	5	6	1	4	2	7	8
Basic disassembly time	3	3.5	3	2.5	6	1.5	2	1.5
Disassembly direction	z+	y+	x−	x−	z+	y+	x+	x+
Disassembly tool	*Tb*	*Ta*	*Td*	*Td*	*Tb*	*Tb*	*Ta*	*Td*
h_{xi}	2	3	3	1	4	2	3	2
End-effector's moving speed	10 cm/s							

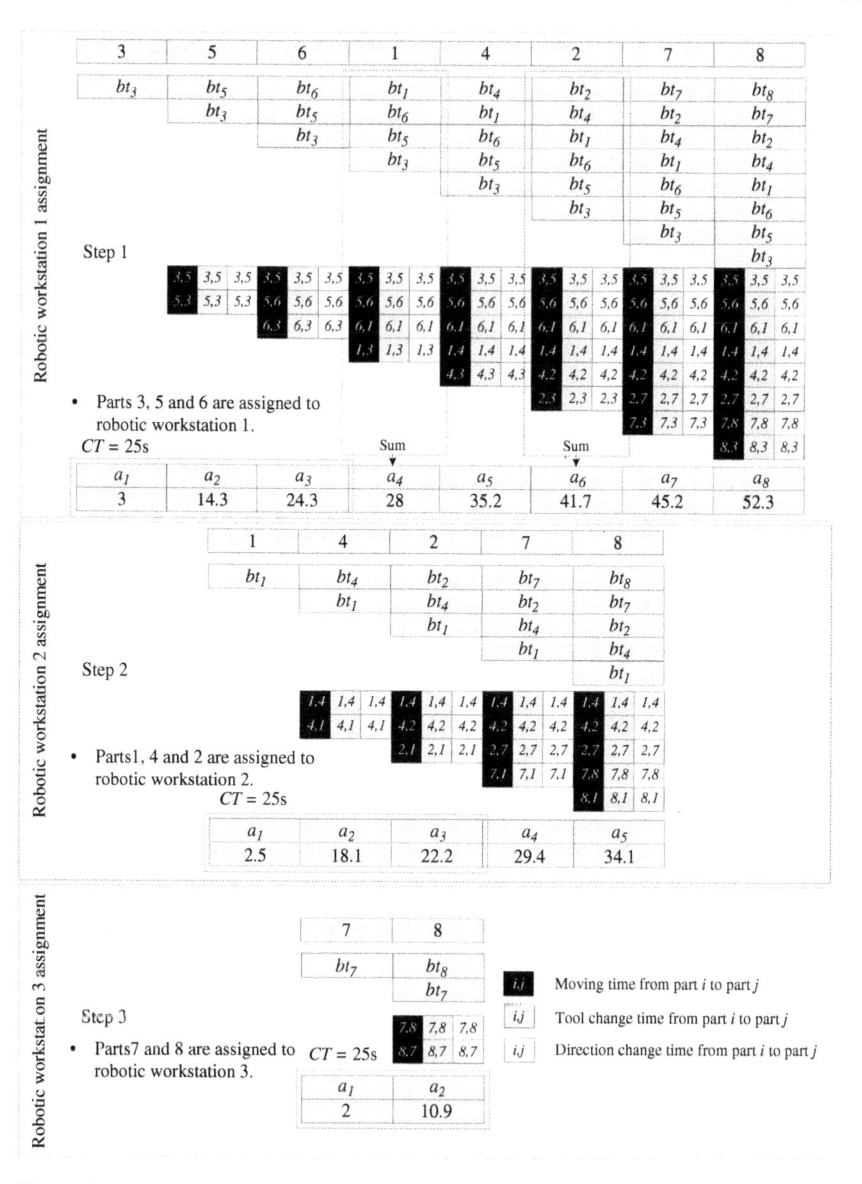

Figure 9.4 The robotic workstation assignment method for RDLBP.

$$\begin{aligned} a_4 &= bt_3 + bt_5 + bt_6 + bt_1 + mt_{3,5} + tt_{3,5} + dt_{3,5} + mt_{5,6} + tt_{5,6} + dt_{5,6} + \\ &\quad mt_{6,1} + tt_{6,1} + dt_{6,1} + mt_{1,3} + tt_{1,3} + dt_{1,3} \\ &= 3 + 3.5 + 3 + 2.5 + 19/10 + 1 + 1 + 25/10 + 2 + 1 + 15/10 + 0 + 0 + \\ &\quad 21/10 + 1 + 2 = 28 \end{aligned} \tag{9.25}$$

$$\begin{aligned} a_6 &= bt_3 + bt_5 + bt_6 + bt_1 + bt_4 + bt_2 + mt_{3,5} + tt_{3,5} + dt_{3,5} + mt_{5,6} + tt_{5,6} + dt_{5,6} + \\ &\quad mt_{6,1} + tt_{6,1} + dt_{6,1} + mt_{1,4} + tt_{1,4} + dt_{1,4} + mt_{4,2} + tt_{4,2} + dt_{4,2} + mt_{2,3} + tt_{2,3} + dt_{2,3} \\ &= 3 + 3.5 + 3 + 2.5 + 6 + 1.5 + 19/10 + 1 + 1 + 25/10 + 1 + 2 + 15/10 + 0 + 0 + \\ &\quad 18/10 + 2 + 1 + 20/10 + 0 + 1 + 25/10 + 0 + 1 = 41.7 \end{aligned} \tag{9.26}$$

$$f_1 = 3 \tag{9.27}$$

$$f_2 = \sum_{j=1}^{m} (CT - t_{j,total})^2 = (25-3)^2 + (25-14.3)^2 + (25-24.3)^2 = 598.98 \tag{9.28}$$

$$f_3 = \sum_{i=1}^{n} i * h_{x_i} = 1\times2 + 2\times3 + 3\times3 + 4\times1 + 5\times4 + 6\times2 + 7\times3 + 8\times2 = 90 \tag{9.29}$$

9.5 Bees Algorithm

9.5.1 The Basic BA

BA mimics honeybees' foraging behavior to get optimal solutions [37]. In basic BA, several parameters are used: number of scout-bees *scoutn*, number of selected sites *m*, number of elite sites *n*, number of follower bees of selected site *mb*, number of follower bees of elite sites *nb*, patch size *ngh* and iteration number *iter*. As shown in Figure 9.5, the solution space is randomly searched by *scoutn* scout bees. The visited sites' fitness values are computed and then sorted in ascending order. The best *m* and *n* sites are regarded as "selected sites" and "elite sites" respectively. The neighborhood space of each elite (non-elites selected) site is searched by *nb* (*mb*) follower bees. If the quality of the newly searched site is better than the original one, the follower bee will replace this scout bee. Otherwise, this scout bee remains unchanged. After that, the remaining *scoutn* – *m* non-selected sites are randomly searched as the global search strategy. All the *scoutn* sites should be sorted by the fitness values in ascending order and the next iteration process starts. When the current iteration number is larger than *iter*, the iteration process stops. Finally, the best bee in the *scoutn* scout Bees is regarded as the best solution.

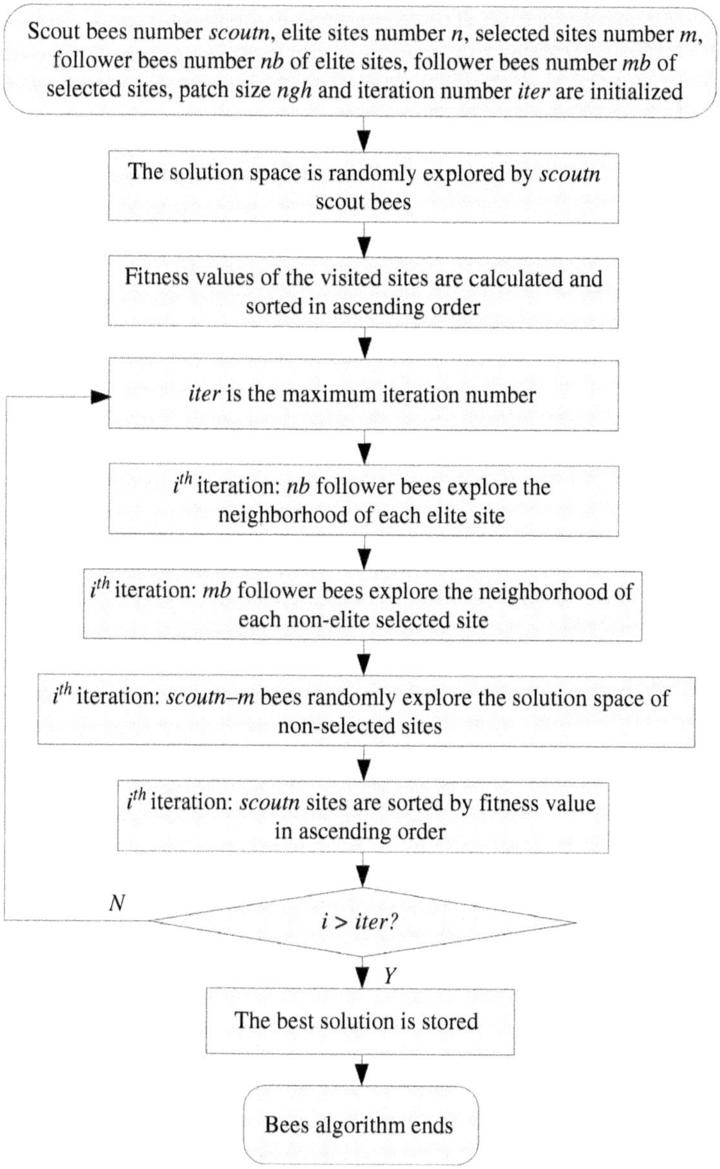

Scout bees number *scoutn*, elite sites number *n*, selected sites number *m*, follower bees number *nb* of elite sites, follower bees number *mb* of selected sites, patch size *ngh* and iteration number *iter* are initialized

The solution space is randomly explored by *scoutn* scout bees

Fitness values of the visited sites are calculated and sorted in ascending order

iter is the maximum iteration number

i^{th} iteration: *nb* follower bees explore the neighborhood of each elite site

i^{th} iteration: *mb* follower bees explore the neighborhood of each non-elite selected site

i^{th} iteration: *scoutn–m* bees randomly explore the solution space of non-selected sites

i^{th} iteration: *scoutn* sites are sorted by fitness value in ascending order

N

$i > iter?$

Y

The best solution is stored

Bees algorithm ends

Figure 9.5 The flowchart of basic BA.

9.5.2 Enhanced Discrete Bees Algorithm

RDSP is a combinatorial optimization problem and solved by enhanced discrete Bees algorithm (EDBA) in this chapter. The parameters *scoutn*, *m*, *b*, *mb*, *nb*, and *iter* of EDBA are similar to those of basic BA. Then, the disassembly model mentioned in Section 9.3 is utilized to generate *scoutn* feasible disassembly sequences which are regarded as *scoutn* scout bees of EDBA in Section 9.5.2.1. The nectar source (sites) are randomly searched and these sites are sorted by fitness values in ascending order. The best *n* and *m* sites are regarded as "elite sites" and "selected sites" respectively. The neighborhood space of each elite (non-elite selected) site is searched by *nb* (*mb*) follower bees using the neighborhood search strategy. The best bee in the *nb* (*mb*) follower bees is selected and it is mutated using the mutation operator, namely, mutated bee. The mutated bee will replace the best bee if it is superior to the best bee. Otherwise, the best bee remains unchanged. The best bee will replace the original bee if the best bee is superior to the original scout bee. Otherwise, the original scout bee remains unchanged. The $scoutn - m$ non-selected sites are replaced by $scoutn - m$ new sites. Then, all the visited sites are sorted by fitness values in ascending order, and the next iteration starts. When the current iteration number is larger than *iter*, the iteration process stops. Figure 9.6 describes the flowchart of EDBA.

9.5.2.1 Representation of Bees

Figure 9.7 represents a bee. The disassembly sequence and disassembly direction arrays are generated using the method mentioned in Section 9.3, and the disassembly tool and disassembly moving time arrays can be determined accordingly. Finally, the fitness value of a bee is calculated using the method mentioned in Section 9.4.2.

9.5.2.2 Neighborhood Search Strategy

The neighborhood search strategy which contains insert and swap operators is described in Figure 9.8.

When the swap operator is used, two integer numbers which indicate the swapping locations of the bee are randomly generated. Then, the corresponding two bits will be exchanged to be a new bee as represented in Figure 9.8a. From Figure 9.8b, one bit is randomly selected and it inserts a random location of a bee for the insert operator.

For both swap and insert operators, the newly generated solutions may be the infeasible disassembly sequence. The new bee's feasibility should be checked after this new bee is generated. If the new solution is infeasible, the original site's neighborhood should be searched until the new solution is feasible.

9.5.2.3 Mutation Operator

A component may be disassembled along several disassembly directions during the disassembly process. As represented in Figure 9.9, a mutation operator is utilized in this chapter to increase the solution diversity. When a mutation operator is used, one bit of disassembly direction array of a bee is randomly selected and it

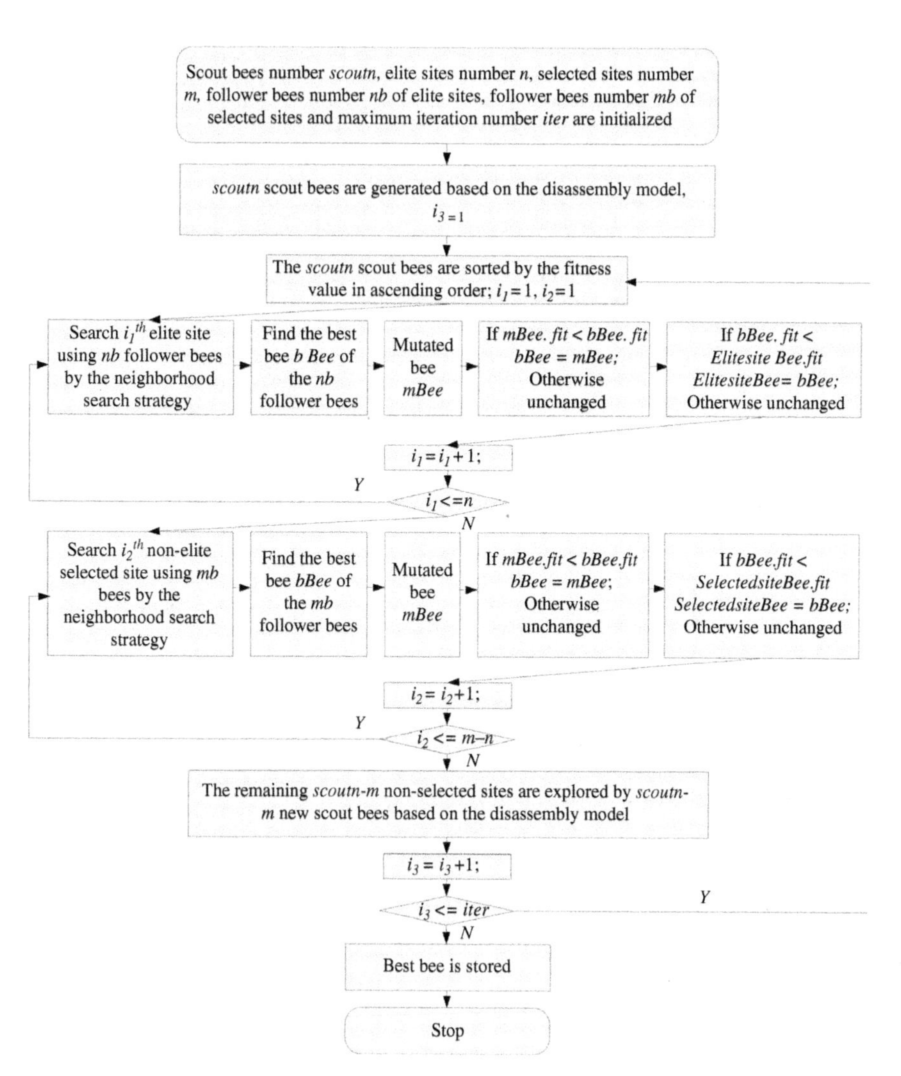

Figure 9.6 The flowchart of EDBA.

Disassembly sequence	3	5	6	1	4	2	7	8
Disassembly direction	+z	+y	−x	−x	+z	+y	+x	+x
Disassembly tool	*Tb*	*Ta*	*Td*	*Td*	*Tb*	*Tb*	*Ta*	*Td*
Disassembly moving time	$mt_{3,5}$	$mt_{5,6}$	$mt_{6,1}$	$mt_{1,4}$	$mt_{4,2}$	$mt_{2,7}$	$mt_{7,8}$	
Fitness value	fit_1							

Figure 9.7 Representation of a bee in EDBA.

(a)

Bee before swapping

3	5	6	1	4	2	7	8
+z	+y	−x	−x	+z	+y	+x	+x
Tb	*Ta*	*Td*	*Td*	*Tb*	*Tb*	*Ta*	*Td*
$mt_{3,5}$	$mt_{5,6}$	$mt_{6,1}$	$mt_{1,4}$	$mt_{4,2}$	$mt_{2,7}$	$mt_{7,8}$	
fit_1							

Bee after swapping

3	5	2	1	4	6	7	8
+z	+y	+y	−x	+z	−x	+x	+x
Tb	*Ta*	*Tb*	*Td*	*Tb*	*Td*	*Ta*	*Td*
$mt_{3,5}$	$mt_{5,2}$	$mt_{2,1}$	$mt_{1,4}$	$mt_{4,6}$	$mt_{6,7}$	$mt_{7,8}$	
fit_2							

(b)

Bee before inserting

3	5	6	1	4	2	7	8
+z	+y	−x	−x	+z	+y	+x	+x
Tb	*Ta*	*Td*	*Td*	*Tb*	*Tb*	*Ta*	*Td*
$mt_{3,5}$	$mt_{5,6}$	$mt_{6,1}$	$mt_{1,4}$	$mt_{4,2}$	$mt_{2,7}$	$mt_{7,8}$	
fit_1							

Bee after inserting

3	5	2	6	1	4	7	8
+z	+y	+y	−x	−x	+z	+x	+x
Tb	*Ta*	*Tb*	*Td*	*Td*	*Tb*	*Ta*	*Td*
$mt_{3,5}$	$mt_{5,2}$	$mt_{2,6}$	$mt_{6,1}$	$mt_{1,4}$	$mt_{4,7}$	$mt_{7,8}$	
fit_3							

Figure 9.8 Neighborhood search strategy of EDBA: (a) swap operator and (b) insert operator.

	3	5	6	1	4	2	7	8
	+z	+y	−x	−x	+z	+y	+x	+x
Bee before mutating	*Tb*	*Ta*	*Td*	*Td*	*Tb*	*Tb*	*Ta*	*Td*
	$mt_{3,5}$	$mt_{5,6}$	$mt_{6,1}$	$mt_{1,4}$	$mt_{4,2}$	$mt_{2,7}$	$mt_{7,8}$	
	fit_1							

	3	5	6	1	4	2	7	8
	+z	+y	−x	−x	+z	−y	+x	+x
Bee after mutating	*Tb*	*Ta*	*Td*	*Td*	*Tb*	*Tb*	*Ta*	*Td*
	$mt_{3,5}$	$mt_{5,6}$	$mt_{6,1}$	$mt_{1,4}$	$mt_{4,2}$	$mt_{2,7}$	$mt_{7,8}$	
	fit_4							

Figure 9.9 Mutation operator.

changes 180° (for example, from +y direction to −y direction). Besides, the new bee's feasibility also needs to be checked.

9.5.2.4 Global Search Strategy

To avoid getting trapped in local optimum, $scoutn - m$ non-selected sites should be replaced by $scoutn - m$ new sites using the method mentioned in Section 9.3.

9.5.3 Improved Multi-Objective Discrete Bees Algorithm

An improved multi-objective discrete Bees algorithm (IMODBA) is utilized to solve RDLBP and its flowchart is represented in Figure 9.10. Firstly, parameters *scoutn*, *m*, *mb*, and *iter* are initialized. Scout bees randomly explore the solution space and multi-objective values of the visited sites are obtained. After that, efficient non-dominated Pareto sorting method (ENS) and crowding-distance

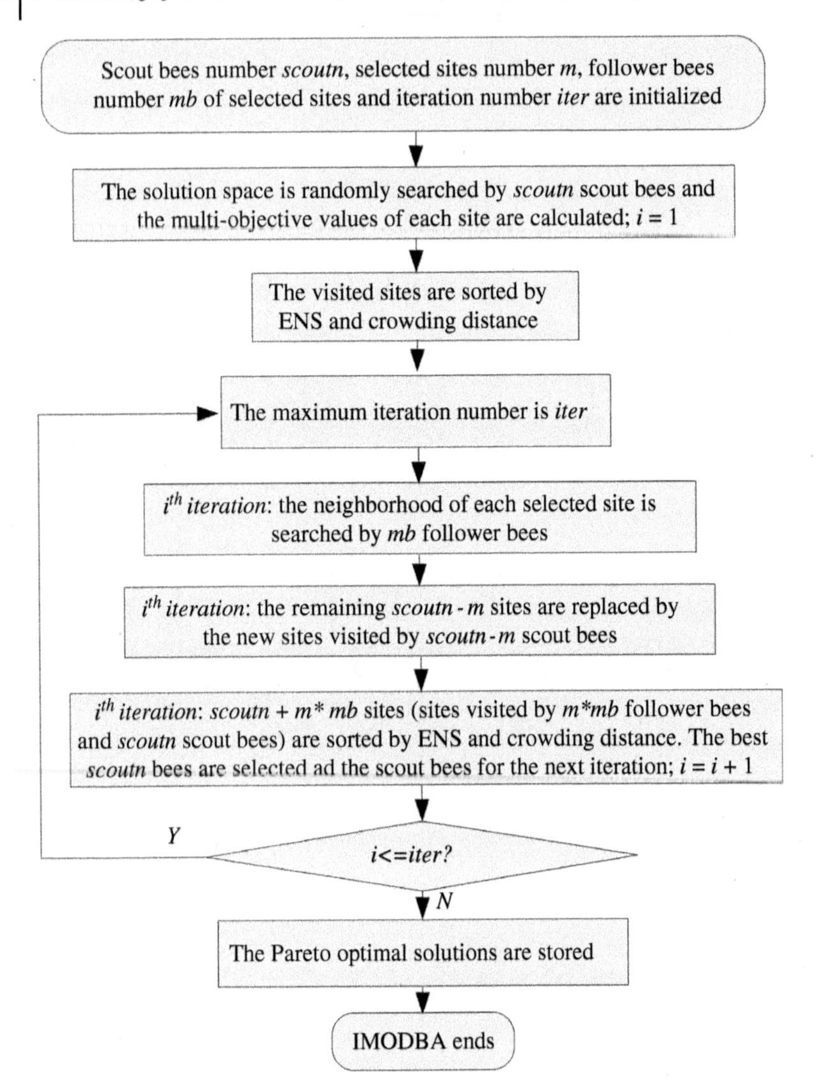

Figure 9.10 The flowchart of IMODBA.

method are utilized to sort the visited sites. Then the best *m* sites are chosen to be the "selected sites," and the neighborhood of each selected site is searched by *mb* follower bees. The remaining *scoutn-m* non-selected sites are replaced by *scoutn-m* new sites. At the end of each iteration process, $scoutn + m^{*}mb$ visited sites are sorted using ENS and crowding distance, the best *scoutn* bees are chosen for the next iteration process. When the current iteration number is larger than *iter*, the iteration process stops.

9.5.3.1 Representation of Bees

Figure 9.11 describes a bee of IMODBA. The disassembly sequence and disassembly direction are obtained in the same manner as in Section 9.5.2.1. The robotic workstation array and multi-objective value array are generated in the same manner as in Section 9.4.3.

9.5.3.2 Pareto Optimal Solution

In this chapter, minimizing the three objectives mentioned in Section 9.4.3 are the optimization objectives of RDLBP as represented by Eq. (9.30).

$$\min F(X) = \left[f_1(X), f_2(X), f_3(X) \right] \quad X \in FX \tag{9.30}$$

Suppose both X_1 and X_2 are feasible disassembly sequences, X_2 is dominated by X_1 if and only if Eq. (9.31) is satisfied:

$$(f_i(X_1) \leq f_i(X_2)) \& \& (f_j(X_1) < f_j(X_2)) \ \forall i = 1,2,3 \ \exists j = 1,2,3 \tag{9.31}$$

Solution X_{dom} is Pareto optimal solution if no solutions dominate solution X_{dom}. The Pareto optimal set is made up of all the Pareto optimal solutions, and the function values of these Pareto optimal solutions are the Pareto optimal front.

9.5.3.3 Pareto Sorting

According to the dominance relationships between different solutions, assigning front ranks to all the solutions is the aim of Pareto sorting. ENS is used to be the Pareto sorting method of IMODBA [38]. The flowchart of ENS is described in Figure 9.12. More details can be found in Appendix A.

Disassembly sequence	3	5	6	1	4	2	7	8
Disassembly direction	+z	+y	−x	−x	+z	+y	+x	+x
Robotic workstations	1	1	1	2	2	2	3	3
Disassembly tool	*Tb*	*Ta*	*Td*	*Td*	*Tb*	*Tb*	*Ta*	*Td*
h_{xi}	2	3	3	1	4	2	3	2
Multi-objective values	3	598.98	90					

Figure 9.11 Representation of a bee in IMODBA.

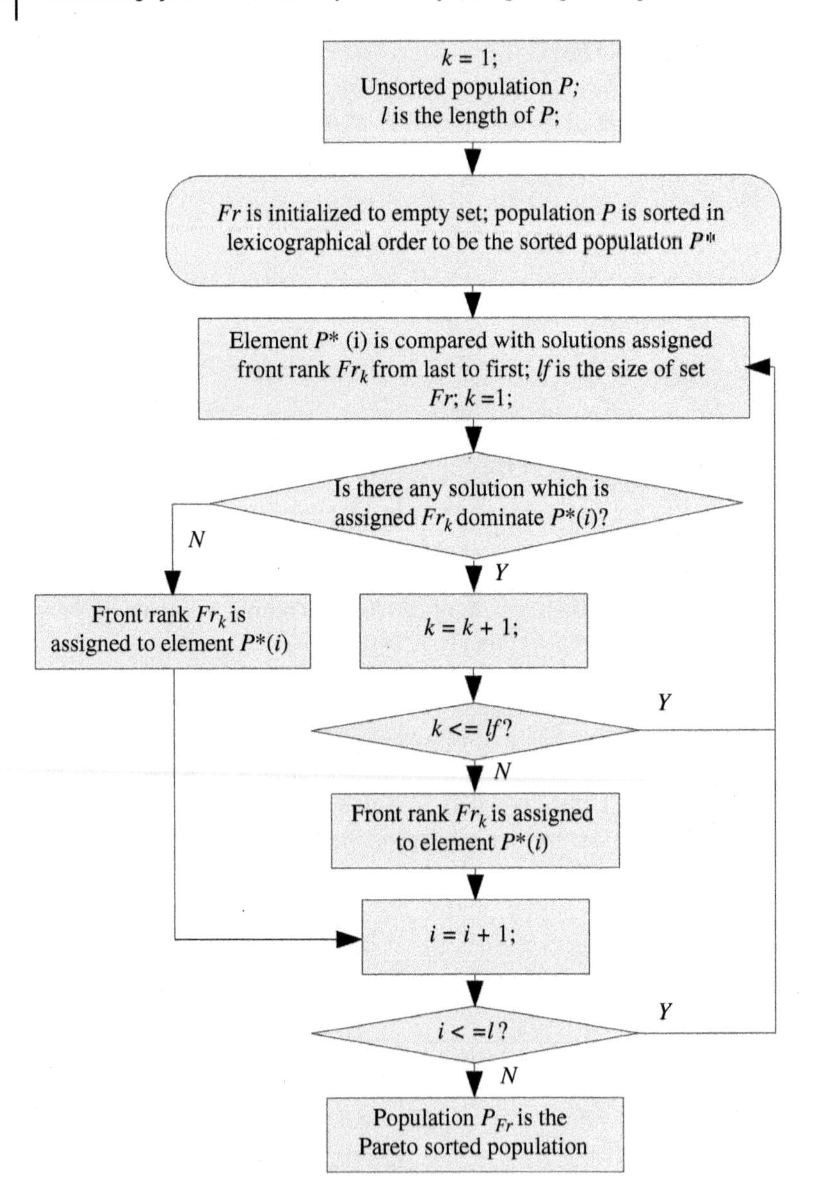

Figure 9.12 The flowchart of ENS.

9.5.3.4 Crowding Distance Computation

To evaluate the solutions with the same front ranks, the crowding distance should be utilized [39]. For the solutions with the same front ranks, the solution with better crowding distance is preferable.

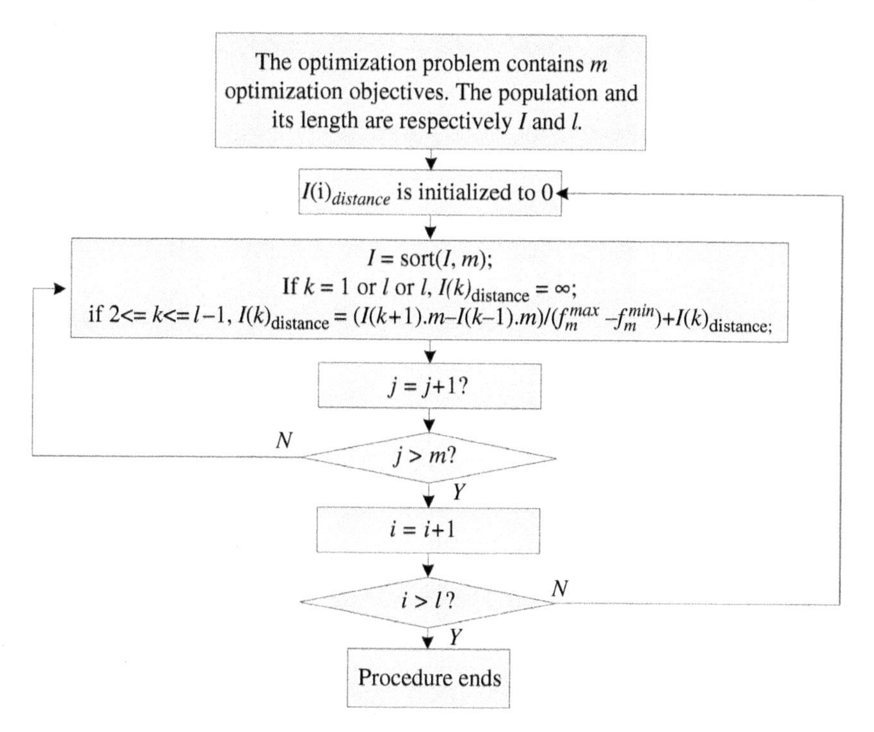

Figure 9.13 Computation process of crowding distance.

As represented in Figure 9.13, when the crowding distance is calculated, the solutions should be sorted according to the objective values in ascending order. For each objective value, the infinite distance value is assigned to the solutions with the largest and smallest objective values. The distance values of the intermediate solutions are calculated using the absolution normalized difference between two adjacent solutions' objective values. The solutions' crowding distance is obtained using the sum of the objectives' distance values.

9.5.3.5 Neighborhood Search Strategy

The sites' neighborhood is randomly explored by the follower bees. For IMODBA, mutation, insert, swap, and inverse operators are used to be the neighborhood search strategy as represented in Figure 9.14. The mutation operator acts on the disassembly direction. A random bit mutates and this bit changes 180° as represented in Figure 9.14a. The insert operator acts on both sequence and direction arrays, and a random bit inserts to a random position. The swap operator acts on both sequence and direction arrays. Two random bits

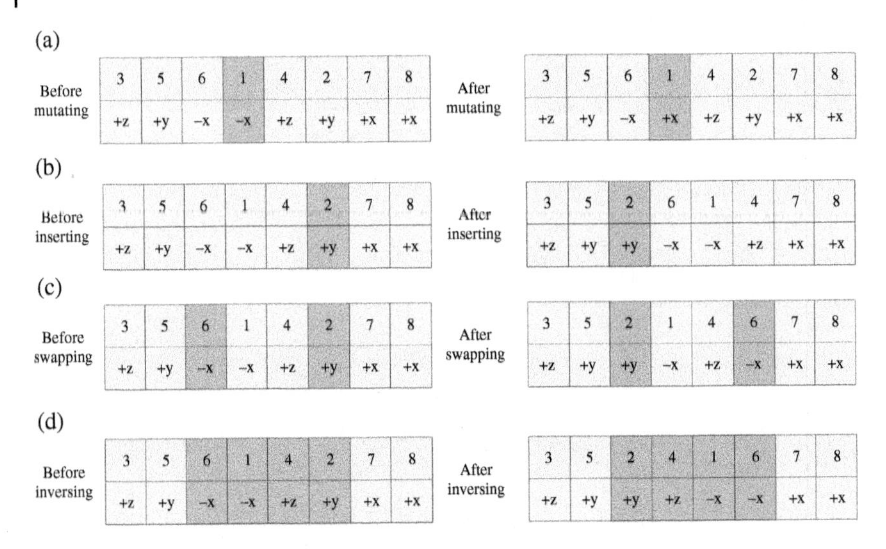

Figure 9.14 Neighborhood search strategy of IMODBA: (a) mutation operator, (b) insert operator, (c) swap operator, and (d) inverse operator.

are exchanged as represented in Figure 9.14c. The inverse operator also acts on both sequence and direction arrays, and a random part is inverted. Similar to the neighborhood search strategy of EDBA, if the new bees are unfeasible, the follower bees continue to search the neighborhood of scout bees until the new bees are feasible.

9.6 Experiments and Results

9.6.1 Case Study

A gear pump is used to verify the proposed methods. The gear pump and its exploded view are described in Figure 9.15. The workflows of RDSP and RDLBP are represented in Figure 9.16. The properties of gear pump's parts are listed in Table 9.2.

The safe distance between the end-effector's moving path and the contour of EOL products is 10 mm. The length matrix MP ($MP = [mp_{ij}]$) is manually calculated. For example, as shown in Figure 9.17, the length of obstacle-avoiding path $mp_{1,13}$ between Bolt A and Shaft B is calculated by Eq. (9.32).

$$\begin{aligned} mp_{1,13} &= DP_1DP_2 + DP_2DP_3 + DP_3DP_4 + DP_4DP_5 + DP_5DP_6 + DP_6DP_7 \\ &= 10 + 18.92 + 72.01 + 127.67 + 20 + 10 = 258.6 \end{aligned} \tag{9.32}$$

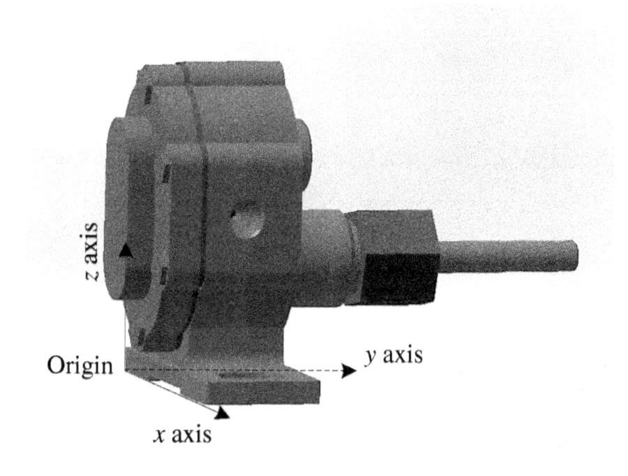

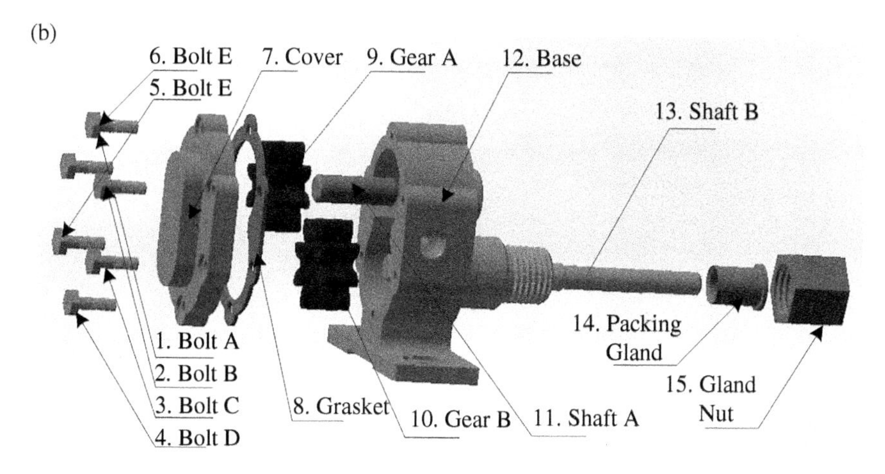

Figure 9.15 The gear pump and its exploded view: (a) the gear pump and (b) exploded view of the gear pump.

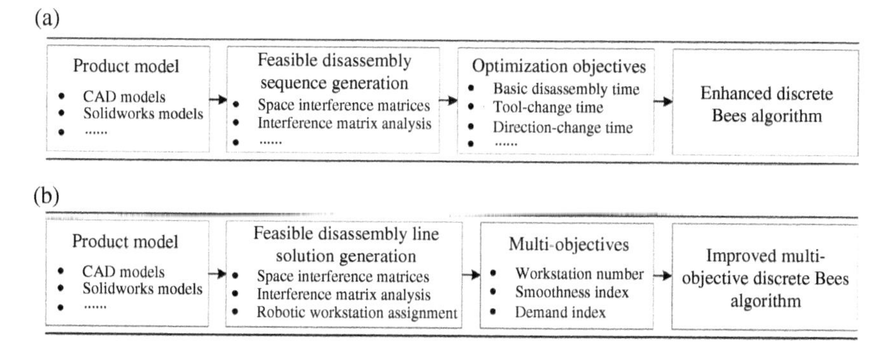

Figure 9.16 The workflow of RDSP and RDLBP.

Table 9.2 Properties of the gear pump's parts.

Parts	Disassembly tasks	Tools	Disassembly point (mm)	h_{xi}	bt/s
1	Unscrewing the Bolt A	Spanner-I	[49.4, −12.6, 105.5]	1	3
2	Unscrewing the Bolt B	Spanner-I	[74.4, −12.6, 81]	1	3
3	Unscrewing the Bolt C	Spanner-I	[74.4, −12.6, 45]	1	3
4	Unscrewing the Bolt D	Spanner-I	[49.4, −12.6, 20.5]	1	3
5	Unscrewing the Bolt E	Spanner-I	[24.4, −12.6, 45]	1	3
6	Unscrewing the Bolt F	Spanner-I	[24.4, −12.6, 81]	1	3
7	Removing the Cover	Gripper-II	[49.4, −20.6, 63]	2	4
8	Removing the Gasket	Gripper-I	[49.4, 1.4, 105.5]	2	3
9	Removing the Gear A	Gripper-I	[49.4, 3.4, 81]	3	6
10	Removing the Gear B	Gripper-I	[49.4, 3.4, 45]	3	6
11	Removing the Shaft A	Gripper-I	[49.4, −7.6, 81]	1	4
12	Removing the Base	Gripper-II	[49.4, 49.4, 81]	4	8
13	Removing the Shaft B	Gripper-I	[49.4, 152.4, 45]	2	4
14	Removing the Packing Gland	Gripper-I	[49.4, 91.4, 45]	2	2
15	Unscrewing the Gland Nut	Spanner-II	[49.4, 96.4, 45]	1	3

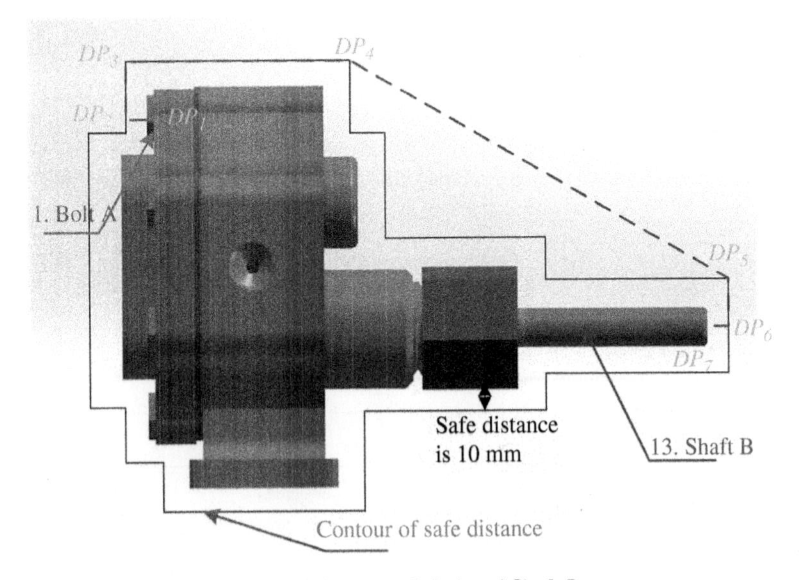

Figure 9.17 The moving path between Bolt A and Shaft B.

9.6.2 Performance Analysis

Simulations were finished on a personal computer with 4 GB memory and 2.3 GHz Intel Core i5-6200U CPU using Matlab 2014b software.

9.6.2.1 RDSP Using EDBA

The performance analysis of RDSP using EDBA contains the following parts: (i) performance analysis with respect to different parameters; (ii) result comparisons using different methods; (iii) comparative analysis using different optimization algorithms.

The parameters *m*, *n*, *mb*, *nb* were respectively 4, 1, 1, and 2. It assumes that the end-effector's moving speed is 12 mm/s. The fitness value and running time under different populations (from 10 to 40) and different iterations (from 100 to 500) are compared. Simulations were run 10 times. It is obvious when the parameter *scoutn* is fixed, the running time improves linearly with the parameter *iter*, as represented in Figure 9.18a. When population and iteration numbers are respectively 10 and 100, EDBA has the worst performance as shown in Figure 9.18b. The reason is that low-quality solutions are obtained using insufficient populations and iterations. However, it is apparent that the solution quality increases with the iterations and populations [8].

Alshibli et al. [40] utilized the Euclidean distance to obtain the moving path length between different DPs. The result (Result 1) obtained using this method is compared with that of the proposed method (Result 2). In this chapter, the near-optimal solutions are compared instead of the real optimal solutions which are difficult to obtain. The parameters *m*, *n*, *mb*, *nb*, *iter*, and *scoutn* were 4, 1, 1, 2, 500, and 40, respectively. The simulation was run 1000 times. In these 1000 solutions, the near-optimal solution is the solution with the minimum fitness values. As represented in Table 9.3 ("1" and "2" respectively represent "y+" and "y−"), Result 2 is apparently different from Result 1, which is generated by the method using the Euclidean distance. For the robot's end-effector, it is not practical to move along the straight line path using the result obtained by the traditional method. Thus, the proposed method is more suitable for use in the robotic disassembly compared with the traditional method.

EDBA is also compared with the other three optimization algorithms. For EDBA and EDBA without mutation operator (EDBA-WMO), parameters *m*, *n*, *mb*, *nb* were respectively 4, 1, 1, and 2. For the genetic algorithm with precedence preserve crossover (GA-PPX) [14], the mutation ratio was 0.1. For the self-adaptive simplified swarm optimization (SASSO), C_p, C_g, and C_w were adaptively determined [41]. Each simulation was run 10 times. For the four optimization algorithms, the population number was 20, the results under different iterations are represented in Figure 9.19a and c. It is apparent that SASSO and GA-PPX

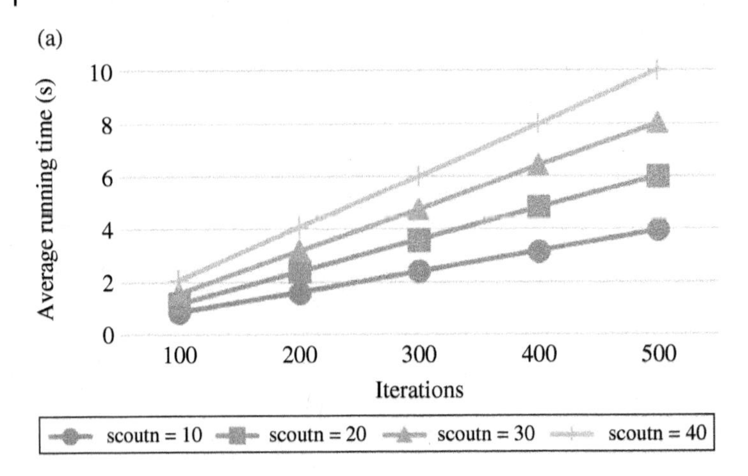

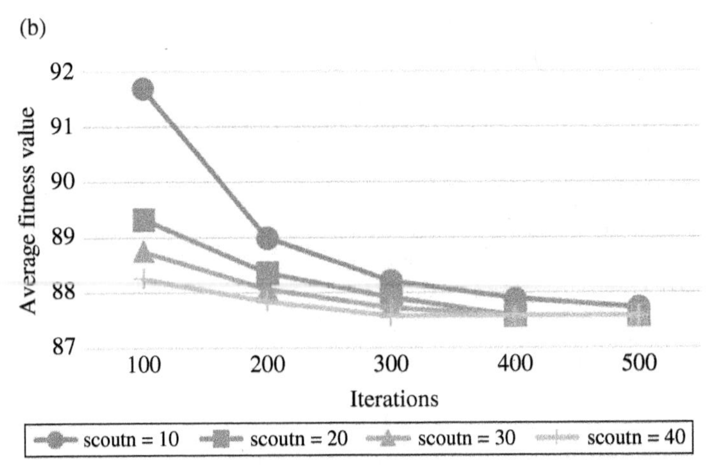

Figure 9.18 Performance analysis of EDBA under different parameters: (a) average running time under different parameters of EDBA and (b) average fitness value under different parameters of EDBA.

Table 9.3 Comparisons of different results.

Result 1	Sequence	5-4-3-2-1-6-7-11-8-9-10-13-15-14-12
	Direction	2-2-2-2-2-2-2-2-2-2-2-2-1-1-1
	Fitness	58.5038
Result 2	Sequence	5-4-3-2-1-6-7-10-9-11-8-13-15-14-12
	Direction	2-2-2-2-2-2-2-2-2-2-2-2-1-1-1
	Fitness	87.5731

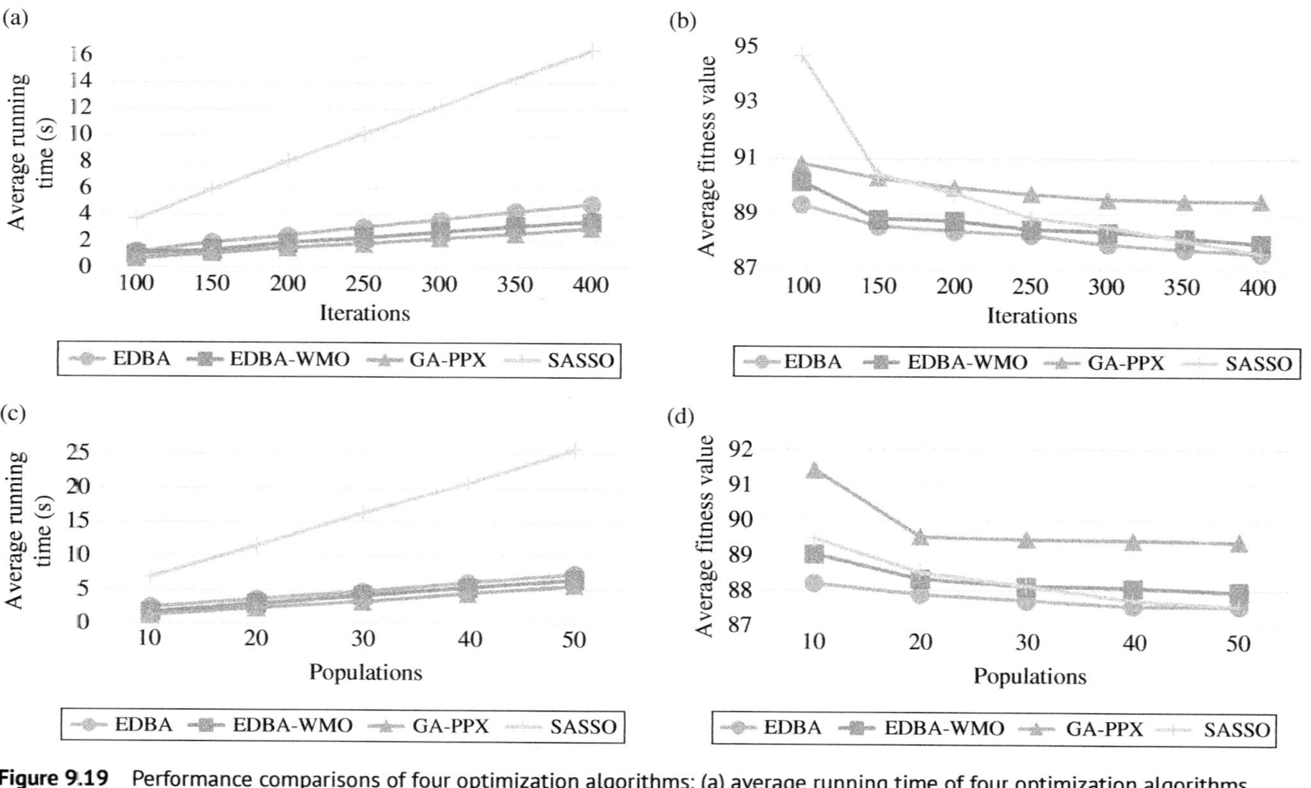

Figure 9.19 Performance comparisons of four optimization algorithms: (a) average running time of four optimization algorithms under different iterations, (b) average fitness value of four optimization algorithms under different iterations, (c) average running time of four optimization algorithms under different populations, and (d) average fitness value of four optimization algorithms under different populations.

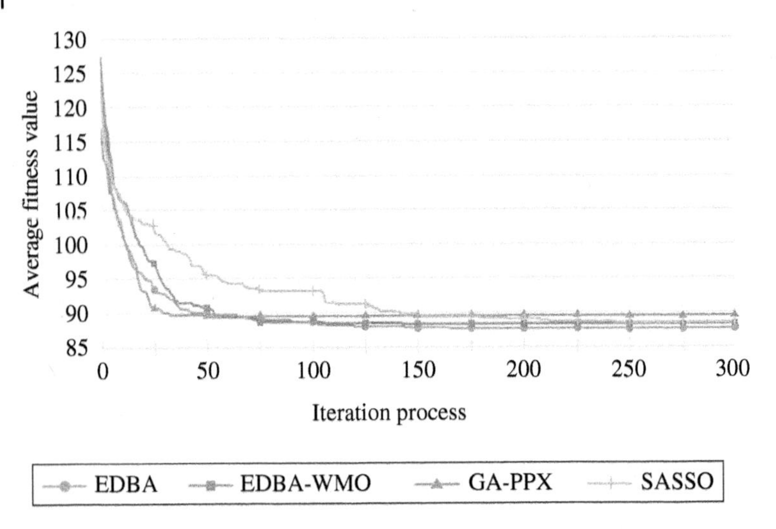

Figure 9.20 Iteration process of the four optimization algorithms.

respectively need the most and the least running time of the four optimization algorithms. EDBA needs more running time compared with EDBA-WMO. The reason is that more time needs to be spent on the mutation operators of EDBA. From Figure 9.19b and d, it is apparent that EDBA can find better solutions than the other three optimization algorithms. For the iteration process, the population and iteration numbers are respectively 20 and 300. From Figure 9.20, it is obvious that EDBA converges to the best solution of the four optimization algorithms.

9.6.2.2 RDLBP Using IMODBA

The performance analysis of IMODBA contains the following parts: (i) performance analysis with respect to different parameters; (ii) comparative analysis of different optimization algorithms; (iii) results comparison of different cases.

For the multi-objective optimization algorithms, auxiliary methods should be utilized to assess the non-dominated solutions' quality. In this chapter, generational distance (GD) [42] and hypervolume indicator (HI) [43] are used. The optimal Pareto solutions used to calculate the GD values are represented in Tables 9.6 and 9.7. The reference point used to calculate the HI values is set to [1.2, 1.2, 1.2]. Generally, lower GD values and greater HI values are preferable.

Before the GD and HI values of non-dominated solutions are calculated, the multi-objective values should be normalized by Eq. (9.33) where $f_{1,min}$, $f_{1,max}$, $f_{2,min}$, $f_{2,max}$, $f_{3,min}$, and $f_{3,max}$ are 3, 5, 0.0548, 765.6372, 228, and 259, respectively.

$$f_{i,norm} = (f_i - f_{i,\min}) / (f_{i,\max} - f_{i,\min}) \quad i = 1,2,3 \tag{9.33}$$

For IMODBA, comparisons with respect to the running time, HI and GD values are made under different parameters. Parameters *m* and *mb* are respectively 15 and 1. It assumes that end-effector's moving speed is 12 cm/s. Each simulation was run 50 times. As represented in Figure 9.21, IMODBA generates the best solutions (HI value is 1.5450 and GD value is 0.0080) when the parameters *iter* and *scoutn* are respectively 800 and 80, but it needs the most running time (67.55 s). When the parameters *iter* and *scoutn* are respectively 100 and 30, IMODBA generates the worst solutions (HI value is 1.4825 and GD value is 0.0149), but it needs the shortest running time (3.26 s) [9].

The comparative analysis of four optimization algorithms is conducted as follows.

- IMODBA: the proposed optimization algorithm in this chapter. The parameters *mb* and *m* are 1 and 15, respectively.
- MODBA: the fast nondominated sorting approach [39] is utilized. The parameters *mb* and *m* are 1 and 15, respectively.
- Multi-objective artificial bees colony (MOABC): the onlooker bees, employed bees, and scout bees are utilized. The probability value of the onlooker bees is the same as the publication [44]. If a food source is not updated over 10 iterations, the employed bee will become the scout bee and new food source will be randomly explored.
- Multi-objective genetic algorithm (MOGA): roulette wheel method, two-point crossover procedure [45], and mutation procedure are utilized. The chromosome selection rate, mutation rate, and crossover rate are respectively 0.1, 0.2, and 0.8.

The results are compared with respect to iterations when parameter *scoutn* is 50 as represented in Figure 9.22. Each simulation was run 50 times. It is apparent that MODBA takes the most running time of the four optimization algorithms as represented in Figure 9.22a. IMODBA takes less running time than MODBA. Table 9.4 describes the running time improvement with respect to iterations. It is apparent that IMODBA and MODBA are superior to MOGA and MOABC in solution quality with respect to iterations, as represented in Figure 9.22c and e. IMODBA generates nearly the same quality of solutions compared with MODBA, the reason is that the solutions sorted by ENS are the same as the traditional methods while they are obtained using less running time. Simulations were conducted under different populations (from 30 to 80) when the iteration number is set to 500. Each simulation was run 50 times. The IMODBA generates optimal solutions using less time compared with MODBA and its running time is comparable with MOGA and MOABC as represented in Figure 9.22b. Table 9.5 describes the running time improvement with respect to populations. From Figure 9.22d and f, it is apparent that solution quality of

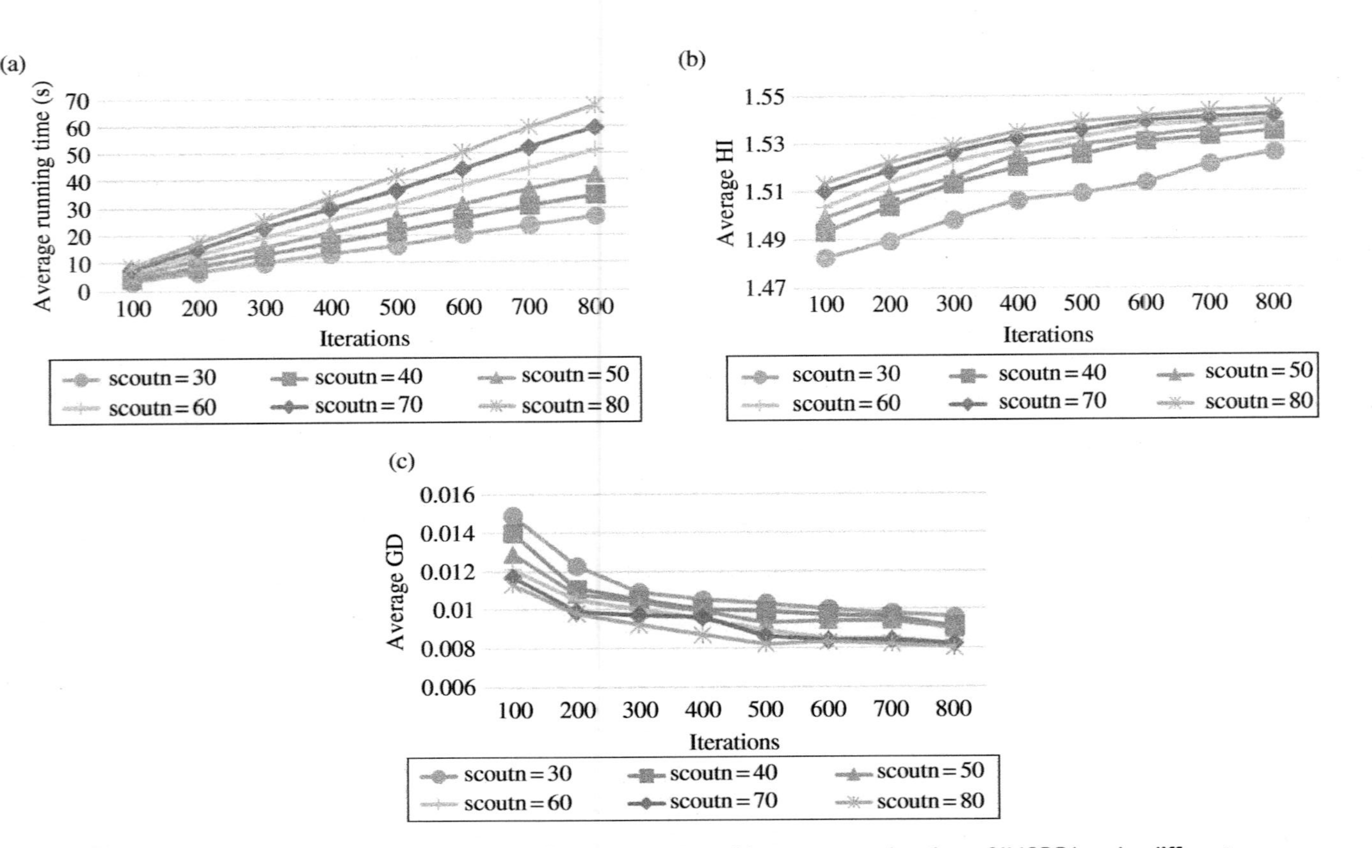

Figure 9.21 Performance analysis of IMODBA under different parameters: (a) average running time of IMODBA under different parameters, (b) average HI values of IMODBA under different parameters, and (c) average GD values of IMODBA under different parameters.

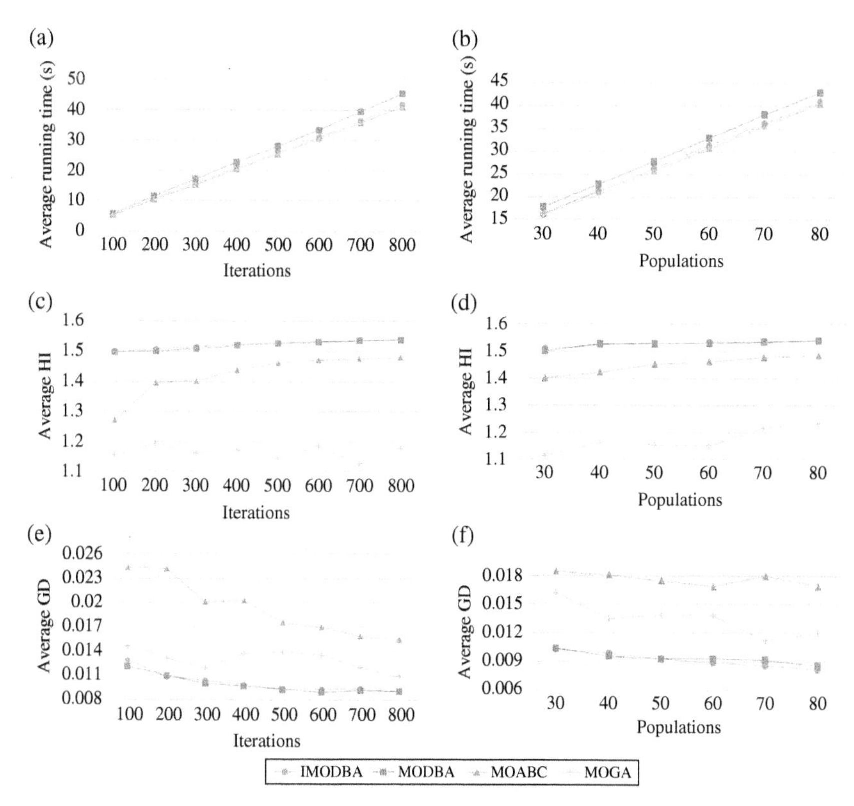

Figure 9.22 Comparative analysis of different optimization algorithms: (a) average running time under different iterations, (b) average running time under different populations, (c) average HI values under different iterations, (d) average HI values under different populations, (e) average GD values under different iterations, and (f) average GD values under different populations.

Table 9.4 Running time improvement with respective to iterations.

Iteration number	100	200	300	400	500	600	700	800
IMODBA	5.31 s	10.77 s	15.71 s	20.98 s	26.18 s	31.16 s	36.55 s	41.86 s
MODBA	5.63 s	11.57 s	17.29 s	22.93 s	28.19 s	33.54 s	39.50s	45.42 s
Improvement	0.32 s	0.80 s	1.58 s	1.95 s	2.01 s	2.38 s	2.95 s	3.56 s
Percentage	5.68%	6.91%	9.14%	8.50%	7.13%	7.10%	7.47%	7.84%

Table 9.5 Running time improvement with respective to populations.

Population number	30	40	50	60	70	80
IMODBA	16.19 s	21.41 s	26.18 s	31.38 s	36.27 s	40.92 s
MODBA	17.83 s	22.85 s	28.19 s	33.02 s	38.14 s	43.03 s
Improvement	1.64 s	1.44 s	2.01 s	1.64 s	1.87 s	2.11 s
Percentage	9.20%	6.30%	7.13%	4.97%	4.90%	4.90%

IMODBA and MODBA is better than MOGA and MOABC under different populations. IMODBA generate nearly the same quality of solutions with MODBA, the reason is also that the solutions sorted by ENS are the same as the traditional methods.

Comparisons are also made with respect to Pareto optimal fronts using different cases as follows.

- Case 1: the basic disassembly time, direction-change time and tool-change time are considered, but the end-effector's moving time is ignored in this case.
- Case 2: the end-effector's moving time, the basic disassembly time, direction-change time, and tool-change time are considered. The end-effector's moving time is calculated using the Euclidean distance.
- Case 3: the proposed method in this chapter is used.

It is hard to exhaust the solutions and obtain real optimal solutions. The near Pareto optimal solutions are generated in the following way. The scout bees number *scoutn*, the iteration number *iter*, the follower bees number *mb* of selected site, and the selected sites number *m* of IMODBA are 80, 800, 1, and 15, respectively. Simulations were run 100 times. The near Pareto optimal solutions are obtained by sorting these 100 groups of non-dominated solutions. Table 9.6 does not list the Pareto fronts obtained by MODBA, the reason is that the solutions obtained by MODBA are the same with IMODBA. Pareto fronts generated by case 3 are apparently different from cases 1 and 2 as represented in Table 9.6. The end-effector's moving time is ignored in case 1 and the obstacle caused by the EOL products is also ignored in case 2. Case 3 should be more applicable for RDLBP compared with cases 1 and 2. In addition, from case 3 in Table 9.6, the fourth and seventh solutions of IMODBA dominates the fourth and eighth solutions of MOABC respectively. The fifth solution obtained by IMODBA dominates the fifth and sixth solutions obtained by MOABC. IMODBA is also superior to MOGA in Pareto fronts number. Thus, IMODBA can generate better Pareto fronts compared with MOGA and MOABC. Table 9.7 lists the Pareto optimal solutions of RDLBP.

Table 9.6 Pareto fronts of three different cases.

Case 1

No.	IMODBA f_1	IMODBA f_2	IMODBA f_3	MOABC f_1	MOABC f_2	MOABC f_3	MOGA f_1	MOGA f_2	MOGA f_3
1	3	1	234	3	1	234	3	1	234
2	3	5	232	3	5	232	3	5	232
3	3	6	230	3	6	230	3	6	230
4	3	10	228	3	10	228	3	10	228
5									
6									
7									
8									
9									
10									
11									
12									
13									

Case 2

No.	IMODBA f_1	IMODBA f_2	IMODBA f_3	MOABC f_1	MOABC f_2	MOABC f_3	MOGA f_1	MOGA f_2	MOGA f_3
1	3	0.5262	238	3	0.5262	238	3	0.5262	238
2	3	2.9193	235	3	2.9193	235	3	2.9193	235
3	3	3.2023	234	3	3.2023	234	3	3.2023	234
4	3	7.5069	233	3	7.5069	233	3	14.3724	232
5	4	14.3724	232	3	14.3724	232	4	198.9339	231
6	4	198.9339	231	4	198.9339	231	4	251.9134	230
7	4	251.9134	230	4	251.9134	230	4	355.8085	229
8	4	355.8085	229	4	355.8085	229	4	448.1330	228
9	4	448.1330	228	4	448.1330	228			
10									
11									
12									
13									

Case 3

No.	IMODBA f_1	IMODBA f_2	IMODBA f_3	MOABC f_1	MOABC f_2	MOABC f_3	MOGA f_1	MOGA f_2	MOGA f_3
1	3	0.0548	246	3	0.0548	246	3	0.0548	246
2	3	0.0860	243	3	0.0860	243	3	0.0860	243
3	3	0.1899	241	3	0.1899	241	3	0.1899	241
4	3	1.0637	239	3	1.5690	240	3	2.0211	236
5	3	2.0211	236	3	2.2897	238	3	3.8500	235
6	3	3.8500	235	3	3.6332	236	3	6.3106	233
7	3	6.3106	233	3	6.3106	235	4	27.0273	232
8	4	27.0273	232	4	26.9303	234	4	27.2973	231
9	4	27.2973	231	4	27.0273	232	4	27.4623	230
10	4	27.4623	230	4	27.2973	231	4	263.5929	229
11	4	263.5929	229	4	27.4623	230	4	284.1421	228
12	4	284.1421	228	4	263.5929	229			
13				4	284.1421	228			

Table 9.7 Pareto optimal solutions.

Algorithms	No.	Sequence	Direction	Robotic workstation	f_1	f_2	f_3
IMODBA	1	4-5-6-2-3-1-7-15-8-14-12-9-13-11-10	2-2-2-2-2-2-1-2-1-1-2-2-1-1	1-1-1-1-1-1-1-2-2-2-2-3-3-3-3	3	0.0548	246
	2	4-2-3-5-6-1-7-15-8-14-12-10-9-13-11	2-2-2-2-2-2-1-2-1-1-2-1-2-2	1-1-1-1-1-1-1-2-2-2-2-3-3-3-3	3	0.0860	243
	3	4-2-3-5-6-1-7-15-14-12-8-10-9-13-11	2-2-2-2-2-2-1-1-1-2-2-1-2-2	1-1-1-1-1-1-1-2-2-2-2-3-3-3-3	3	0.1899	241
	4	15-14-3-5-4-6-1-2-7-9-8-12-13-10-11	1-1-2-2-2-2-2-2-2-2-1-1-1-1	1-1-1-1-1-2-2-2-2-2-2-3-3-3-3	3	1.0637	239
	5	15-14-13-2-6-3-5-1-4-7-8-12-10-9-11	1-1-1-2-2-2-2-2-2-2-2-1-1-1-1	1-1-1-1-1-2-2-2-2-2-2-3-3-3-3	3	2.0211	236
	6	1-6-5-3-2-4-7-10-9-15-14-12-8-13-11	2-2-2-2-2-2-2-2-2-1-1-1-1-1-1	1-1-1-1-1-1-1-2-2-2-2-3-3-3-3	3	3.8500	235
	7	15-14-13-5-6-1-2-3-4-7-9-12-10-8-11	1-1-1-2-2-2-2-2-2-2-2-1-1-1-1	1-1-1-1-1-2-2-2-2-2-2-3-3-3-3	3	6.3106	233
	8	15-14-3-6-13-5-1-4-2-12-9-7-8-10-11	1-1-2-2-1-2-2-2-2-1-1-2-1-2-1	1-1-1-1-2-2-2-2-3-3-3-4-4-4-4	4	27.0273	232
	9	15-14-3-6-13-4-1-5-2-12-9-7-10-8-11	1-1-2-2-1-2-2-2-2-1-1-2-1-2-1	1-1-1-1-2-2-2-2-3-3-3-4-4-4-4	4	27.2973	231
	10	15-14-5-2-13-3-1-4-6-12-9-7-8-11-10	1-1-2-2-1-2-2-2-2-1-1-2-1-2-1	1-1-1-1-2-2-2-2-3-3-3-4-4-4-4	4	27.4623	230
	11	15-14-13-3-4-1-5-2-6-12-9-7-10-8-11	1-1-1-2-2-2-2-2-2-1-1-2-1-2-1	1-1-1-1-2-2-2-2-2-3-3-3-4-4-4	4	263.5929	229
	12	15-14-13-3-4-1-5-2-6-12-9-10-7-8-11	1-1-1-2-2-2-2-2-2-1-1-1-2-2-1	1-1-1-1-2-2-2-2-2-3-3-3-4-4-4	4	284.1421	228
MOABC	1	15-14-13-6-2-3-5-1-4-7-11-12-9-10-8	1-1-1-2-2-2-2-2-2-2-2-1-1-1-1	1-1-1-1-1-2-2-2-2-2-2-3-3-3-3	3	1.5690	240
	2	15-14-13-2-3-5-1-4-6-7-8-12-10-11-9	1-1-1-2-2-2-2-2-2-2-2-1-1-1-1	1-1-1-1-1-2-2-2-2-2-2-3-3-3-3	3	2.2897	238
	3	15-14-13-2-3-5-6-1-4-7-8-12-10-9-11	1-1-1-2-2-2-2-2-2-2-2-1-1-1-1	1-1-1-1-1-2-2-2-2-2-2-3-3-3-3	3	3.6332	236
	4	15-14-3-6-13-4-1-5-2-12-9-8-7-11-10	1-1-2-2-1-2-2-2-2-1-1-1-2-1-2	1-1-1-1-2-2-2-2-3-3-3-4-4-4-4	4	26.9303	234
	5–13	The 1st ~ 3rd and 7th ~ 12th solutions of IMODBA.					
MOGA	1–11	The 1st ~ 3rd and 5th ~ 12th solutions of IMODBA.					

9.7 Conclusion

Scheduling of robotic disassembly in remanufacturing mainly contains RDSP and RDLBP, which were respectively solved by EDBA and IMODBA. Disassembly precedence relationships of products were represented by a disassembly model. Including the safe distance, the moving time between different DPs was generated using the obstacle-avoiding path length and the end-effector's moving speed. After that, the optimization objectives of RDSP and RDLBP were also introduced. Then, two types of BA were developed to solve RDSP and RDLBP, namely, EDBA and IMODBA. Finally, a gear pump was utilized to verify the proposed methods. The results show that the proposed methods are more applicable to be used in the scheduling of robotic disassembly compared with the traditional methods. EDBA is superior to EDBA-WMO, GA-PPX, and SASSO in solution quality, while IMODBA is also superior to MOGA and MOABC in solution quality. The practical implication of this chapter is to provide guidance to generate the optimal disassembly solutions of disassembling EOL products for both RDCs and RDLs. However, there are also some limitations of the proposed methods. First, the disassembly model used in this chapter can only generate orthogonal disassembly directions for each component while in the practical robotic disassembly process, the disassembly directions of components should be more complex. Besides, the obstacle-avoiding path length is manually obtained, it is time-consuming and cumbersome for manual labor. In the future, methods of efficiently obtaining the obstacle-avoiding path length should be studied. Besides, the disassembly process is full of uncertainties. This chapter assumes that disassembly time is constant and deterministic. In the future, we will study more practical methods which are applicable to handle the uncertainties in the robotic disassembly process.

Acknowledgments

This work was supported by National Natural Science Foundation of China (Grant Nos. 51775399 and 51475343), Engineering and Physical Sciences Research Council (EPSRC), UK (Grant No. EP/N018524/1), and the China Scholarship Council (CSC).

A. Appendix A

An example of ENS is illustrated here as follows. As shown in Figure 9.A1, for the given population P, the sorted population P^* is obtained according to lexicographical order. In step 1, the front rank Fr_1 is assigned to the 1st solution (solution 3) of the

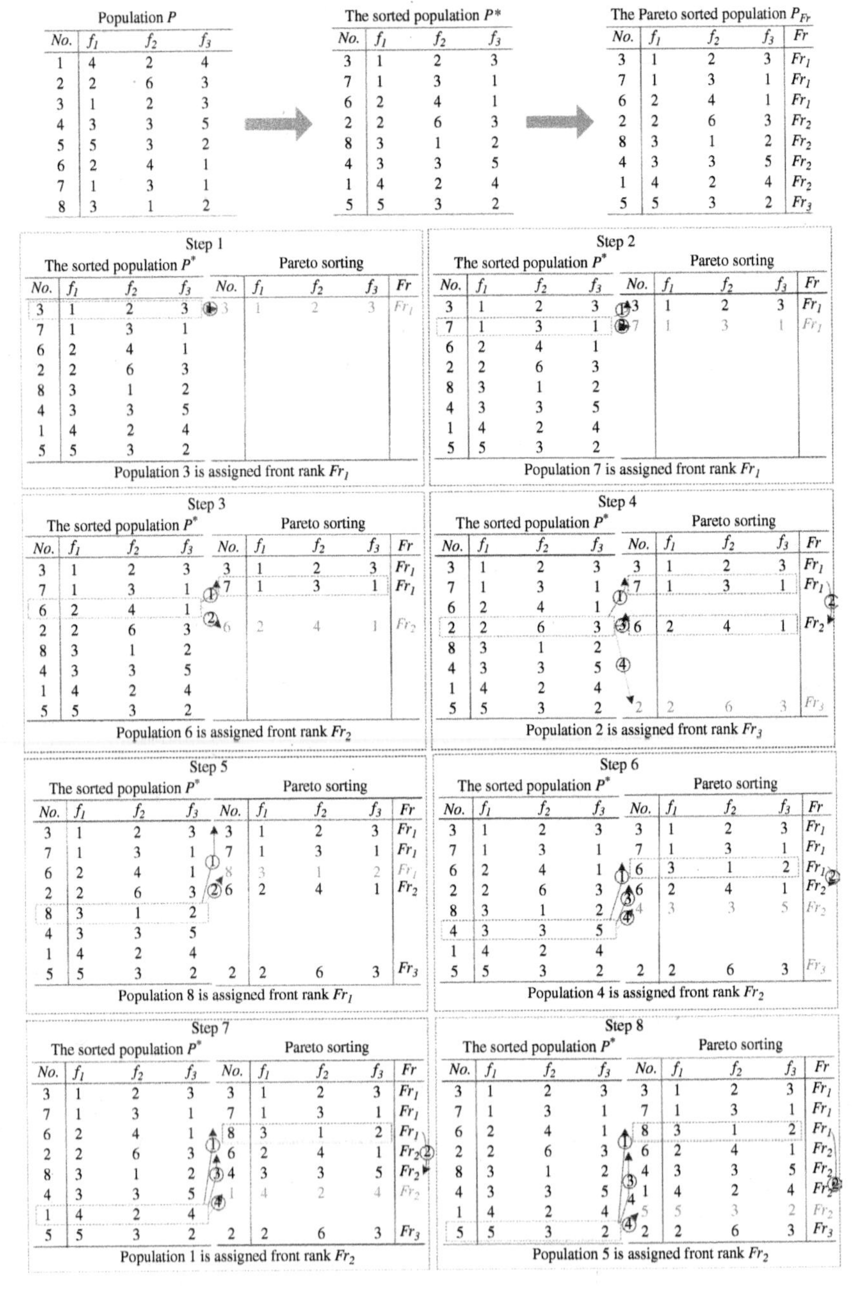

Figure 9.A1 An example to illustrate the process of ENS.

sorted population P^*. In step 2, the 2nd solution of population P^* is compared with the solutions with front rank Fr_1 from the last to the first. It is obvious that solution 7 is not dominated by solution 3 and front rank Fr_1 is assigned to solution 7. Then, comparisons are made between solution 6 and the solutions with front rank Fr_1 from the last to the first. It is apparent that solution 7 dominates solution 6 and front rank Fr_2 does not exist in set Fr. Thus, front rank Fr_2 is created and it is assigned to the solution 6. Based on the same rules, through steps 4–8, all the solutions in sorted population P^* are assigned the front ranks as shown in Figure 9.A1.

References

1 Xu, B.S. (2010). State of the art and future development in remanufacturing engineering. *Trans. Mater. Heat Treatment* 31 (1): 10–14.

2 Ren, Y.P., Yu, D.Y., Zhang, C.Y. et al. (2017). An improved gravitational search algorithm for profit-oriented partial disassembly line balancing problem. *Int. J. Prod. Res.* 55 (24): 7302–7316.

3 Diallo, C., Venkatadri, U., Khatab, A., and Bhakthavatchalam, S. (2017). State of the art review of quality, reliability and maintenance issues in closed-loop supply chains with remanufacturing. *Int. J. Prod. Res.* 55 (5): 1277–1296.

4 Guide, V.D.R. (2000). Production planning and control for remanufacturing: industry practice and research needs. *J. Oper. Manage.* 18 (4): 467–483.

5 Vongbunyong, S., Kara, S., and Pagnucco, M. (2012). A framework for using cognitive robotics in disassembly automation. Proceedings of the 19th CIRP International Conference on Life Cycle Engineering, 173–178.

6 Vongbunyong, S., Kara, S., and Pagnucco, M. (2013). Basic behaviour control of the vision-based cognitive robotic disassembly automation. *Assem. Autom.* 33 (1): 38–56.

7 Vongbunyong, S., Kara, S., and Pagnucco, M. (2015). Learning and revision in cognitive robotics disassembly automation. *Rob. Comput.-Integr. Manuf.* 34: 79–94.

8 Liu, J.Y., Zhou, Z.D., Pham, D.T. et al. (2018). Robotic disassembly sequence planning using enhanced discrete bees algorithm in remanufacturing. *Int. J. Prod. Res.* 56 (9): 3134–3151.

9 Liu, J.Y., Zhou, Z.D., Pham, D.T. et al. (2018). An improved multi-objective discrete bees algorithm for robotic disassembly line balancing problem in remanufacturing. *Int. J. Adv. Manuf. Technol.* 97 (9–12): 3937–3962.

10 Xu, W.J., Tian, S.S., Liu, Q. et al. (2016). An improved discrete bees algorithm for correlation-aware service aggregation optimization in cloud manufacturing. *Int. J. Adv. Manuf. Technol.* 84 (1–4): 17–28.

11 Hussein, W.A., Sahran, S., and Abdullah, S.N.H.S. (2016). A fast scheme for multilevel thresholding based on a modified bees algorithm. *Knowl.-Based Syst.* 101: 114–134.

12 Oztemel, E. and Selam, A.A. (2017). Bees algorithm for multi-mode, resource-constrained project scheduling in molding industry. *Comput. Ind. Eng.* 112: 187–196.

13 Xing, Y.F., Wang, C.E., and Liu, Q. (2012). Disassembly sequence planning based on Pareto ant Colony algorithm. *J. Mech. Eng.* 48 (9): 186–192.

14 Kheder, M., Trigui, M., and Aifaoui, N. (2015). Disassembly sequence planning based on a genetic algorithm. *Proc. Inst. Mech. Eng., Part C: J. Mech. Eng. Sci.* 229 (12): 2281–2290.

15 Jin, G.Q., Li, W.D., Wang, S., and Gao, S.M. (2017). A systematic selective disassembly approach for waste electrical and electronic equipment with case study on liquid crystal display televisions. *Proc. Inst. Mech. Eng., Part B: J. Eng. Manuf.* 231 (13): 2261–2278.

16 Meng, K., Lou, P.H., Peng, X.H., and Prybutok, V. (2016). An improved co-evolutionary algorithm for green manufacturing by integration of recovery option selection and disassembly planning for end-of-life products. *Int. J. Prod. Res.* 54 (18): 5567–5593.

17 Kim, H.W. and Lee, D.H. (2017). An optimal algorithm for selective disassembly sequencing with sequence-dependent set-ups in parallel disassembly environment. *Int. J. Prod. Res.* 55 (24): 7317–7333.

18 ElSayed, A., Kongar, E., Gupta, S., and Sobh, T. (2011). An online genetic algorithm for automated disassembly sequence generation. In: ASME 2011 International Design Engineering Technical Conferences and Computers and Information in Engineering Conference, 657–664. Washington: American Society of Mechanical Engineers.

19 ElSayed, A., Kongar, E., and Gupta, S.M. (2010). A genetic algorithm approach to end-of-life disassembly sequencing for robotic disassembly. Proceedings of the 2010 Northeast Decision Sciences Institute Conference, Alexandria, VA, USA (26–28 March), 402–408.

20 ElSayed, A., Kongar, E., Gupta, S.M., and Sobh, T. (2012). A robotic-driven disassembly sequence generator for end-of-life electronic products. *J. Intell. Rob. Syst.* 68 (1): 43–52.

21 McGovern, S.M. and Gupta, S.M. (2007). A balancing method and genetic algorithm for disassembly line balancing. *Eur. J. Oper. Res.* 179 (3): 692–708.

22 Ilgin, M.A., Akcay, H., and Araz, C. (2017). Disassembly line balancing using linear physical programming. *Int. J. Prod. Res.* 55 (20): 6108–6119.

23 Ding, L.P., Feng, Y.X., Tan, J.R., and Gao, Y.C. (2010). A new multi-objective ant colony algorithm for solving the disassembly line balancing problem. *Int. J. Adv. Manuf. Technol.* 48 (5–8): 761–771.

24 Aydemir-Karadag, A. and Turkbey, O. (2013). Multi-objective optimization of stochastic disassembly line balancing with station paralleling. *Comput. Ind. Eng.* 65 (3): 413–425.

25 Bentaha, M.L., Battaia, O., and Dolgui, A. (2014). A sample average approximation method for disassembly line balancing problem under uncertainty. *Comput. Oper. Res.* 51: 111–122.

26 Mete, S., Cil, Z.A., Agpak, K. et al. (2016). A solution approach based on beam search algorithm for disassembly line balancing problem. *J. Manuf. Syst.* 41: 188–200.

27 Kalayci, C.B. and Gupta, S.M. (2013). Ant colony optimization for sequence-dependent disassembly line balancing problem. *J. Manuf. Technol. Manage.* 24 (3): 413–427.

28 Kalayci, C.B. and Gupta, S.M. (2013). A particle swarm optimization algorithm with neighborhood-based mutation for sequence-dependent disassembly line balancing problem. *Int. J. Adv. Manuf. Technol.* 69 (1–4): 197–209.

29 Liu, J. and Wang, S.W. (2017). Balancing disassembly line in product recovery to promote the coordinated development of economy and environment. *Sustainability* 9 (2): 309–323.

30 Kalayci, C.B., Polat, O., and Gupta, S.M. (2016). A hybrid genetic algorithm for sequence-dependent disassembly line balancing problem. *Ann. Oper. Res.* 242 (2): 321–354.

31 Kalayci, C.B. and Gupta, S.M. (2014). A tabu search algorithm for balancing a sequence-dependent disassembly line. *Prod. Plann. Control* 25 (2): 149–160.

32 Xu, J., Zhang, S.Y., and Fei, S.M. (2011). Product remanufacture disassembly planning based on adaptive particle swarm optimization algorithm. *J. Zhejiang Univ. Eng. Sci.* 45 (10): 1746–1752.

33 Luo, Y., Peng, Q., and Gu, P. (2016). Integrated multi-layer representation and ant colony search for product selective disassembly planning. *Comput. Ind.* 75: 13–26.

34 Agrawal, S. and Tiwari, M.K. (2008). A collaborative ant colony algorithm to stochastic mixed-model U-shaped disassembly line balancing and sequencing problem. *Int. J. Prod. Res.* 46 (6): 1405–1429.

35 Kalaycilar, E.G., Azizoglu, M., and Yeralan, S. (2016). A disassembly line balancing problem with fixed number of workstations. *Eur. J. Oper. Res.* 249 (2): 592–604.

36 Constantinescu, D. and Croft, E.A. (2000). Smooth and time-optimal trajectory planning for industrial manipulators along specified path. *J. Rob. Syst.* 17 (5): 233–249.

37 Pham, Q.T., Pham, D.T., and Castellani, M. (2012). A modified bees algorithm and a statistics-based method for tuning its parameters. *Proc. Inst. Mech. Eng., Part I: J. Syst. Control Eng.* 226 (3): 287–301.

38 Zhang, X.Y., Tian, Y., Cheng, R., and Jin, Y.C. (2015). An efficient approach to nondominated sorting for evolutionary multiobjective optimization. *IEEE Trans. Evol. Comput.* 19 (2): 201–213.

39 Deb, K., Pratap, A., Agarwal, S., and Meyarivan, T. (2002). A fast and elitist multiobjective genetic algorithm: NSGA-II. *IEEE Trans. Evol. Comput.* 6 (2): 182–197.

40 Alshibli, M., ElSayed, A., Kongar, E. et al. (2015). Disassembly sequencing using tabu search. *J. Intell. Rob. Syst.* 82 (1): 1–11.

41 Yeh, W.C. (2012). Optimization of the disassembly sequencing problem on the basis of self-adaptive simplified swarm optimization. *IEEE Trans. Syst., Man, Cybern. – Part A: Syst. Humans* 42 (1): 250–261.

42 Okabe, T., Jin, Y., and Sendhoff, B. (2003). A critical survey of performance indices for multi-objective optimisation. Proceedings of 2003 Congress on IEEE on Evolutionary Computation, Canberra, Australia (8–12 December), 878–885.

43 Zitzler, E., and Thiele, L. (1998). Multiobjective optimization using evolutionary algorithms—a comparative case study. Proceedings of 1998 International Conference on Parallel Problem Solving from Nature, Amsterdam, Netherlands (27–30 September), 292–301.

44 Akbari, R., Hedayatzadeh, R., Ziarati, K., and Hassanizadeh, B. (2012). A multi-objective artificial bee colony algorithm. *Swarm Evol. Comput.* 2: 39–52.

45 Akpınar, S. and Bayhan, G.M. (2011). A hybrid genetic algorithm for mixed model assembly line balancing problem with parallel workstations and zoning constraints. *Eng. Appl. Artif. Intell.* 24 (3): 449–457.

10

A Modified Fireworks Algorithm to Solve the Heat and Power Generation Scheduling Problem in Power System Studies

Mohammad Sadegh Javadi[1], Ali Esmaeel Nezhad[2], Seyed-Ehsan Razavi[3], Abdollah Ahmadi[4], and João P.S. Catalão[5]*

[1] *Department of Electrical Engineering, Shiraz Branch, Islamic Azad University, Shiraz, Iran*
[2] *Department of Electrical, Electronic, and Information Engineering, University of Bologna, Bologna, Italy*
[3] *School of engineering and IT, Murdoch University, Perth, Australia*
[4] *School of Electrical Engineering and Telecommunications, University of New South Wales, Sydney, NSW, Australia*
[5] *INESC TEC and Faculty of Engineering of the University of Porto, Porto, Portugal*

10.1 Introduction

Combined Heat and Power (CHP) production has been introduced as a much more efficient scheme to generate power in plants and heat in boilers, offering higher energy efficiency and far lower fuel consumption [1]. Following the increasing growth of CHP units over the last decade, the conventional Economic Load Dispatch (ELD), which determined operating points of thermal plants in power-only production optimization problems, is transformed to Combined Heat and Power Scheduling (CHPS). Obviously, considering a primary objective function that is minimizing the operational cost, more constraints involved in the CHPS make it more difficult to solve compared to the ELD. As the most severe constraint to be satisfied, load balance including both power and heat can be pointed out. The authors in [2] have also considered the valve point effect in the CHPS problem that adds a sinusoidal term to the quadratic cost function of conventional thermal units. Although this modeling presents a more realistic viewpoint, it increases the nonlinearity order of the problem. To include transmission losses in the CHPS problem, two different paths can be followed. The first and the

* *corresponding author*

Evolutionary Computation in Scheduling, First Edition. Edited by Amir H. Gandomi, Ali Emrouznejad, Mo M. Jamshidi, Kalyanmoy Deb, and Iman Rahimi.

most accurate one is to model transmission lines and other network details such as bus voltages in the context of Power Flow (PF) or Optimal Power Flow (OPF) [3]. However, as the second solution, some attempts have been made to include transmission losses using Kron's loss formula – an equivalent loss matrix – into the CHPS without considering the voltage of the network busses [4, 5]. Another consideration that turns the CHPS problem into a complex optimization problem is the issue of Feasible Operating Region (FOR) owning to CHP units. Different types of CHP units have different FORs determined by CHP units' manufacturer. In reality, these FORs are non-convex [6–11]. However, in some studies, they have been approximated by convex ones with the aim of simplicity [12, 13].

Before moving on to the further operational step, it is worth mentioning that ELD presents an optimal operating point associated with only one snapshot of the power system, while System Operators (SOs) need a wider time resolution for the operational goals and this is where the Short-term CHPS arises. The CHPS is mainly discussed over 24 hours, or 168 hours, that is daily or weekly, and determines the optimal operating points of each generating unit in order to minimize the operating cost. Needless to say, considering CHP units in the short-term scheduling problem is of high importance for SOs. In this respect, the authors in [14–16] have presented CHP Unit Commitment (CHPUC) models in which minimum up/down time limitations of units are taken into consideration. Anand et al. in [14] have considered the flexibility of CHP units taking advantage of replacing single-shaft turbines with multi-shaft ones to provide a variable ratio of heat and power output, called dual-mode CHP. To investigate the daily generation scheduling of CHP in various viewpoints and with different considerations, plenty of research has been carried out [13, 17–22]. PF constraints have been included in [17] while the model proposed in [18] has been further elaborated to include security constraints of the power system. Furthermore, the inclusion of an Electric Storage System (ESS), Thermal Storage System (TSS), and industrial customers have been taken into account in the CHPS problem [13, 20]. Minimizing toxic emissions such as CO_2 is the subsidiary goal in the CHPS. In this regard, an optimal model desirable for Generation Companies (GENCOs) has been proposed in [21] to minimize the cost associated with both operational and environmental CO_2 emissions. To present an accurate model, the valve-point effect of thermal units and spinning reserve market has also been seen in [21]. In [22], short-term participation of CHP units is investigated in the presence of demand response programs while a heat buffer tank is incorporated into the framework to store heat. Over a wider time span, Majić et al. in [19] studied a 48-hour CHP scheduling considering energy storage, while most of the aforementioned studies in the literature have carried out day-ahead scheduling.

Achieving optimal solutions in the CHPS problem has always been a challenging and attractive issue for power SOs and researchers. In between, evolutionary algorithms have always played a pivotal role in reaching optimal solutions. Similar to many other engineering optimization problems, the basic Genetic Algorithm (GA) or some modified versions are the most favorable tools for the proposed problem. The Improved Genetic Algorithm with Multiple Updating (IGA-MU) is used to solve the CHPS problem in [6]. The IGA-MU takes advantage of the Improved Evolutionary Direction Operator (IEDO), which effectively searches for solutions and a Multiplier Updating (MU) tool to avoid deforming the augmented Lagrange function. The authors in [10] have employed a Self-Adaptive Real-Coded GA (SARGA) for the CHPS, by which the exploration capability of the basic GA is considerably improved. In this technique, a selection tournament is created using a Simulated Binary Crossover (SBX) between two solutions and eventually the better one is considered to be placed in a mating pool. GA can also act as a part of a heuristic approach in the CHPS problem. In [21], the proposed model is divided into two loops, one loop for thermal units and the one for demand, and the former is handled by GA. Particle swarm optimization (PSO) is another powerful tool to cope with the problem of CHPS. Indeed, PSO with selective operators was conducted on the CHPS problem [11] when it was, later on, revealed that the results might be either infeasible or trapped into local optimums. To overcome this shortage, Time-Varying Acceleration Coefficients PSO (TVAC-PSO) is proposed and has been tested on the CHPS problem [5]. Indeed, TVAC-PSO benefits from the adaptive coefficients in the PSO that can vary during iterations. Moreover, a heuristic PSO-based optimization, called Civilized Swarm Optimization (CSO), has been used to solve daily scheduling of CHP units in [14]. In this algorithm, the local search is done by means of binary successive approximation which iteratively updates the unit status. Apart from GA and PSO, other evolutionary algorithms have found their paths to CHP generation scheduling. Vasebi et al. in [7] applied Harmony Search Algorithm (HSA) to the CHPS problem. However, the solutions were infeasible in some cases. It is worth mentioning that an analogy made in [23] between HSA and the Lagrangian Relaxation (LR) method demonstrated the effectiveness of the HSA in large-scale networks. [9] showed that the Ant Colony Search Algorithm (ACSA) itself has some difficulties in terms of CHPS constraints handling and convergence, and therefore attempted to bridge the gap with the help of Tuba Search (TS) and GA incorporated into the ACSA. To reduce the computational burden of the CHPS problem, the authors in [24] utilized the Bee Colony Optimization Algorithm (BCOA) while its performance was validated by comparing the obtained results with those of Real Coded Genetic Algorithm (RCGA) and PSO. Also, Rong et al. in [16] introduced a new dynamic programing approach for the CHPUC named Dynamic Regrouping-based Dynamic Programing-Relaxation and Sequential Commitment (DRDP-RSC) which investigates the generation scheduling of CHP on a daily, weekly, and monthly basis. To see more applications of heuristic and evolutionary algorithms in solving Combined Heat and Economic Dispatch (CHPED), a comprehensive review has been carried out in [25].

10.1.1 Statistics Related to Documents

This section reports statistics related to the published documents in the field of ELD and Combined Heat and Power Scheduling (CHPS) using the Scopus database. The search resulted in 274 documents.

Figure 10.1 shows the number of documents published each year from 1990 and Figure 10.2 depicts the journals with the highest number of publications in the area of CHPS, where *Energy* is the most interesting journal. The authors and

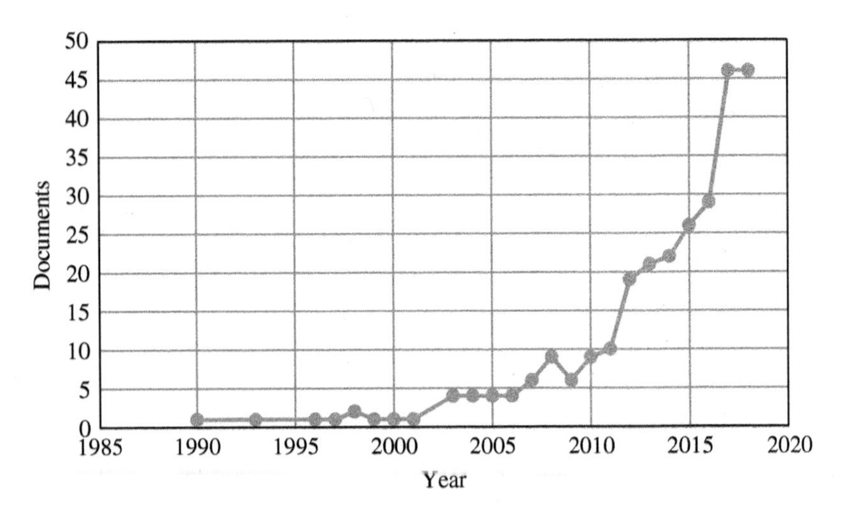

Figure 10.1 Documents by year.

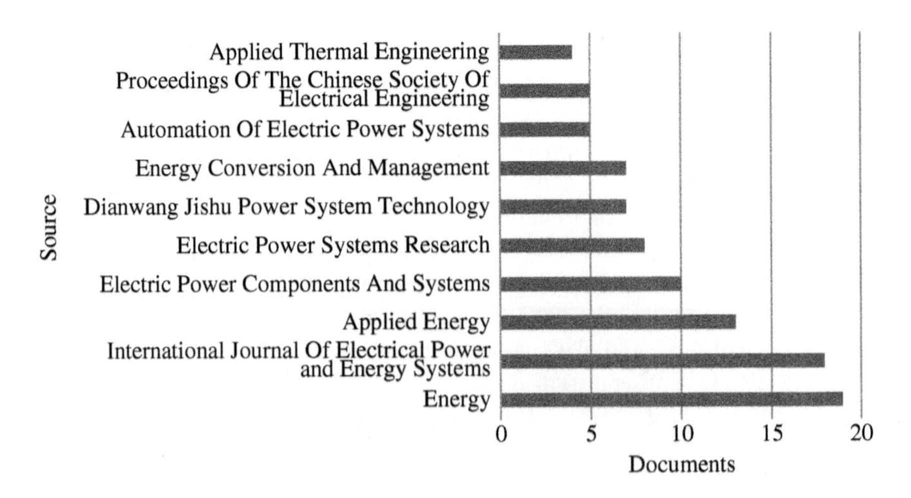

Figure 10.2 Documents by source.

the universities with the highest number of publications in the area of CHPS have been sorted in Figures 10.3 and 10.4 where M. Basu and his affiliated university, Jadavpur University, have been ranked first.

Figure 10.1 depicts the number of publications per year.

Figure 10.2 depicts the number of publications by source and Figure 10.3 depicts the number of publications by author.

Besides, Figure 10.4 depicts the number of publications by affiliation and Figure 10.5 indicates the number of publications by country.

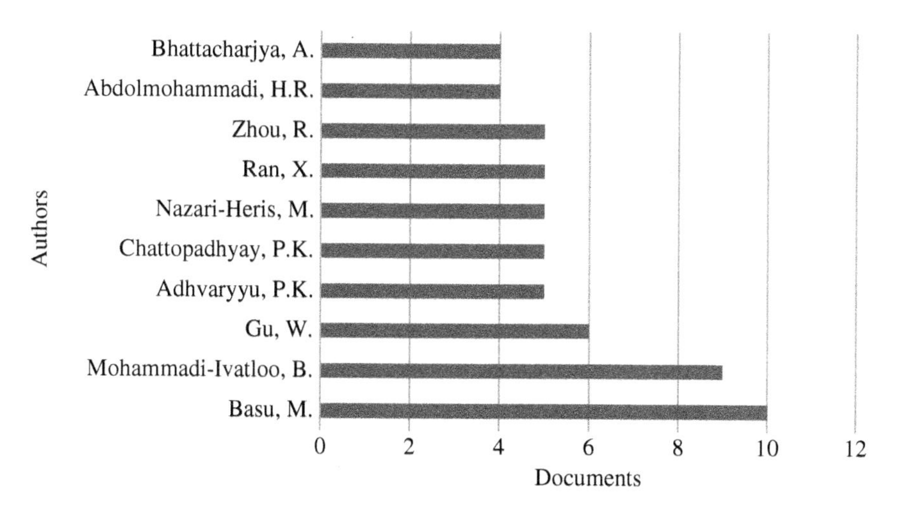

Figure 10.3 Documents by author.

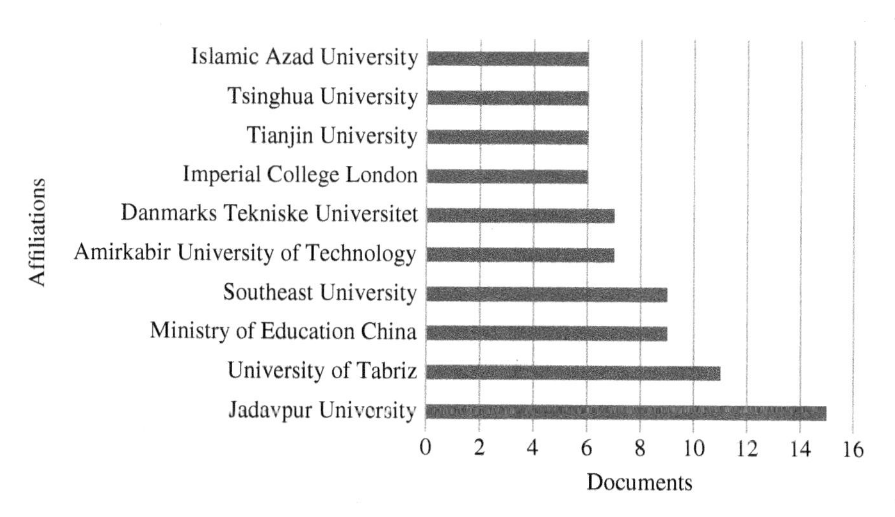

Figure 10.4 Documents by affiliation.

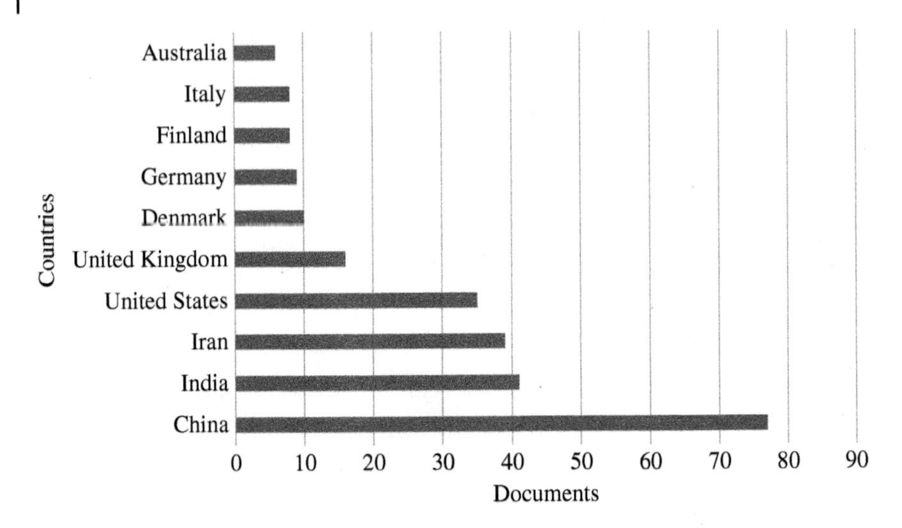

Figure 10.5 Documents by country.

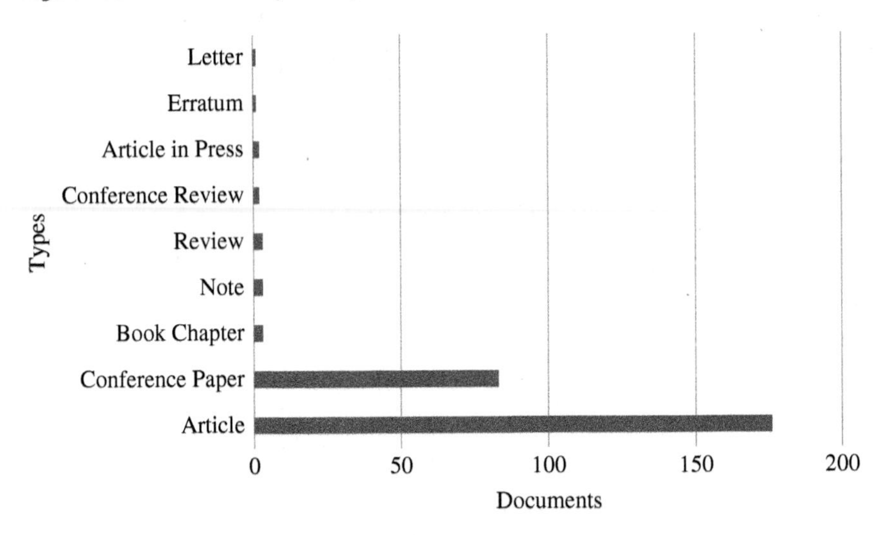

Figure 10.6 Documents by type.

Figure 10.6 illustrates the numbers of publications by type while Figure 10.7 shows the number of publications by subject area.

In this study, the problem of short-term CHPS is investigated using the modified fireworks algorithm [26] as an optimization tool. The modified fireworks algorithm is one of the emerging heuristic population-based algorithms that show a strong performance in complex optimization problems. Moreover, the proposed problem is simulated in two phases, which are ELD and CHPS. In the first phase,

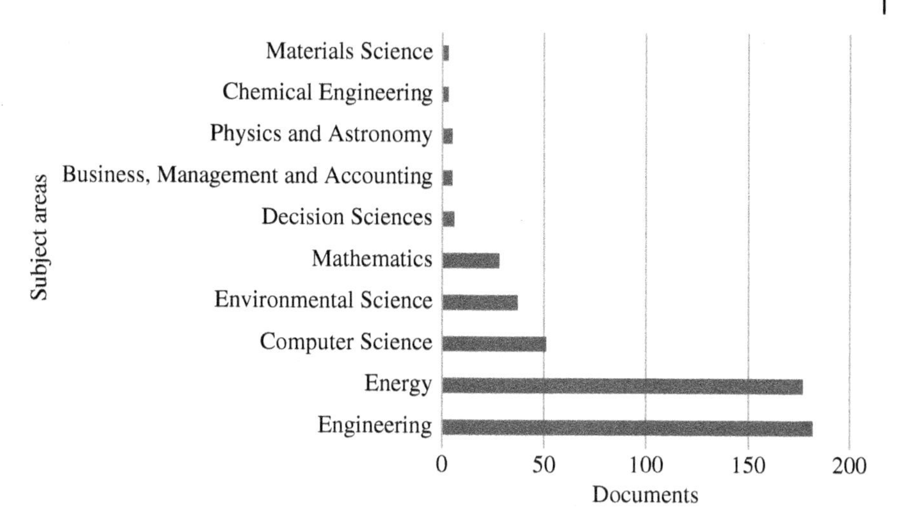

Figure 10.7 Documents by subject area.

aiming at demonstrating the effectiveness of the modified fireworks algorithm in terms of optimality and feasibility, the obtained results are compared with those of existing evolutionary algorithms. In the second stage, the daily CHP scheduling is studied. The results illustrate that the modified fireworks algorithm is well capable of handling the proposed problem and guaranteeing the optimal solution.

The remaining sections are structured as follows. Section 10.2 formulates the objective function and constraints of the problem while Section 10.3 explains the modified fireworks algorithm. Section 10.4 is devoted to the numerical simulations and, finally, Section 10.5 draws some conclusions.

10.2 Modeling

The CHPS problem is a non-linear optimization problem, mainly due to the fuel cost function of the generating units. First, the simpler form of the CHPS problem is proposed in the context of CHPED problem, where the scheduling period is one hour. It is worth mentioning that the CHPS problem is defined for more than one hour and usually up to 168 hours on an hourly basis. The basic economic dispatch problems are only proposed to determine the operating points of the thermal generating units, while it has been more completed by taking into consideration the valve-point effect and the transmission system losses. The CHPS problem has been introduced to power system studies in recent years, and different solution methods have been used. The CHPED problem is proposed to determine the optimal operating point of those units in service such that the electrical and heat load demand is fully met. This problem is generally formulated as an optimization problem with the cost function to be minimized as:

$$Min \quad TC = \sum_{i=1}^{N_T} F_i^T(PG_i^T) + \sum_{j=1}^{N_B} F_j^B(PH_j^B) + \sum_{k=1}^{N_{CHP}} F_k^{CHP}(PG_k^{CHP}, PH_k^{CHP}) \tag{10.1}$$

In the objective function, TC stands for the total cost, which includes the fuel cost of thermal generating units denoted by $F_{j,t}^B(PH_{j,t}^B)$, the fuel cost of heat-only units denoted by $F_{j,t}^B(PH_{j,t}^B)$, as well as the fuel cost of CHP units indicated by $F_{k,t}^{CHP}(PG_{k,t}^{CHP}, PH_{k,t}^{CHP})$, as respectively presented in Eqs. (10.2)–(10.4).

$$F_i^T(PG_i^T) = \alpha_i (PG_i^T)^2 + \beta_i PG_i^T + \gamma_i + \left|\lambda_i \sin(\rho_i (PG_i^{T,Min} - PG_i^T))\right| \tag{10.2}$$

$$F_j^B(PH_j^B) = a_j (PH_j^B)^2 + b_j PH_j^B + c_j \tag{10.3}$$

$$\begin{aligned} F_k^{CHP}(PG_k^{CHP}, PH_k^{CHP}) = {} & a_k (PG_k^{CHP})^2 + b_k PG_k^{CHP} + c_k + d_k (PH_k^{CHP})^2 \\ & + e_k PH_k^{CHP} + f_k PG_k^{CHP} PH_k^{CHP} \end{aligned} \tag{10.4}$$

The expression presented for modeling the fuel cost of thermal generating units includes a sinusoidal term added to a polynomial function as shown in Figure 10.8. In this regard, the conventional cost function of thermal generating units is modeled using a polynomial function of the second order while assigning the impact of the valve-point effect causes a sinusoidal effect, shown by the second term. The fuel cost function of the heat-only units has been represented in Eq. (10.3) while Eq. (10.4) represents the cost function of CHP units. It is noteworthy that the problem presented includes several constraints.

$$\sum_{j=1}^{N_B} PH_j^B + \sum_{k=1}^{N_{CHP}} PH_k^{CHP} = PH_L \tag{10.5}$$

$$\sum_{i=1}^{N_T} PG_i^T + \sum_{k=1}^{N_{CHP}} PG_k^{CHP} = PG_L \tag{10.6}$$

$$PG_i^{T,Min} \le PG_i^T \le PG_i^{T,Max} \tag{10.7}$$

$$PH_j^{B,Min} \le PH_j^B \le PH_j^{B,Max} \tag{10.8}$$

$$\left\{PG_k^{CHP}, PH_k^{CHP}\right\} \in FOR_k^{CHP} \tag{10.9}$$

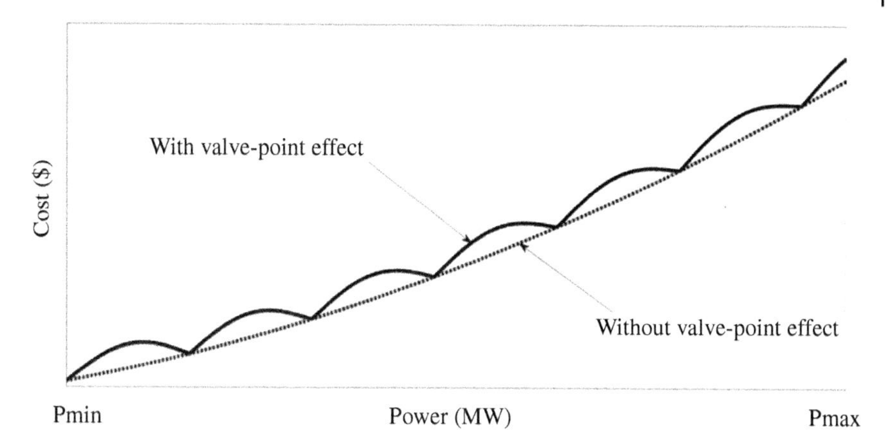

Figure 10.8 The fuel cost function of thermal generating units.

The power and heat balance equations are regarded as the most significant limitations in power systems so that the generated heat and electrical power should be exactly equal to the heat and electrical load demands as Eqs. (10.5) and (10.6), respectively. The power generation limits of thermal, heat-only, and CHP units are indicated through Eqs. (10.7)–(10.9), respectively. As shown in Eq. (10.9), the CHP should operate in the permitted range, which is a 2D closed surface characterized by the generated heat and power known as Feasible Operating Region (FOR). This chapter takes into account three types of FORs as demonstrated in Figures 10.9–10.11.

The FOR of the CHP units can be either convex known as "Type 1" or non-convex known as "Type 2" or "Type 3." Generally, specifying the optimal operating point within Type 2 and Type 3 FORs would be a difficult task which may be intractable on several occasions. Stepping up the generated electrical power in Type 1 CHPs causes the generated heat to reduce. Stepping up the generated heat also leads to electrical power reduction. Such circumstances are regarded as common operation of CHP units. As Figure 10.9 shows, the angles of Type 1 FOR are all less than 180°, while this is not true for Type 2 and Type 3 CHPs, as Figures 10.10 and 10.11 indicate.

10.2.1 Modeling Using a Heuristic Approach

Using heuristic methods for solving the presented method, the objective function and the constraints should be stated properly. First, the variables of the problem should be specified. For the ELD problem in which the operating point of all assets should be determined, one variable should be defined for each of the heat-only units, one variable for each thermal unit, and two variables for each CHP

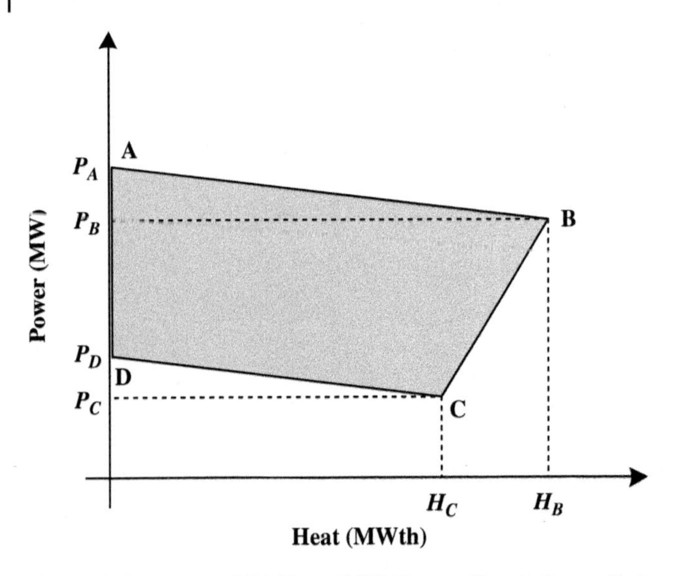

Figure 10.9 Convex FOR (Type: 1) [5] (*Source:* Electric Power Systems Research Journal with permission from Elsevier).

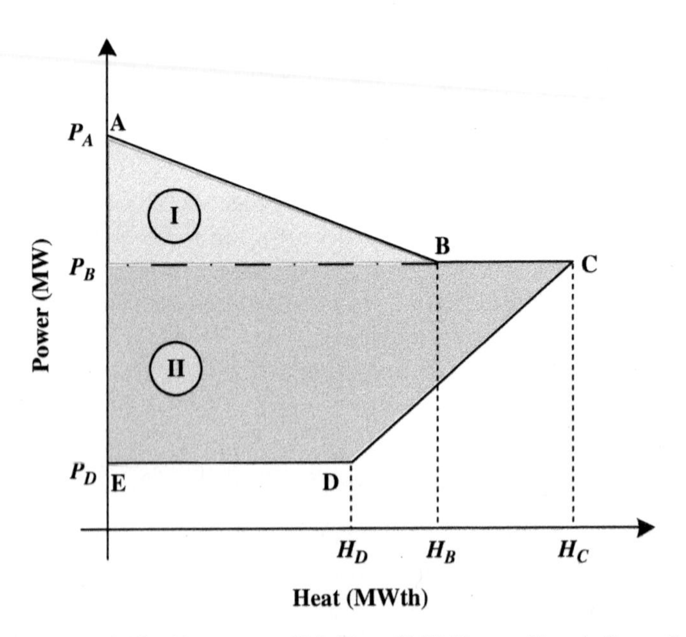

Figure 10.10 Non-convex FOR (Type: 2) [5] (*Source:* Electric Power Systems Research Journal with permission from Elsevier).

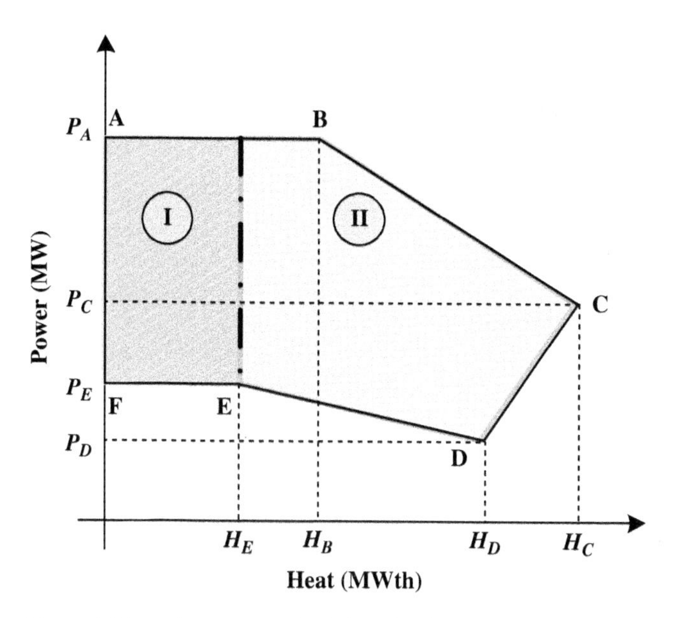

Figure 10.11 Non-convex FOR (Type: 3) [5] (*Source:* Electric Power Systems Research Journal with permission from Elsevier).

unit. It should be noted that these operating points must be within the FOR of the CHP unit. In this respect, N_T, N_B, and N_{CHP} variable matrices should be defined as follows proportionally to the number of thermal units, heat-only units, and CHP units, respectively:

$$Var = \left[\underbrace{pg_2^T, \ldots, pg_{N_T}^T}_{\text{Thermal Units}-1} ; \underbrace{ph_2^B, \ldots, ph_{N_B}^T}_{\text{Boilers}-1} ; \underbrace{pg_1^{CHP}, \ldots, pg_{N_{CHP}}^{CHP} ; ph_1^{CHP}, \ldots, ph_{N_{CHP}}^{CHP}}_{\text{CHP Units}} \right] \tag{10.10}$$

In general, optimization problems constrained by lower and upper limits on a single variable can be modeled using a simple mapping of the stochastic space [0,1] to the permitted range. For instance, the power generation of thermal units and heat generation of heat-only units are stated as follows:

The one-dimensional mapping is used to determine the permitted operating range of the thermal and heat-only units as stated in Eqs. (10.11)–(10.12).

$$PG_i^T = PG_i^{T,Min} + pg_i^T \left(PG_i^{T,Max} - PG_i^{T,Min}\right) \tag{10.11}$$

$$PH_j^B = PH_j^{B,Min} + ph_j^B \left(PH_j^{B,Max} - PH_j^{B,Min}\right) \tag{10.12}$$

The two-dimensional mapping should be used for the CHP units as the electricity and heat generation are mutually related. In this respect, the relationship between the FOR and the stochastic variables should be optimally implemented. For instance, for the convex FOR shown in Figure 10.9, it can be easily modeled. The mapped two-dimensional FOR, which is a rectangle, is depicted in Figure 10.12.

Accordingly, the primary mapping is obtained as follows:

$$PG_k^{CHP} = PG_k^{CHP,Min} + pg_k^{CHP}(PG_k^{CHP,Max} - PG_k^{CHP,Min}) \tag{10.13}$$

$$PH_k^{CHP} = PH_k^{CHP,Min} + ph_k^{CHP}(PH_k^{CHP,Max} - PH_k^{CHP,Min}) \tag{10.14}$$

After the two-dimensional mapping, it is necessary to apply the CHP-related constraints. If the heat generation is between 0 and H_C, the generated power must be within the area surrounded by the line connecting points $(0,P_A)$ and (H_B,P_B) and the line connecting points $(0,P_D)$ and (H_C,P_C). For example, if the electricity generation is above the first line or below the second line, the electric power generation must be fixed on the corresponding value on the mentioned lines. For the state in which the heat generation is between Hc and H_B, the electrical power generation must within the area surrounded by the line connecting points (H_B,P_B) and (H_C,P_C) and line connecting points $(0,P_A)$ and (H_B,P_B). If the electrical power generation is outside this area, the value of electrical power generation must be fixed on the line.

If the FOR is non-convex, using the above linear equations would make a part of the search space unreachable. This issue has been depicted in Figure 10.13.

In this regard, one solution would be dividing the search space into multiple convex sets. If one of the vertices of the FOR is greater than 180°, the plane

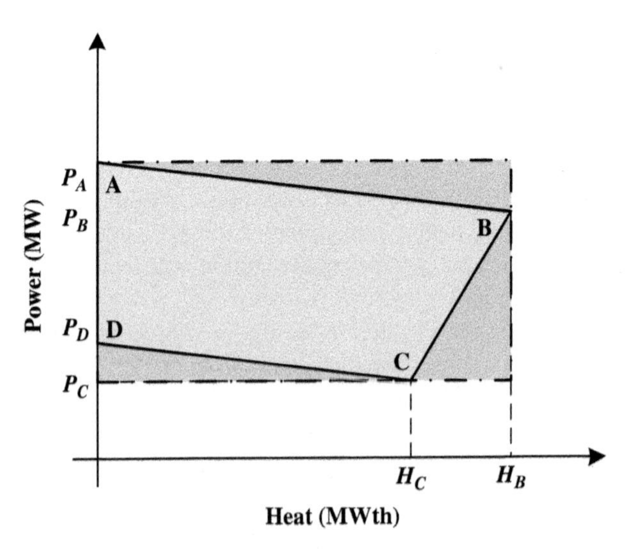

Figure 10.12 Two-dimensional mapping of the convex FOR.

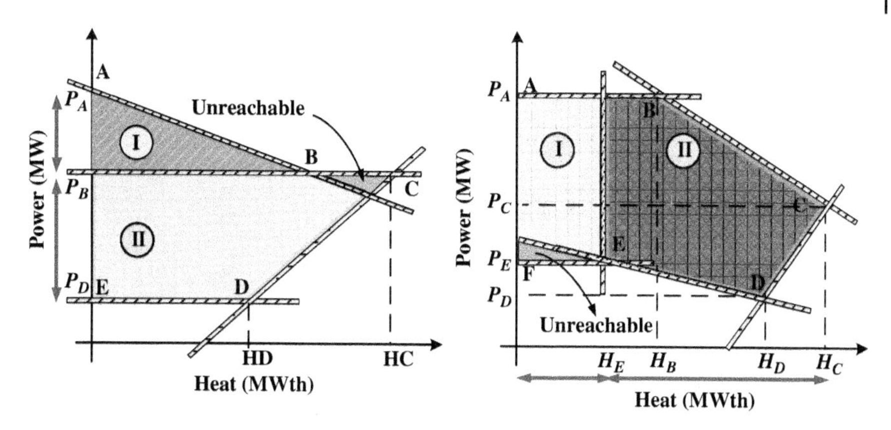

Figure 10.13 Unreachable areas in non-convex FORs.

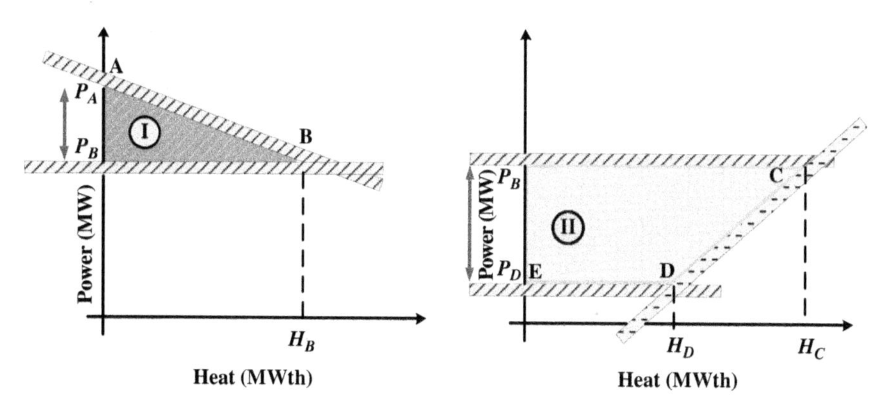

Figure 10.14 Decomposition technique for the Type 2 FOR.

mentioned would be non-convex. Thus the plane should be decomposed to several planes with all vertices less than 180°. Accordingly, the continuous problem will be converted into a discrete optimization problem. Using the above mentioned technique, the linear equations proposed for the convex FOR can be used for each of the decomposed planes.

For the units with non-convex FOR, first the FOR is decomposed into two convex regions. Afterwards, using the technique mentioned, the operating points are determined. Figures 10.14 and 10.15 illustrate the decomposition technique used for non-convex FORs.

For the Type 2 FOR, in Figure 10.14, if the electrical power generation is between A and B, the equations corresponding to the region I are used. If the electrical power generation is between B and D, the equations corresponding to region 2 are used. In Figure 10.15 the regions would be selected after determining the heat power

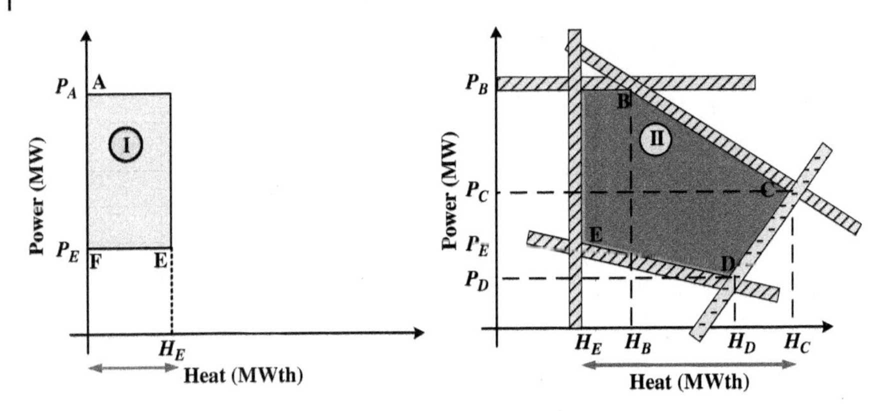

Figure 10.15 Decomposition technique for the Type 3 FOR.

generation. If the heat power is less than H_E, the equations corresponding to the region I are used. If the heat power generation is more than H_E, the equations corresponding to region II are used. After determining the power generation of the CHP units, other constraints of the units are applied. The most important constraint is the electrical and heat power balance equations. As mentioned in Eq. (10.10), the number of thermal units and heat-only units are N_T-*1* and N_B-*1*, respectively. It is noteworthy that the thermal unit 1 and the heat-only unit 1 are considered as the slack units. These units are supposed to ensure the electrical and heat power balance. As a result, the power balance constraint can be simply satisfied. However, it is possible that the power generation in slack units may be out of the feasible range of such units. If so, it should be assigned to the objective function using a penalty factor. The power generation equations of the slack thermal and heat-only units are stated as follows:

$$PG_1^T = PG_L - \sum_{i=2}^{N_T} PG_i^T - \sum_{k=1}^{N_{CHP}} PG_k^{CHP} \tag{10.15}$$

$$PH_j^B = PH_L - \sum_{j=2}^{N_B} PH_j^B - \sum_{k=1}^{N_{CHP}} PH_k^{CHP} \tag{10.16}$$

Finally, the objective function of the problem is reformulated as follows:

$$\begin{aligned} &Min \\ &TC = \sum_{i=1}^{N_T} F_i^T(PG_i^T) + \sum_{j=1}^{N_B} F_j^B(PH_j^B) + \sum_{k=1}^{N_{CHP}} F_k^{CHP}(PG_k^{CHP}, PH_k^{CHP}) \\ &+ \lambda_1^T \langle PG_1^T - PG_1^{T,Min} \rangle + \langle \lambda_1^T PG_1^{T,Max} - PG_1^T \rangle \\ &+ \lambda_1^B \langle PH_1^B - PH_1^{B,Min} \rangle + \langle \lambda_1^B PH_1^{B,Max} - PH_1^B \rangle \end{aligned} \tag{10.17}$$

The bracket-operator <> denotes the absolute value of the operand if the operand is negative; and if the operand is non-negative, it returns zero [5].

The simulation results obtained from mathematical and heuristic methods show that in cases where the problem and the FOR constraint has not been properly modeled, the operating points may fall out of the FORs. Moreover, reaching the global optimum would be very difficult due to the non-linear behavior of the problem. A section is presented in the simulation results to address this issue.

10.2.2 Expanding the ELD Problem to the Generation Scheduling Problem

As the chapter focuses on solving the CHPS problem, the ELD problem has been extended to the generation scheduling problem. Hence, the scheduling horizon will be more than one hour. The generation scheduling problem is generally defined over 24 hours, i.e. 1 day, or 168 hours, i.e. 1 week. Due to the fluctuating load demand over different hours, optimal operating points of the units would change from one hour to another. It is noteworthy that this change in the power and heat generation is limited due to technical and thermodynamic constraints. Such limitations should therefore be applied to the problem. The objective function and the constraints of the CHPS problem are presented as follows:

$$Min$$

$$TC = \sum_{t=1}^{N_H}\sum_{i=1}^{N_T} F_{i,t}^{T}(PG_{i,t}^{T}) + \sum_{t=1}^{N_H}\sum_{j=1}^{N_B} F_{j,t}^{B}(PH_{j,t}^{B}) + \sum_{t=1}^{N_H}\sum_{k=1}^{N_{CHP}} F_{k,t}^{CHP}(PG_{k,t}^{CHP}, PH_{k,t}^{CHP}) \tag{10.18}$$

$$F_{i,t}^{T}(PG_{i,t}^{T}) = \alpha_i (PG_{i,t}^{T})^2 + \beta_i PG_{i,t}^{T} + \gamma_i + \left|\lambda_i \sin(\rho_i (PG_i^{T,Min} - PG_{i,t}^{T}))\right| \tag{10.19}$$

$$F_{j,t}^{B}(PH_{j,t}^{B}) = a_j (PH_{j,t}^{B})^2 + b_j PH_{j,t}^{B} + c_j \tag{10.20}$$

$$\begin{aligned} F_{k,t}^{CHP}(PG_{k,t}^{CHP}, PH_{k,t}^{CHP}) = {} & a_k (PG_{k,t}^{CHP})^2 + b_k PG_{k,t}^{CHP} + c_k + d_k (PH_{k,t}^{CHP})^2 \\ & + e_k PH_{k,t}^{CHP} + f_k PG_{k,t}^{CHP} PH_{k,t}^{CHP} \end{aligned} \tag{10.21}$$

$$\sum_{j=1}^{N_B} PH_{j,t}^{B} + \sum_{k=1}^{N_{CHP}} PH_{k,t}^{CHP} = PH_{L,t} \tag{10.22}$$

$$\sum_{i=1}^{N_T} PG_{i,t}^{T} + \sum_{k=1}^{N_{CHP}} PG_{k,t}^{CHP} = PG_{L,t} \tag{10.23}$$

$$PG_i^{T,Min} \leq PG_{i,t}^T \leq PG_i^{T,Max} \tag{10.24}$$

$$PH_j^{B,Min} \leq PH_{j,t}^B \leq PH_j^{B,Max} \tag{10.25}$$

$$\left\{PG_{k,t}^{CHP}, PH_{k,t}^{CHP}\right\} \in FOR_k^{CHP} \tag{10.26}$$

$$PG_{i,t}^T - PG_{i,t-1}^T \leq RUPG_i^T \tag{10.27}$$

$$PG_{i,t-1}^T - PG_{i,t}^T \leq RDPG_i^T \tag{10.28}$$

$$PH_{j,t}^B - PH_{j,t-1}^B \leq RUPH_j^B \tag{10.29}$$

$$PH_{j,t-1}^B - PH_{j,t}^B \leq RDPH_j^B \tag{10.30}$$

$$PG_{k,t}^{CHP} - PG_{k,t-1}^{CHP} \leq RUPG_k^{CHP} \tag{10.31}$$

$$PG_{k,t-1}^{CHP} - PG_{k,t}^{CHP} \leq RDPG_k^{CHP} \tag{10.32}$$

$$PH_{k,t}^{CHP} - PH_{k,t-1}^{CHP} \leq RUPH_k^{CHP} \tag{10.33}$$

$$PH_{k,t-1}^{CHP} - PH_{k,t}^{CHP} \leq RDPH_k^{CHP} \tag{10.34}$$

The objective function of the scheduling problem is defined as mitigating the total operating cost of the system while satisfying the electrical and heat load demand. The scheduling period of the problem is N_H hours while t is the index of time (one hour). In other words, the optimal operating points of the units are determined on an hourly basis. In addition to the constraints defined for the ELD problem, there are also other constraints that should be taken into consideration relating to the thermodynamic considerations of the units. One of these constraints is the ramping rate of the generating units which is defined as the maximum amount of power generation that can change from one hour to another. In this respect, the Ramp-Up (RU) and Ramp-Down (RD) limits have been considered for the units. The constraints of the scheduling problem which are in common with the ELD problem are modeled similarly, but the preliminary population matrix should be extended to the scheduling period, i.e. NH. Besides, the following framework is used to apply the RU and RD constraints. Assume that the power generation limitation is as follows:

$$P_x^{Min} \leq P_{x,t} \leq P_x^{Max} \tag{10.35}$$

This constraint is modeled as below for hour 2 to hour 24:

$$P_{x,t} = P_{x,t}^{Min} + p_{x,t}\left(P_{x,t}^{Max} - P_{x,t}^{Min}\right) \qquad t > 1 \tag{10.36}$$

where the upper and lower bounds vary on an hourly basis. These bounds are stated as follows:

$$\hat{P}_{x,t}^{Min} = Max\left\{P_x^{Min}, P_{x,t-1} - RDP_x\right\} \tag{10.37}$$

$$\hat{P}_{x,t}^{Max} = Min\left\{P_x^{Max}, P_{x,t-1} + RUP_x\right\} \tag{10.38}$$

These bounds also depend upon the power generation of the units at the previous hour. Accordingly, the RU and RD constraints can be easily modeled. It should be noted that power generation at the first hour can be determined randomly using the preliminary matrix or using the power generation for the previous hour.

In the next section the proposed model is implemented on the ELD problem using the modified fireworks algorithm, and the results are verified. Moreover, some comparisons have been made with other references. Afterwards, the CHPS problem is simulated and the obtained results are discussed.

As mentioned in the ELD problem, determining the optimal operating point of the CHP unit is much more difficult compared to thermal and heat-only units. However, utilizing the suggested technique, the operating points can be optimally and simply determined.

10.3 Fireworks Algorithm

Fireworks optimization method was first introduced in 2010 for dealing with global optimization problems [27]. This algorithm comprises three main parts similar to other optimization algorithms as (i) initialization, (ii) local search, and (iii) selection.

10.3.1 Initialization

The first step is to randomly choose *N* solutions from the search space as fireworks are exploded at different points in the sky.

10.3.2 Local Search

This algorithm includes two types of fireworks: good and bad fireworks. As is obvious from the name, the good fireworks result in a huge population of sparks in a narrow range while the other type leads to a small population of sparks in a wide range. It is noteworthy that the sparks and range respectively denote the number of sampled solutions and the distance from the central solution. If the problem is formulated as a minimization problem, the radius of the explosion and the number of sparks relating to each firework can be stated as below:

$$A_i = \hat{A} \times \frac{f(x_i) - y_{\min} + \varepsilon}{\sum_{j=1}^{N} (f(x_j) - y_{\min}) + \varepsilon} \tag{10.39}$$

$$N_i = \hat{N} \times \frac{y_{\max} - f(x_i) + \varepsilon}{\sum_{j=1}^{N} (y_{\max} - f(x_j)) + \varepsilon} \tag{10.40}$$

In which A shows the radius of the explosion. A denotes the largest radius, N_i indicates the produced sparks by the firework i while the total number of the sparks has been represented by N. Also, it is worth mentioning that the fitness of x_i is shown by $f(x_i)$ while $y_{max} = max(f(x_i))$; $y_{min} = min(f(x_i))$. The very low number utilized to avoid the denominator turning to zero is indicated by ε. Moreover, two types of sparks are generally produced as explosion sparks and mutation sparks where the neighborhood search is done using the first one and the second type would be used to raise the variety of the population. For solution x_i associated with d dimensions, $z(z<d)$ are determined on a random basis while the value of the dimension k changes as:

$$x_{ik} = x_{ik} + deviation \tag{10.41}$$

In the case of explosion sparks, deviation $= Ai \times U(a, b)$, where $U(a, b)$ a number uniformly chosen from [a, b]. In case of the second-type sparks, i.e. mutation sparks, *deviation* $= Ai \times N(\mu, \sigma^2)$, where $N(\mu, \sigma^2)$ indicates a number chosen from a Gaussian distribution with μ and σ^2 as mean as variance.

10.3.3 Selection

The fireworks algorithm allocates its reproductive trials with the disparity between one solution and the remaining, while the best solution is returned to the successive generation because of the principle of holding the fittest individual. The individuals having a longer distance from others are more likely to be selected, disregarding fitness. If K is the candidate set, the roulette wheel mechanism is employed to select the rest n-1 solutions. The probability of selecting a candidate solution x_i is determined as follows:

$$p(x_i) = \frac{R(x_i)}{\sum_{x_j \in K} R(x_j)} \tag{10.42}$$

where $R(x_i) = \sum_{x_j \in k} \| x_i - x_j \|$.

10.3.4 Modified Fireworks Algorithm

A novel neighborhood construction method and the disparity metric are introduced in this section.

10.3.5 Local Search

The cause, beyond leading to good solutions using the Genitor algorithm, relates to its neighborhood construction method, which is also used in the modified fireworks algorithm, where the technique to produce the explosion sparks is revised.

10.3.6 Crossover Operator

It should be noted that two solutions are selected within the solution space and saved as Parent 1 and Parent 2 while the new solution would be Child. First, the crossover operator chooses K random positions in Parent 1. Then the related operator elements in Parent 2 are positioned. Elements in Parent 1 not belonging to the selected elements are directly reversed to Child with identical positions. Eventually, the free slots are allocated to the positioned elements in Parent 2 in the same order.

10.3.7 Explosion Sparks

Using the fireworks algorithm, once the spark number and the radius of the explosion are specified, all solutions are produced in its neighborhood, while no other solution is included. Nevertheless, a child solution is produced for a crossover operator using two parents, i.e. an extra solution must be included in addition to S_i. If S_{best} and S_i represent the ongoing best solution and the ongoing solution respectively, explosion sparks are produced as follows. In case $S_i = S_{best}$, S_{best} is selected as Parent 1 and another solution is selected in random as Parent 2. In case $S_i \neq S_{best}$, S_i is picked as Parent 1 and S_{best} as Parent 2. Moreover, parameter k would be equal to the explosion radius. Therefore, a child solution will be produced once k elements are chosen and the crossover operator is assigned to Parents 1 and 2.

10.3.8 Mutation Sparks

This type of spark is utilized to keep the diversity of solution and improve the local search capability. Thus the method must totally differ from the crossover operator. Three operators are proposed in [28] and the interchange operator is found to be the best. The mutation sparks are produced in this chapter with the interchange operator.

Assume Π represents the set of all permutations of the number {1,2,...,n}, and π is one of the permutations, where $\pi \in \Pi$ as follows:

$$\pi = \{\pi_1, \pi_2, \ldots, \pi_n\} \tag{10.43}$$

π_i indicates the element i in π where $\pi_i \neq \pi_j \quad (i \neq j)$.

The interchange operator selects two elements π_i and π_j in random and their locations are exchanged. π' is obtained as below provided that $i > j$.

$$\pi' = \{\pi_1, \ldots, \pi_{j-1}, \pi_i, \pi_{j+1}, \ldots, \pi_{i-1}, \pi_j, \pi_{i+1}, \ldots, \pi_n\} \tag{10.44}$$

10.3.9 Selection

The fireworks algorithm allocates the reproductive trials with the Euclidean distance between a solution and all the remaining. A proper distance between two permutations is obtained using the number of different arcs [29] provided that $\pi = \{\pi_1, \pi_2, \ldots, \pi_n\}$, and $\sigma = \{\sigma_1, \sigma_2, \ldots, \sigma_n\}$. π and σ should be transformed as follows to determine the distance.

$$\pi_P = \{(\pi_1, \pi_2), (\pi_2, \pi_3), \ldots, (\pi(n-2), \pi(n-1)), (\pi(n-1)), \pi_n)\} \tag{10.45}$$

$$\sigma_P = \{(\sigma_1, \sigma_2), (\sigma_2, \sigma_3), \ldots, (\sigma(n-2), \sigma(n-1)), (\sigma(n-1)), \sigma_n)\} \tag{10.46}$$

The distance between π and σ is determined as:

$$D(\pi, \sigma) = n - |\pi_P \cap \sigma_P| \tag{10.47}$$

10.4 Simulation Results

The presented models for the ELD and CHPS are solved using the modified fireworks algorithm, and the results are discussed in this section. The proposed framework has been coded in Matlab software run on a Core i7 Laptop with 12 GB RAM.

10.4.1 Case 1: ELD

The proposed ELD model has been implemented as a power system with a considerable number of generating units and solved using the modified fireworks algorithm. There are 13 thermal generating units operating, as well as 6 CHP units and 5 generating units. The thermal generating units and CHP units (Types 10.1–10.3) are denoted by T_1–T_{13} and CHP_{14}–CHP_{19}, while the heat-only units are indicated by B_{20}–B_{24}. The data of the generating units and the load demand data is the same as [30].

Table 10.1 represents the obtained results using the proposed method beside those reported by other references using other optimization algorithms. Although Teaching Learning Based Optimization (TLBO) and Oppositional Teaching Learning Based Optimization (OTLBO) gave better results compared to the modified fireworks algorithm, the reported operating points are infeasible, as the power generated by CHP units falls out of the related FOR. Thus, the results are not valid. The modified fireworks algorithm has reached the best solution at which the total cost of the system for satisfying the load demand is $57 963.9.

10.4.2 Case 2: CHPS

This case study uses the same data as the ED case, and it is solved using the modified fireworks algorithm. Only thermal units T_1–T_3 and T_{10}–T_{11} are used. Other units are decommitted for maintenance. Furthermore, the electrical and heat load demand patterns over the 24 hours are demonstrated in Figure 10.16.

The electrical peak load occurs at hour 18, and the heat peak load occurs at hour 1, while the electrical load demand increases over the day while the heat load demand decreases during the day. The optimal operating points of the generating units over the 24 hours are determined to take into account the thermal constraints of the thermal units, i.e. RU and RD constraints. Figure 10.17 shows the convergence of the modified fireworks algorithm for 10 runs.

It is worth mentioning that a sensitivity analysis has also been done to assess the total cost versus different load demand from 0.75 of the original load to 1.25 times the original load demand. Table 10.2 represents the results obtained.

A sensitivity analysis has been carried out to evaluate the performance modified fireworks algorithm in reaching the optimal solutions. To this end, the program has been run 10 times, taking into consideration different number sparks from 10 to 30 and three different scenarios. If the number of sparks is 10, the optimal solution would be obtained only in one of the 10 runs. In other words, the rate of achieving the optimal solution is 10%. By increasing the number of sparks to 20, this value increases to 40%. Finally, if 30 sparks are considered, all optimal solutions are obtained. The average convergence time has also experienced an incremental trend. Table 10.3 represents the brief results.

10.5 Conclusion

This chapter investigated the optimal short-term CHP scheduling problem. The proposed power system includes different generation technologies such as thermal generating units, CHP units, as well as heat-only units. The CHP units, unlike other units, are associated with much more complexity regarding their FOR,

Table 10.1 Simulation results obtained by different techniques for case study 4.

Output	TLBO[a] [31]	OTLBO[a] [31]	CPSO [5]	TVAC-PSO [5]	GSA [4]	Modified fireworks
PG1T	628.324	538.5656	680.0000	538.5587	538.5150	628.3185
PG2T	227.3588	299.2123	0.0000	224.4608	224.4727	143.1414
PG3T	225.9347	299.1220	0.0000	224.4608	224.4611	143.1414
PG4T	110.3721	109.9920	180.0000	109.8666	109.8666	159.7331
PG5T	110.2461	109.9545	180.0000	109.8666	109.8666	159.7331
PG6T	160.1761	110.4042	180.0000	109.8666	109.9008	159.7331
PG7T	108.3552	109.8045	180.0000	109.8666	109.8666	159.7331
PG8T	110.5379	109.6862	180.0000	109.8666	109.8666	159.7331
PG9T	110.5672	109.8992	180.0000	109.8666	109.8666	159.7331
PG10T	75.7562	77.3992	50.5304	77.5210	77.5210	40.0000
PG11T	41.8698	77.8364	50.5304	77.5210	77.5341	40.0000
PG12T	92.4789	55.2225	55.0000	120.0000	120.0000	55.0000
PG13T	57.5140	55.0861	55.0000	120.0000	120.0000	55.0000
PG14CHP	82.5628	81.7524	117.4854	88.3514	92.5632	81.0000
PG15CHP	41.4891	41.7615	45.9281	40.5611	40.0050	40.0000
PG16CHP	84.7710	82.2730	117.4854	88.3514	84.4916	81.0000
PG17CHP	40.5874	40.5599	45.9281	40.5611	40.0079	40.0000
PG18CHP	10.0010	10.0002	10.0013	10.0245	10.0000	10.0000
PG19CHP	31.0978	31.4679	42.1109	40.4288	41.1998	35.0000

Output	TLBO[a] [31]	OTLBO[a] [31]	CPSO [5]	TVAC-PSO [5]	GSA [4]	Modified fireworks
PH14CHP	105.9125	105.2219	125.2754	108.9256	111.2790	104.8000
PH15CHP	76.2843	76.5205	80.1175	75.4844	74.9980	75.0000
PH16CHP	106.9125	105.5142	125.2754	108.9256	106.7495	104.8000
PH17CHP	75.5061	75.4833	80.1174	75.4840	74.9978	75.0000
PH18CHP	39.9986	39.9999	40.0005	40.0104	40.0000	40.0000
PH19CHP	18.2205	18.3944	23.2322	22.4676	22.8181	20.0000
PH20B	468.2278	468.9043	415.9815	458.7020	458.8811	470.4000
PH21B	59.9867	59.9994	60.0000	60.0000	60.0000	60.0000
PH22B	59.9814	59.9999	60.0000	60.0000	60.0000	60.0000
PH23B	119.6074	119.9854	120.0000	120.0000	120.0000	120.0000
PH24B	119.6030	119.9768	120.0000	120.0000	120.0000	120.0000
TC ($)	58006.9992	57856.2676	59736.2635	58122.7460	58121.8640	57983.9000

[a] Infeasible.

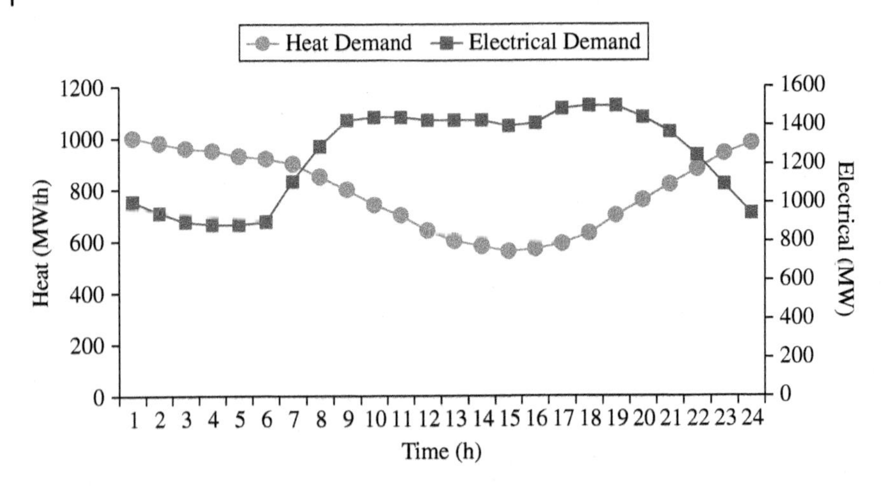

Figure 10.16 Daily electrical and heat load demand.

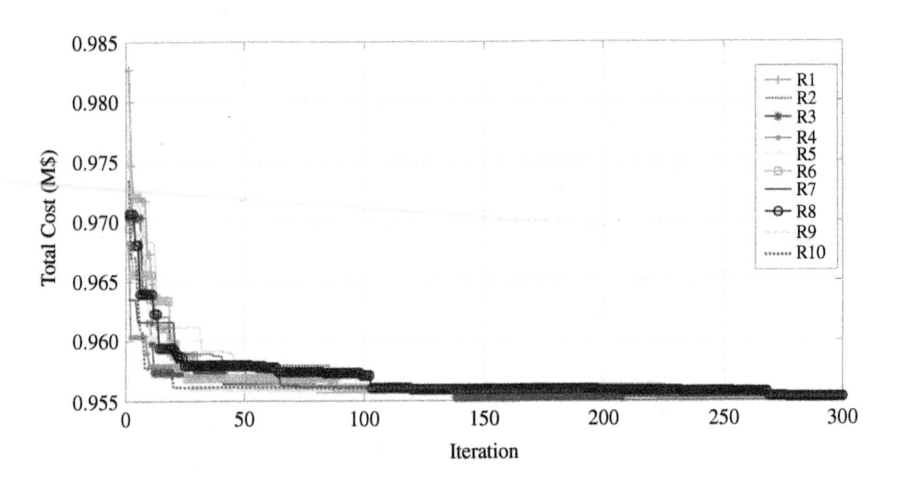

Figure 10.17 Convergence trends of the modified fireworks algorithm for 10 runs.

which can be either a convex or a non-convex plane. This chapter proposed a decomposition technique for decomposing the non-convex FORs to multiple convex FORs. The modified fireworks algorithm was also used to solve the problem presented and its performance was verified through a sensitivity analysis and by comparing the results obtained with those derived by other optimization algorithms.

Table 10.2 Sensitivity analysis results.

Load	Total Cost ($)
0.75	838 087.2092
0.80	856 810.6816
0.85	881 030.8993
0.90	905 478.8464
0.95	925 915.8465
1.00	955 226.5588
1.05	981 142.8492
1.10	1 006 999.430
1.15	1 032 972.547
1.20	1 067 069.007
1.25	1 092 448.790

Table 10.3 Brief results of the sensitivity analysis.

Number of sparks	Minimum cost ($)	Maximum cost ($)	Average cost ($)	Rate of achieving the optimal solution	Average solution time (s)
10	955 226.5588	957 383.9578	956 248.2390	10	96.65
20	955 226.5588	957 301.9621	955 974.6812	40	184.61
30	955 226.5588	955 226.5588	955 226.5588	100	295.98

Acknowledgment

J.P.S. Catalão acknowledges the support by FEDER funds through COMPETE 2020 and by Portuguese funds through FCT, under POCI-01-0145-FEDER-029803 (02/SAICT/2017) and POCI-01-0145-FEDER-006961 (UID/EEA/50014/2019).

References

1 Rong, A. and Lahdelma, R. (2007). An effective heuristic for combined heat-and-power production planning with power ramp constraints. *Appl. Energy* 84 (3): 307–325.

2 Basu, M. (2010). Combined heat and power economic dispatch by using differential evolution. *Electr. Power Compon. Syst.* 38 (8): 996–1004.

3 Adhvaryyu, P.K., Chattopadhyay, P.K., and Bhattacharya, A. (2017). Dynamic optimal power flow of combined heat and power system with valve-point effect using krill herd algorithm. *Energy* 127: 756–767.

4 Beigvand, S.D., Abdi, H., and La Scala, M. (2016). Combined heat and power economic dispatch problem using gravitational search algorithm. *Electr. Pow. Syst. Res.* 133: 160–172.

5 Mohammadi-Ivatloo, B., Moradi-Dalvand, M., and Rabiee, A. (2013). Combined heat and power economic dispatch problem solution using particle swarm optimization with time varying acceleration coefficients. *Electr. Pow. Syst. Res.* 95: 9–18.

6 Su, C.-T. and Chiang, C.-L. (2004). An incorporated algorithm for combined heat and power economic dispatch. *Electr. Pow. Syst. Res.* 69 (2): 187–195.

7 Vasebi, A., Fesanghary, M., and Bathaee, S.M.T. (2007). Combined heat and power economic dispatch by harmony search algorithm. *Int. J. Electr. Power Energy Syst.* 29 (10): 713–719.

8 Khorram, E. and Jaberipour, M. (2011). Harmony search algorithm for solving combined heat and power economic dispatch problems. *Energ. Conver. Manage.* 52 (2): 1550–1554.

9 Song, Y.H., Chou, C.S., and Stonham, T.J. (1999). Combined heat and power economic dispatch by improved ant colony search algorithm. *Electr. Pow. Syst. Res.* 52 (2): 115–121.

10 Subbaraj, P., Rengaraj, R., and Salivahanan, S. (2009). Enhancement of combined heat and power economic dispatch using self adaptive real-coded genetic algorithm. *Appl. Energy* 86 (6): 915–921.

11 Ramesh, V., Jayabarathi, T., Shrivastava, N., and Baska, A. (2009). A novel selective particle swarm optimization approach for combined heat and power economic dispatch. *Electr. Power Compon. Syst.* 37 (11)): 1231–1240.

12 Kia, M., Nazar, M.S., Sepasian, M.S. et al. (2017). An efficient linear model for optimal day ahead scheduling of CHP units in active distribution networks considering load commitment programs. *Energy* 139: 798–817.

13 Kia, M., Nazar, M.S., Sepasian, M.S. et al. (2017). New framework for optimal scheduling of combined heat and power with electric and thermal storage systems considering industrial customers inter-zonal power exchanges. *Energy* 138: 1006–1015.

14 Anand, H., Narang, N., and Dhillon, J.S. (2018). Unit commitment considering dual-mode combined heat and power generating units using integrated optimization technique. *Energ. Conver. Manage.* 171: 984–1001.

15 Zugno, M., Morales, J.M., and Madsen, H. (2016). Commitment and dispatch of heat and power units via affinely adjustable robust optimization. *Comput. Oper. Res.* 75: 191–201.

16 Rong, A., Hakonen, H., and Lahdelma, R. (2009). A dynamic regrouping based sequential dynamic programming algorithm for unit commitment of combined heat and power systems. *Energ. Conver. Manage.* 50 (4): 1108–1115.
17 Alipour, M., Zare, K., and Seyedi, H. (2017). Power flow constrained short-term scheduling of CHP units. In: *Sustainable Development in Energy Systems* (ed. B. Azzopardi), 147–165. Cham: Springer International Publishing.
18 Kia, M., Nazar, M.S., Sepasian, M.S. et al. (2017). Optimal day ahead scheduling of combined heat and power units with electrical and thermal storage considering security constraint of power system. *Energy* 120: 241–252.
19 Majić, L., Krželj, I., and Delimar, M. (2013). Optimal scheduling of a CHP system with energy storage. 36th International Convention on Information and Communication Technology, Electronics and Microelectronics (MIPRO), 1253–1257.
20 Kia, M., Nazar, M.S., Sepasian, M.S. et al. (2017). Coordination of heat and power scheduling in micro-grid considering inter-zonal power exchanges. *Energy* 141: 519–536.
21 Nazari, M.E. and Ardehali, M.M. (2017). Profit-based unit commitment of integrated CHP-thermal-heat only units in energy and spinning reserve markets with considerations for environmental CO_2 emission cost and valve-point effects. *Energy* 133: 621–635.
22 Alipour, M., Zare, K., and Mohammadi-Ivatloo, B. (2014). Short-term scheduling of combined heat and power generation units in the presence of demand response programs. *Energy* 71: 289–301.
23 Javadi, M.S., Esmaeel Nezhad, A., and Sabramooz, S. (2012). Economic heat and power dispatch in modern power system harmony search algorithm versus analytical solution. *Sci. Iran.* 19 (6): 1820–1828.
24 Basu, M. (2011). Bee colony optimization for combined heat and power economic dispatch. *Expert Syst. Appl.* 38 (11): 13527–13531.
25 Nazari-Heris, M., Mohammadi-Ivatloo, B., and Gharehpetian, G.B. (2018). A comprehensive review of heuristic optimization algorithms for optimal combined heat and power dispatch from economic and environmental perspectives. *Renew. Sustain. Energy Rev.* 81: 2128–2143.
26 Liu, Z., Feng, Z., and Ke, L. (2016) A modified fireworks algorithm for the multi-resource range scheduling problem. International Conference on Swarm Intelligence (ICSI 2016): Advances in Swarm Intelligence, 535–543. Cham: Springer International Publishing.
27 Tan, Y. and Zhu, Y. (2010). *Fireworks Algorithm for Optimization*, 355–364. Berlin, Heidelberg: Springer.
28 Liu, Z., Feng, Z., and Ke, L. (2015). Fireworks algorithm for the multi-satellite control resource scheduling problem. 2015 IEEE Congress on Evolutionary Computation (CEC): 1280–1286.

29 Jones, T. and Forrest, S. (1995). Fitness distance correlation as a measure of problem difficulty for genetic algorithms. Presented at the Proceedings of the 6th International Conference on Genetic Algorithms.

30 Razavi, S.-E., Javadi, M.S., and Esmaeel Nezhad, A. (2014). Mixed-integer nonlinear programming framework for combined heat and power units with nonconvex feasible operating region: feasibility, optimality, and flexibility evaluation. *Int. Trans. Electr. Energy Syst.* 29 (3): e2767.

31 Roy, P.K., Paul, C., and Sultana, S. (2014). Oppositional teaching learning based optimization approach for combined heat and power dispatch. *Int. J. Electr. Power Energy Syst.* 57: 392–403.

Index

Evolutionary Computation in Scheduling, First Edition. Edited by Amir H. Gandomi, Ali Emrouznejad, Mo M. Jamshidi, Kalyanmoy Deb, and Iman Rahimi.

d

S